Data Feminism

**&lt;strong&gt; Ideas Series**

Edited by David Weinberger

The &lt;strong&gt; Ideas Series explores the latest ideas about how technology is affecting culture, business, science, and everyday life. Written for general readers by leading technology thinkers and makers, books in this series advance provocative hypotheses about the meaning of new technologies for contemporary society.

The &lt;strong&gt; Ideas Series is published with the generous support of the MIT Libraries.

*Hacking Life: Systematized Living and Its Discontents*, Joseph M. Reagle Jr.

*The Smart Enough City: Putting Technology in Its Place to Reclaim Our Urban Future*, Ben Green

*Sharenthood: Why We Should Think before We Post about Our Kids*, Leah A. Plunkett

*Data Feminism*, Catherine D'Ignazio and Lauren F. Klein

# Data Feminism

Catherine D'Ignazio and Lauren F. Klein

The MIT Press
Cambridge, Massachusetts
London, England

First MIT Press paperback edition, 2023

This book was set in ITC Stone Serif Std and ITC Stone Sans Std by Toppan Best-set Premedia Limited. Printed and bound in the United States of America.

Library of Congress Cataloging-in-Publication Data

Names: Kanarinka, author. | Klein, Lauren F., author.
Title: Data feminism / Catherine D'Ignazio and Lauren F. Klein.
Description: Cambridge, Massachusetts : The MIT Press, [2020] | Series: <Strong> ideas series | Includes bibliographical references and index.
Identifiers: LCCN 2019036137 | ISBN 9780262044004 (hardcover)—9780262547185 (paperback)
Subjects: LCSH: Feminism. | Feminism and science. | Big data—Social aspects. | Quantitative research—Methodology—Social aspects. | Power (Social sciences)
Classification: LCC HQ1190 .K375 2020 | DDC 305.42—dc23
LC record available at https://lccn.loc.gov/2019036137

10  9  8

This book is dedicated to our children: Gastón, Orlando, Margarita, Loie, and Aurora. You inspire us to live our feminism every day.

# Contents

# Acknowledgments

It's often said that a book is the project of many hands. This saying is certainly true for this book. A debt of gratitude is owed to Patsy Baudoin, who first wrote to put the two of us in touch. Without her email message, this project would not exist.

This project also would not exist without all the activists, journalists, artists, designers, engineers, scientists, scholars, and teachers whose work we describe in this book, as well as the numerous additional projects that we discussed during the writing process, but ultimately did not have the space to include. Your work is an inspiration.

We remain ever appreciative of the dozens of people who reviewed and commented on our online manuscript draft. As we said when we posted the draft, we believe that their direct and critical words were tokens of generosity and votes of confidence in our ability to hear and be transformed by our readers. Among these readers were Alexandrina Agloro, Kecia Ali, Jessica Bellamy, Ruha Benjamin, Rahul Bhargava, Rena Bivens, Alison Booth, Carol Chiodo, Patricio Davila, Michelle Doerr, Jordan Ellenberg, Jack Gieseking, Ben Green, Ksenia Gueletina, Oliver Haimson, Jaron Heard, Anna Lauren Hoffman, Hannah House, Pratyusha Kalluri, Os Keyes, David Kim, Heather Krause, Nick Lally, Christopher Linzy, Liz Losh, Momin Malik, Laura Mandell, Surya Mattu, Elijah Meeks, Rebecca Michelson, Susana Morris, Ron Morrison, Maria Munir, Bethany Nowviskie, Yoehan Oh, Thomas Padilla, Jonas Parnow, Margaret Pearce, Firaz Peer, Tawana Petty, Anne Pollock, Miriam Posner, Zara Rahman, Gabriela Rodriguez Beron, James Scott-Brown, Nicole Siggins, Erik Simpson, Nikki Stevens, Zach Van Stanley, Annette Vee, Fernanda Viégas, Lee Vinsel, Maya Wagoner, Jacque Wernimont, and Sarah Yerima. We hope that the revised version of this book has demonstrated how much we have valued and learned from your thoughts.

In particular, we would like to thank Marian Dörk, Aristea Fotopoulou, Shannon Mattern, Yanni Loukissas, and Seeta Peña Gangadharan in their capacity as official online peer reviewers. They demonstrated the best of academic generosity in their willingness to join our experimental review process and make their impressively comprehensive comments in full public view.

We would also like to thank Sasha Costanza-Chock, Alexis Lothian, Zara Rahman, and the other anonymous peer reviewers of our book proposal and manuscript draft. Your generous and generative comments pushed us to clarify each and every one of the ideas we express in these pages.

Our editors, Patsy Baudoin and David Lobenstine, helped provide shape and cohesion to this book at crucial phases of the writing process.

Our student research assistants, Isabel Carter at Emerson College and Zoe Wangstrom at Georgia Tech, both contributed countless hours to ensuring the images, references, and other details in this book are accurate and complete.

At the MIT Press, Gita Manaktala remained steadfast in her support of this project, always responding to our queries with attention and detail. Catherine Ahearn devoted countless hours to creating the online version of the manuscript draft. Jessica Lipton, Nhora Lucia Serrano, and Kyle Gipson provided invaluable editorial assistance. Melinda Rankin carefully copyedited each line of this book. Mandy Keifetz created the detailed index that concludes the volume.

We would also like to thank David Weinberger, who first solicited the project, enthusiastically backed it from the first iteration, and reviewed multiple drafts of our book proposal; Mushon Zer-Aviv, who believed in and supported this work in its earliest and most nascent stages, and even used his own invited talk at Data & Society to uplift this work; Alison Booth and Liz Losh, whose early endorsement of the project enabled it to receive crucial external funding; and our contacts, colleagues, and friends who have read and commented on portions of this manuscript: Mariel García-Montes, Regina Larrea, Sarah Blackwood, Nihad Farooq, Matthew K. Gold, Yanni Loukissas, Miriam Posner, Kyla Schuller, Karen Weingarten, and Jacqueline Wernimont.

In addition, we would like to thank those who have directly aided us to have the courage to do the hard, uncomfortable, and urgent work of accounting for how structural forces manifest themselves in our own identities and everyday experiences, especially Kimberly Seals Allers, Jenn Roberts of Versed Education, and Chris Miller, as well as our interlocutors in our scholarly communities and online. You have taught us that anti-oppression work is never finished, but also that there is joy in the struggle, even when so much of the struggle is to take responsibility for one's own position within the matrix of domination.

We would like to thank the people and institutions that have extended us invitations to talk about this project and offered their own thoughts and ideas in return. These include Eric Gordon and the Engagement Lab at Emerson College, Ethan

Zuckerman and the MIT Center for Civic Media, Dietmar Offenhuber at Northeastern University, Elizabeth Maddock Dillon and the NULab at Northeastern University, Kevin Hamilton and the University of Illinois Champaign-Urbana, Meghan Kelly and the University of Wisconsin Geography Department, Pablo Rey and Think Commons, Mariana Santos and Chicas Poderosas, Mushon Zer-Aviv and ISVIS, Isabelle Mireilles and the Information+ Conference, Enrico Bertini and Mortiz Stefaner of the Data Stories podcast, Jonathan Schwabisch of the PolicyViz podcast, the American Association of Geographers, the University of Southern Maine, Winnie Poster and the Labor Tech reading group, Cecilia Balbin and the Universidad Católica de Argentina, Paul Benzon and Skidmore College, Rebecca Munson and the Princeton Center for Digital Humanities, the Data Justice Conference, Thai Jungpanich and Civichub, Spencer Keralis and the rest of the organizing team for Digital Frontiers 2018, Tarez Graban and the Digital Scholars Discussion Group at the University of Florida, Sara Ortolani at the London College of Communication, Jonathan Gray and the Department of Digital Humanities at King's College London, Diane Jakacki and the Program in Comparative Humanities at Bucknell University, Mimi Onuoha and the eyeo 2019 conference, the Data Power 2019 conference, and the organizers of the IEEE VIS 2016 Workshop on Visualization for the Digital Humanities, which gave us the first opportunity to write together.

We would also like to thank our own institutions. By the time this book is published, we will have each moved on to new positions—Catherine in the Department of Urban Studies and Planning at MIT, and Lauren in the Departments of English and Quantitative Theory & Methods at Emory University. But when we began this project, Catherine worked in the Journalism Department at Emerson College and Lauren worked in the School of Literature, Media, and Communication at Georgia Tech. We will remain forever appreciative of the support provided by our colleagues at our first institutional homes: Miranda Banks, André Brock, Betsy DiSalvo, Carl DiSalvo, Nihad Farooq, Lina Maria Giraldo, Eric Gordon, Narin Hassan, Janet Kolodzy, Chris Le Dantec, Todd Michney, Paul Mihailidis, Susana Morris, Janet Murray, Brad Rittenhouse, Jacqueline Jones Royster, Richard Utz, Lauren Wilcox, Joycelyn Wilson, and Greg Zinman. Emerson students in the Data Visualization classes and Women in Journalism class taught by Diane Mermigas offered excellent feedback on chapter drafts. The institutions of Georgia Tech and Emerson College should also be thanked for their financial support of this book.

We would also like to thank Dr. Norman Stearns and Irma Mann Stearns, who supported this project with the Emerson College Mann Stearns Award, as well as the American Council of Learned Societies, which supported this project with a 2019–2020 ACLS

Collaborative Fellowship. The year of research leave made possible by the ACLS fellowship enabled this project to advance at a crucial moment in our careers.

Last but not least, we are deeply grateful to our partners, Dave Raymond and Greg Zinman, for their intellectual contributions, as well as their (significant) household and family contributions that made this project possible. We are also grateful to our parents, Janet and Fred D'Ignazio and Diane and Francis Klein, for teaching us to love learning and do work that makes a difference in the world. We also thank our partners, our parents, and our children's additional caregivers, grandparents Mike and Susan Raymond, Cuban grandparents Mayda and Coco Marrero, María Lopez Rodas, Milagros Banciella-Dickie and Melissa Gault, for giving us the time to write.

# Introduction: Why Data Science Needs Feminism

Christine Mann Darden first passed through the gates of NASA's Langley Research Center in Hampton, Virginia, in the summer of 1967. Her newly minted master's degree in applied math had earned her a position as a data analyst there. In the city of Hampton, and across the United States, tensions were running high. In Los Angeles, a massive protest against the Vietnam War ended only when more than a thousand armed police officers attacked the peaceful protestors. One month later, even more violence engulfed the city of Detroit after a police raid spiraled out of control. The 1967 Detroit Riot, or the 1967 Detroit Rebellion, as it is increasingly known, resulted in over forty deaths and one thousand injuries.

The gates of Langley might have shielded Darden from those physical confrontations, but her work there was no less removed from the national stage. By 1967, the space race was well underway, and the United States was losing. The Soviet Union had already sent a man into space and a rocket to the moon. The only thing standing in the way of a Soviet victory was to put those two pieces together. Meanwhile, the United States had suffered a series of defeats—and, in January of that year, an outright disaster, when a sudden fire during a launch test of the *Apollo 1* spacecraft killed all three astronauts on board.

While the nation mourned, everyone at NASA threw themselves back into their work—including Darden, who began her data analyst job in the immediate aftermath of the *Apollo 1* disaster. Two years later, it would be her precise analysis of the physics of rocket reentry that would help to ensure the successful return of the *Apollo 11* mission from the moon, effectively winning the space race for the United States. But it would be Darden herself, as a Black woman with technical expertise, working at a federal agency in which sexism and racism openly prevailed, who demonstrated that the ideological mission of the United States—as a land based on the ideals of liberty, equality, and opportunity for all—was far from accomplished (figure 0.1).

**Figure 0.1**
Christine Darden in the control room of the Unitary Plan Wind Tunnel at NASA's Langley
Research Center in 1975. Courtesy of NASA.

The 1960s, after all, were years of social protest and transformation as well as exploration into outer space. Darden herself had participated in several lunch-counter sit-
ins at the Hampton Institute, the historically Black college that she attended for her
undergraduate studies.[1] By the time that Darden joined NASA, its Virginia facility had
been officially desegregated for several years. But it had yet to reckon with the unof-
ficial segregation of both race and gender that remained in place—particularly among
its women employees known as *computers*.

Darden's arrival at Langley coincided with the early days of digital computing.
Although Langley could claim one of the most advanced computing systems of the
time—an IBM 704, the first computer to support floating-point math—its resources
were still limited. For most data analysis tasks, Langley's Advanced Computing Division

relied upon human computers like Darden herself. These computers were all women, trained in math or a related field, and tasked with performing the calculations that determined everything from the best wing shape for an airplane, to the best flight path to the moon. But despite the crucial roles they played in advancing this and other NASA research, they were treated like unskilled temporary workers. They were brought into research groups on a project-by-project basis, often without even being told anything about the source of the data they were asked to analyze. Most of the engineers, who were predominantly men, never even bothered to learn the computers' names.

These women computers have only recently begun to receive credit for their crucial work, thanks to scholars of the history of computing—and to journalists like Margot Lee Shetterly, whose book, *Hidden Figures: The American Dream and the Untold Story of the Black Women Who Helped Win the Space Race*, along with its film adaptation, is responsible for bringing Christine Darden's story into the public eye.[2] Her story, like those of her colleagues, is one of hard work under discriminatory conditions. Each of these women computers was required to advocate for herself—and some, like Darden, chose also to advocate for others. It is because of both her contributions to data science and her advocacy for women that we have chosen to begin our book, *Data Feminism*, with Darden's story. For *feminism* begins with a belief in the "political, social, and economic equality of the sexes," as the *Merriam-Webster Dictionary* defines the term—as does, for the record, Beyoncé.[3] And any definition of feminism also necessarily includes the activist work that is required to turn that belief into reality. In *Data Feminism*, we bring these two aspects of feminism together, demonstrating a way of thinking about data, their analysis, and their display, that is informed by this tradition of feminist activism as well as the legacy of feminist critical thought.

As for Darden, she did not only apply her skills of data analysis to spaceflight trajectories; she also applied them to her own career path. After working at Langley for a number of years, she began to notice two distinct patterns in her workplace: men with math credentials were placed in engineering positions, where they could be promoted through the ranks of the civil service, while women with the same degrees were sent to the computing pools, where they languished until they retired or quit. She did not want to become one of those women, nor did she want others to experience the same fate. So she gathered up her courage and decided to approach the chief of her division to ask him why. As Darden, now seventy-five, told Shetterly in an interview for *Hidden Figures*, his response was sobering: "Well, nobody's ever complained," he told Darden. "The women seem to be happy doing that, so that's just what they do."

In today's world, Darden might have gotten her boss fired—or at least served with an Equal Employment Opportunity Commission complaint. But at the time that

Darden posed her question, stereotypical remarks about "what women do" were par for the course. In fact, challenging assumptions about what women could or couldn't do—especially in the workplace—was the central subject of Betty Friedan's best-selling book, *The Feminine Mystique*. Published in 1963, *The Feminine Mystique* is often credited with starting feminism's so-called second wave.[4] Fed up with the enforced return to domesticity following the end of World War II, and inspired by the national conversation about equality of opportunity prompted by the civil rights movement, women across the United States began to organize around a wide range of issues, including reproductive rights and domestic violence, as well as the workplace inequality and restrictive gender roles that Darden faced at Langley.

That said, Darden's specific experience as a Black woman with a full-time job was quite different than that of a white suburban housewife—the central focus of *The Feminine Mystique*. And when critics rightly called out Friedan for failing to acknowledge the range of experiences of women in the United States (and abroad), it was women like Darden, among many others, whom they had in mind. In *Feminist Theory: From Margin to Center*, another landmark feminist book published in 1984, bell hooks puts it plainly: "[Friedan] did not discuss who would be called in to take care of the children and maintain the home if more women like herself were freed from their house labor and given equal access with white men to the professions. She did not speak of the needs of women without men, without children, without homes. She ignored the existence of all non-white women and poor white women. She did not tell readers whether it was more fulfilling to be a maid, a babysitter, a factory worker, a clerk, or a prostitute than to be a leisure-class housewife."[5]

In other words, Friedan had failed to consider how those additional dimensions of individual and group identity—like race and class, not to mention sexuality, ability, age, religion, and geography, among many others—intersect with each other to determine one's experience in the world. Although this concept—*intersectionality*—did not have a name when hooks described it, the idea that these dimensions cannot be examined in isolation from each other has a much longer intellectual history.[6] Then, as now, key scholars and activists were deeply attuned to how the racism embedded in US culture, coupled with many other forms of oppression, made it impossible to claim a common experience—or a common movement—for all women everywhere. Instead, what was needed was "the development of integrated analysis and practice based upon the fact that the major systems of oppression are interlocking."[7] These words are from the Combahee River Collective Statement, written in 1978 by the famed Black feminist activist group out of Boston. In this book, we draw heavily from intersectionality and other concepts developed through the work of *Black feminist* scholars and activists because they offer some of the best ways for negotiating this multidimensional terrain.

Indeed, feminism must be intersectional if it seeks to address the challenges of the present moment. We write as two straight, white women based in the United States, with four advanced degrees and five kids between us. We identify as middle-class and *cisgender*—meaning that our gender identity matches the sex that we were assigned at birth. We have experienced sexism in various ways at different points of our lives— being women in tech and academia, birthing and breastfeeding babies, and trying to advocate for ourselves and our bodies in a male-dominated health care system. But we haven't experienced sexism in ways that other women certainly have or that nonbinary people have, for there are many dimensions of our shared identity, as the authors of this book, that align with dominant group positions. This fact makes it impossible for us to speak *from experience* about some oppressive forces—racism, for example. But it doesn't make it impossible for us to educate ourselves and then speak about racism and the role that white people play in upholding it. Or to challenge ableism and the role that abled people play in upholding it. Or to speak about class and wealth inequalities and the role that well-educated, well-off people play in maintaining those. Or to believe in the logic of *co-liberation*. Or to advocate for justice through equity. Indeed, a central aim of this book is to describe a form of intersectional feminism that takes the inequities of the present moment as its starting point and begins its own work by asking: How can we use data to remake the world?[8]

This is a complex and weighty task, and it will necessarily remain unfinished. But its size and scope need not stop us—or you, the readers of this book—from taking additional steps toward justice. Consider Christine Darden, who, after speaking up to her division chief, heard nothing from him but radio silence. But then, two weeks later, she was indeed promoted and transferred to a group focused on sonic boom research. In her new position, Darden was able to begin directing her own research projects and collaborate with colleagues of all genders as a peer. Her self-advocacy serves as a model: a sustained attention to how systems of oppression intersect with each other, informed by the knowledge that comes from direct experience. It offers a guide for challenging power and working toward justice.

## What Is Data Feminism?

Christine Darden would go on to conduct groundbreaking research on sonic boom minimization techniques, author more than sixty scientific papers in the field of computational fluid dynamics, and earn her PhD in mechanical engineering—all while "juggling the duties of Girl Scout mom, Sunday school teacher, trips to music lessons, and homemaker," Shetterly reports. But even as she ascended the professional ranks, she could tell that her scientific accomplishments were still not being recognized as readily as those of her male counterparts; the men, it seemed, received promotions far more quickly.

Darden consulted with Langley's Equal Opportunity Office, where a white woman by the name of Gloria Champine had been compiling a set of statistics about gender and rank. The data confirmed Darden's direct experience: that women and men—even those with identical academic credentials, publication records, and performance reviews—were promoted at vastly different rates. Champine recognized that her data could support Darden in her pursuit of a promotion and, furthermore, that these data could help communicate the systemic nature of the problem at hand. Champine visualized the data in the form of a bar chart, and presented the chart to the director of Darden's division.[9] He was "shocked at the disparity," Shetterly reports, and Darden received the promotion she had long deserved.[10] Darden would advance to the top rank in the federal civil service, the first Black woman at Langley to do so. By the time that she retired from NASA, in 2007, Darden was a director herself.[11]

Although Darden's rise into the leadership ranks at NASA was largely the result of her own knowledge, experience, and grit, her story is one that we can only tell as a result of the past several decades of feminist activism and critical thought. It was a national feminist movement that brought women's issues to the forefront of US cultural politics, and the changes brought about by that movement were vast. They included both the shifting gender roles that pointed Darden in the direction of employment at NASA and the creation of reporting mechanisms like the one that enabled her to continue her professional rise. But Darden's success in the workplace was also, presumably, the result of many unnamed colleagues and friends who may or may not have considered themselves feminists. These were the people who provided her with community and support—and likely a not insignificant number of casserole dinners—as she ascended the government ranks. These types of collective efforts have been made increasingly legible, in turn, because of the feminist scholars and activists whose decades of work have enabled us to recognize that labor—emotional as much as physical—as such today.

As should already be apparent, feminism has been defined and used in many ways. Here and throughout the book, we employ the term *feminism* as a shorthand for the diverse and wide-ranging projects that name and challenge sexism and other forces of oppression, as well as those which seek to create more just, equitable, and livable futures. Because of this broadness, some scholars prefer to use the term *feminisms*, which clearly signals the range of—and, at times, the incompatibilities among—these various strains of feminist activism and political thought. For reasons of readability, we choose to use the term *feminism* here, but our feminism is intended to be just as expansive. It includes the work of regular folks like Darden and Champine, public intellectuals like Betty Friedan and bell hooks, and organizing groups like the Combahee River Collective, which have taken direct action to achieve the equality of the sexes. It also includes the work of scholars and other cultural critics—like Kimberlé Crenshaw and Margot Lee Shetterly, among many more—who have used writing to explore the social, political, historical, and conceptual reasons behind the inequality of the sexes that we face today.

In the process, these writers and activists have given voice to the many ways in which today's status quo is unjust.[12] These injustices are often the result of historical and contemporary differentials of power, including those among men, women, and nonbinary people, as well as those among white women and Black women, academic researchers and Indigenous communities, and people in the Global North and the Global South. Feminists analyze these power differentials so that they can change them. Such a broad focus—one that incorporates race, class, ability, and more—would have sounded strange to Friedan or to the white women largely credited for leading the fight for women's suffrage in the nineteenth century.[13] But the reality is that women of color have long insisted that any movement for gender equality must also consider the ways in which privilege and oppression are intersectional.

Because the concept of intersectionality is essential for this whole book, let's get a bit more specific. The term was coined by legal theorist Kimberlé Crenshaw in the late 1980s.[14] In law school, Crenshaw had come across the antidiscrimination case of *DeGraffenreid v. General Motors*. Emma DeGraffenreid was a Black working mother who had sought a job at a General Motors factory in her town. She was not hired and sued GM for discrimination. The factory did have a history of hiring Black people: many Black men worked in industrial and maintenance jobs there. They also had a history of hiring women: many white women worked there as secretaries. These two pieces of evidence provided the rationale for the judge to throw out the case. Because the company did hire Black people and did hire women, it could not be discriminating based on race or gender. But, Crenshaw wanted to know, what about discrimination on the basis of race *and* gender together? This was something different, it was real, and it needed to be named. Crenshaw not only named the concept, but would go on to explain and elaborate the idea of intersectionality in award-winning books, papers, and talks.[15]

Key to the idea of *intersectionality* is that it does not only describe the intersecting aspects of any particular person's identity (or *positionalities*, as they are sometimes termed).[16] It also describes the intersecting forces of privilege and oppression at work in a given society. *Oppression* involves the systematic mistreatment of certain groups of people by other groups. It happens when power is not distributed equally—when one group controls the institutions of law, education, and culture, and uses its power to systematically exclude other groups while giving its own group unfair advantages (or simply maintaining the status quo).[17] In the case of gender oppression, we can point to the sexism, cissexism, and patriarchy that is evident in everything from political representation to the wage gap to who speaks more often (or more loudly) in a meeting.[18] In the case of racial oppression, this takes the form of racism and white supremacy. Other forms of oppression include ableism, colonialism, and classism. Each has its particular history and manifests differently in different cultures and contexts, but all involve a dominant group that accrues power and privilege at the expense of others. Moreover, these forces of power and privilege on the one hand and oppression on the other mesh together in ways that multiply their effects.

The effects of privilege and oppression are not distributed evenly across all individuals and groups, however. For some, they become an obvious and unavoidable part of daily life, particularly for women and people of color and queer people and immigrants: the list goes on. If you are a member of any or all of these (or other) minoritized groups, you experience their effects everywhere, shaping the choices you make (or don't get to make) each day. These systems of power are as real as rain. But forces of oppression can be difficult to detect when you benefit from them (we call this a *privilege hazard* later in the book). And this is where data come in: it was a set of intersecting systems of power and privilege that Darden was intent on exposing when she posed her initial question to her division chief. And it was that same set of intersecting systems of power and privilege that Darden sought to challenge when she approached Champine. Darden herself didn't need any more evidence of the problem she faced; she was already living it every day.[19] But when her experience was recorded as data and aggregated with others' experiences, it could be used to challenge institutional systems of power and have far broader impact than on her career trajectory alone.

In this way, Darden models what we call *data feminism*: a way of thinking about data, both their uses and their limits, that is informed by direct experience, by a commitment to action, and by intersectional feminist thought. The starting point for data feminism is something that goes mostly unacknowledged in data science: power is not distributed equally in the world. Those who wield power are disproportionately elite, straight, white, able-bodied, cisgender men from the Global North.[20] The work of data feminism is first to tune into how standard practices in data science serve to reinforce these existing inequalities and second to use data science to challenge and change the distribution of power.[21] Underlying data feminism is a belief in and commitment to *co-liberation*: the idea that oppressive systems of power harm all of us, that they undermine the quality and validity of our work, and that they hinder us from creating true and lasting social impact with data science.

We wrote this book because we are data scientists and data feminists. Although we speak as a "we" in this book, and share certain identities, experiences, and skills, we have distinct life trajectories and motivations for our work on this project. If we were sitting with you right now, we would each introduce ourselves by answering the question: What brings you here today? Placing ourselves in that scenario, here is what we would have to say.

**Catherine:** I am a hacker mama. I spent fifteen years as a freelance software developer and experimental artist, now professor, working on projects ranging from serendipitous news-recommendation systems to countercartography to civic data literacy to making breast pumps not suck. I'm here writing this book because, for one, the hype around big data and AI is deafeningly male and white and technoheroic and the time is now to reframe that world with a feminist lens. The second reason I'm here is that my recent experience running a large, equity-focused hackathon taught me just how much people like me—basically, well-meaning liberal white people—are part of the problem in struggling for social justice. This book is one attempt to expose such workings of power, which are inside us as much as outside in the world.[22]

**Lauren:** I often describe myself as a professional nerd. I worked in software development before going to grad school to study English, with a particular focus on early American literature and culture. (Early means very early—like, the eighteenth century.) As a professor at an engineering school, I now work on research projects that translate this history into contemporary contexts. For instance, I'm writing a book about the history of data visualization, employing machine-learning techniques to analyze abolitionist newspapers, and designing a haptic recreation of a hundred-year-old visualization scheme that looks like a quilt. Through projects like these, I show how the rise of the concept of "data" (which, as it turns out, really took off in the eighteenth century) is closely connected to the rise of our current concepts of gender and race. So one of my reasons for writing this book is to show how the issues of racism and sexism that we see in data science today are by no means new. The other reason is to help translate humanistic thinking into practice and, in so doing, create more opportunities for humanities scholars to engage with activists, organizers, and communities.[23]

We both strongly believe that data can do good in the world. But for it to do so, we must explicitly acknowledge that a key way that power and privilege operate in the world today has to do with the word *data* itself. The word dates to the mid-seventeenth century, when it was introduced to supplement existing terms such as *evidence* and *fact*. Identifying information as data, rather than as either of those other two terms, served a rhetorical purpose.[24] It converted otherwise debatable information into the solid basis for subsequent claims. But what information needs to become data before it can be trusted? Or, more precisely, whose information needs to become data before it can be considered as fact and acted upon?[25] Data feminism must answer these questions, too.

The story that begins with Christine Darden entering the gates of Langley, passes through her sustained efforts to confront the structural oppression she encountered there, and concludes with her impressive array of life achievements, is a story about the power of data. Throughout her career, in ways large and small, Darden used data to make arguments and transform lives. But that's not all. Darden's feel-good biography is just as much a story about the larger systems of power that required data—rather than the belief in her lived experience—to perform that transformative work. An institutional mistrust of Darden's experiential knowledge was almost certainly a factor in Champine's decision to create her bar chart. Champine likely recognized, as did Darden herself, that she would need the bar chart to be believed.

In this way, the alliance between Darden and Champine, and their work together, underscores the flaws and compromises that are inherent in any data-driven project. The process of converting life experience into data always necessarily entails a reduction of that experience—along with the historical and conceptual burdens of the term. That Darden and Champine were able to view their work as a success despite these inherent constraints underscores even more the importance of listening to and learning from people whose lives and voices are behind the numbers. No dataset or analysis or visualization or model or algorithm is the result of one person working alone. Data feminism can help to remind us that before there are data, there are people—people who offer up their experience to be counted and analyzed, people who perform that counting and analysis, people who visualize the data and promote the findings of any particular project, and people who use the product in the end. There are also, always, people who go uncounted—for better or for worse. And there are problems that cannot be represented—or addressed—by data alone. And so data feminism, like justice, must remain both a goal and a process, one that guides our thoughts and our actions as we move forward toward our goal of remaking the world.

## Data and Power

It took five state-of-the-art IBM System/360 Model 75 machines to guide the *Apollo 11* astronauts to the moon. Each was the size of a car and cost $3.5 million dollars. Fast forward to the present. We now have computers in the form of phones that fit in our pockets and—in the case of the 2019 Apple iPhone XR—can perform more than 140 million more instructions per second than a standard IBM System/360.[26] That rate of change is astounding; it represents an exponential growth in computing capacity (figure 0.2a). We've witnessed an equally exponential growth in our ability to collect and record information in digital form—and in the ability to have information collected about us (figure 0.2b).

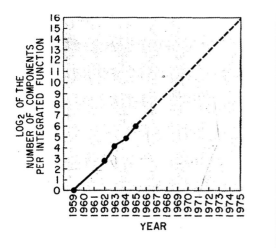

**Figure 0.2**
(a) The time-series chart included in the original paper on Moore's law, published in 1965, which posited that the number of transistors that could fit on an integrated circuit (and therefore contribute to computing capacity) would double every year. Courtesy of Gordon Moore. (b) Several years ago, researchers concluded that transistors were approaching their smallest size and that Moore's law would not hold. Nevertheless, today's computing power is what enabled Dr. Katie Bouman, a postdoctoral fellow at MIT, to contribute to a project that involved processing and compositing approximately five petabytes of data captured by the Event Horizon Telescope to create the first ever image of a black hole. After the publication of this photo in April 2019 showing her excitement—as one of the scientists on the large team that worked for years to capture the image—Bouman was subsequently trolled and harassed online. Courtesy of Tamy Emma Pepin/ Twitter.

But the act of collecting and recording data about people is not new at all. From the registers of the dead that were published by church officials in the early modern era to the counts of Indigenous populations that appeared in colonial accounts of the Americas, data collection has long been employed as a technique of consolidating knowledge about the people whose data are collected, and therefore consolidating power over their lives.[27] The close relationship between data and power is perhaps most clearly visible in the historical arc that begins with the logs of people captured and placed aboard slave ships, reducing richly lived lives to numbers and names. It passes through the eugenics movement, in the late nineteenth and early twentieth centuries, which sought to employ data to quantify the superiority of white people over all others. It continues today in the proliferation of biometrics technologies that, as sociologist Simone Browne has shown, are disproportionately deployed to surveil Black bodies.[28]

When Edward Snowden, the former US National Security Agency contractor, leaked his cache of classified documents to the press in 2013, he revealed the degree to which the federal government routinely collects data on its citizens—often with minimal regard to legality or ethics.[29] At the municipal level, too, governments are starting to collect data on everything from traffic movement to facial expressions in the interests of making cities "smarter."[30] This often translates to reinscribing traditional urban patterns of power such as segregation, the overpolicing of communities of color, and the rationing of ever-scarcer city services.[31]

But the government is not alone in these data-collection efforts; corporations do it too—with profit as their guide. The words and phrases we search for on Google, the times of day we are most active on Facebook, and the number of items we add to our Amazon carts are all tracked and stored as data—data that are then converted into corporate financial gain. The most trivial of everyday actions—searching for a way around traffic, liking a friend's cat video, or even stepping out of our front doors in the morning—are now hot commodities. This is not because any of these actions are exceptionally interesting (although we do make an exception for Catherine's cats) but because these tiny actions can be combined with other tiny actions to generate targeted advertisements and personalized recommendations—in other words, to give us more things to click on, like, or buy.[32]

This is the data economy, and corporations, often aided by academic researchers, are currently scrambling to see what behaviors—both online and off—remain to be turned into data and then monetized. Nothing is outside of *datafication*, as this process is sometimes termed—not your search history, or Catherine's cats, or the butt that Lauren is currently using to sit in her seat. To wit: Shigeomi Koshimizu, a Tokyo-based professor of engineering, has been designing matrices of sensors that collect data at 360 different positions around a rear end while it is comfortably ensconced in a chair.[33] He proposes that people have unique butt signatures, as unique as their fingerprints. In the future, he suggests, our cars could be outfitted with butt-scanners instead of keys or car alarms to identify the driver.

Although datafication may occasionally verge into the realm of the absurd, it remains a very serious issue. Decisions of civic, economic, and individual importance are already and increasingly being made by automated systems sifting through large amounts of data. For example, PredPol, a so-called predictive policing company founded in 2012 by an anthropology professor at the University of California, Los Angeles, has been employed by the City of Los Angeles for nearly a decade to determine which neighborhoods to patrol more heavily, and which neighborhoods to (mostly) ignore. But because PredPol is based on historical crime data and US policing practices have always disproportionately surveilled and patrolled neighborhoods of color, the predictions of where crime will happen in the future look a lot like the racist practices of the past.[34] These systems create what mathematician and writer Cathy O'Neil, in *Weapons of Math Destruction: How Big Data Increases Inequality and Threatens Democracy*, calls a "pernicious feedback loop," amplifying the effects of racial bias and of the criminalization of poverty that are already endemic to the United States.

O'Neil's solution is to open up the computational systems that produce these racist results. Only by knowing what goes in, she argues, can we understand what comes out. This is a key step in the project of mitigating the effects of biased data. Data feminism additionally requires that we trace those biased data back to their source. PredPol and the "three most objective data points" that it employs certainly amplify existing biases, but they are not the root cause.[35] The cause, rather, is the long history of the criminalization of Blackness in the United States, which produces biased policing practices, which produce biased historical data, which are then used to develop risk models for the future.[36] Tracing these links to historical and ongoing forces of oppression can help us answer the ethical question, Should this system exist?[37] In the case of PredPol, the answer is a resounding no.

Understanding this long and complicated chain reaction is what has motivated Yeshimabeit Milner, along with Boston-based activists, organizers, and mathematicians, to found Data for Black Lives, an organization dedicated to "using data science to create concrete and measurable change in the lives of Black communities."[38] Groups like the Stop LAPD Spying coalition are using explicitly feminist and antiracist methods to quantify and challenge invasive data collection by law enforcement.[39] Data journalists are reverse-engineering algorithms and collecting qualitative data at scale about maternal harm.[40] Artists are inviting participants to perform ecological maps and using AI for making intergenerational family memoirs (figure 0.3a).[41]

All these projects are *data science*. Many people think of data as numbers alone, but data can also consist of words or stories, colors or sounds, or any type of information that is systematically collected, organized, and analyzed (figures 0.3b, 0.3c).[42] The *science* in data science simply implies a commitment to systematic methods of observation and experiment. Throughout this book, we deliberately place diverse data science examples alongside each other. They come from individuals and small groups, and from across academic, artistic, nonprofit, journalistic, community-based, and for-profit organizations. This is due to our belief in a capacious definition of data science, one that seeks to include rather than exclude and does not erect barriers based on formal credentials, professional affiliation, size of data, complexity of technical methods, or other external markers of expertise. Such markers, after all, have long been used to prevent women from fully engaging in any number of professional fields, even as those fields—which include data science and computer science, among many others—were largely built on the knowledge that women were required to teach themselves.[43] An attempt to push back against this gendered history is foundational to data feminism, too.

Throughout its own history, feminism has consistently had to work to convince the world that it is relevant to people of all genders. We make the same argument: that data feminism is for everybody. (And here we borrow a line from bell hooks.)[44] You will notice that the examples we use are not only about women, nor are they created only by women. That's because *data feminism isn't only about women*. It takes more than one gender to have gender inequality and more than one gender to work toward justice. Likewise, *data feminism isn't only for women*. Men, nonbinary, and genderqueer people are proud to call themselves feminists and use feminist thought in their work. Moreover, *data feminism isn't only about gender*. Intersectional feminists have keyed us into how race, class, sexuality, ability, age, religion, geography, and more are factors that together influence each person's experience and opportunities in the world. Finally, *data feminism is about power—about who has it and who doesn't*. Intersectional feminism examines unequal power. And in our contemporary world, data is power too. Because the power of data is wielded unjustly, it must be challenged and changed.

## Data Feminism in Action

Data is a double-edged sword. In a very real sense, data have been used as a weapon by those in power to consolidate their control—over places and things, as well as

**Figure 0.3**
We define data science expansively in this book—here are three examples. (a) *Not the Only One* by Stephanie Dinkins (2017), is a sculpture that features a Black family through the use of artificial intelligence. The AI is trained and taught by the underrepresented voices of Black and brown individuals in the tech sector. (b) Researcher Margaret Mitchell and colleagues, in "Seeing through the Human Reporting Bias" (2016), have worked on systems to infer what is *not said* in human speech for the purposes of image classification. For example, people say "green bananas" but not "yellow bananas" because yellow is implied as the default color of the banana. Similarly, people say "woman doctor" but do not say "man doctor," so it is the words that are *not spoken* that encode the bias. (c) A gender analysis of Hollywood film dialogue, "Film Dialogue from 2,000 Screenplays Broken Down by Gender and Age," by Hanah Anderson and Matt Daniels, created for The Pudding, a data journalism start-up (2017).

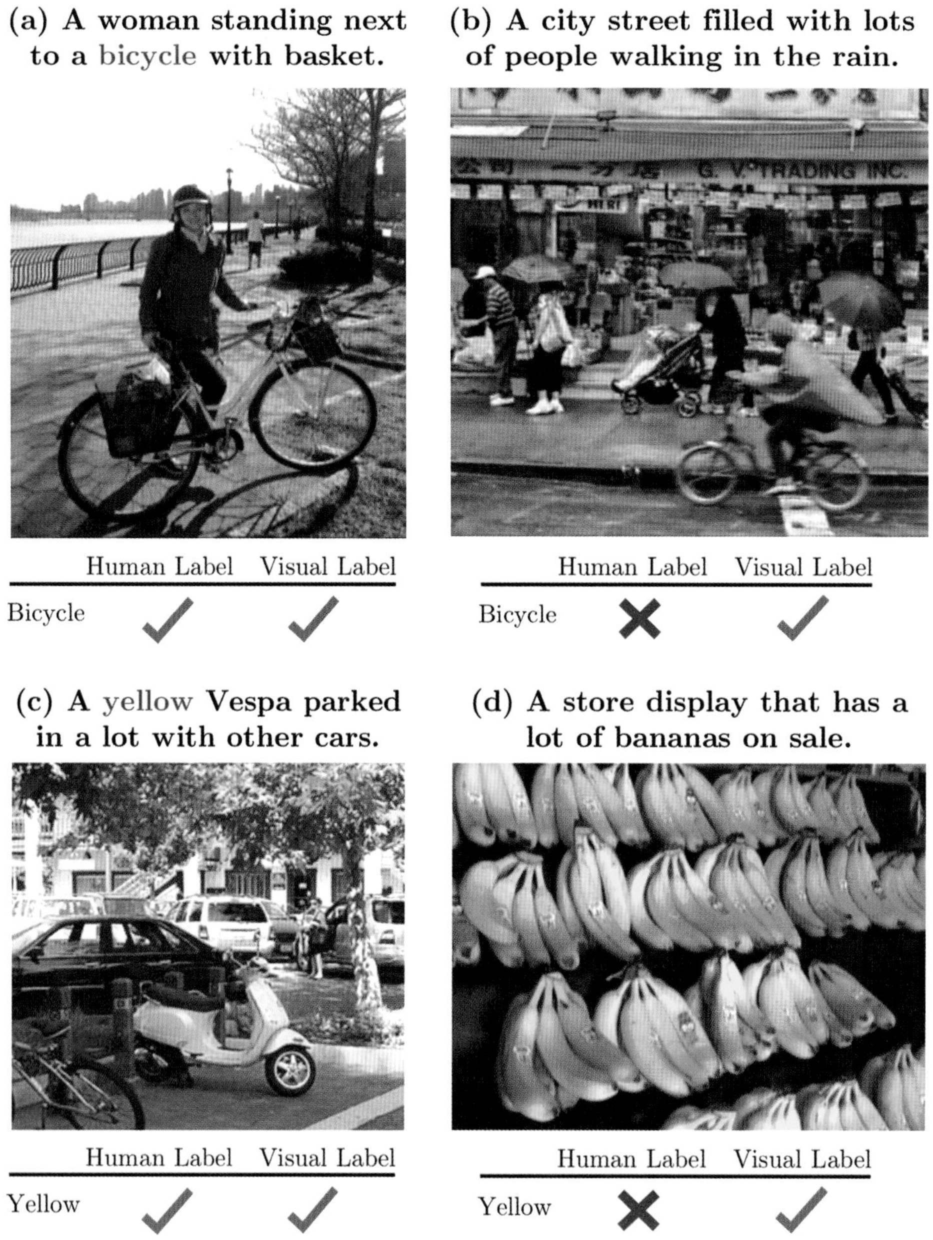

(a) A woman standing next
to a bicycle with basket.

| | Human Label | Visual Label |
|---|---|---|
| Bicycle | ✓ | ✓ |

(b) A city street filled with lots
of people walking in the rain.

| | Human Label | Visual Label |
|---|---|---|
| Bicycle | ✗ | ✓ |

(c) A yellow Vespa parked
in a lot with other cars.

| | Human Label | Visual Label |
|---|---|---|
| Yellow | ✓ | ✓ |

(d) A store display that has a
lot of bananas on sale.

| | Human Label | Visual Label |
|---|---|---|
| Yellow | ✗ | ✓ |

**Figure 0.3** (continued)

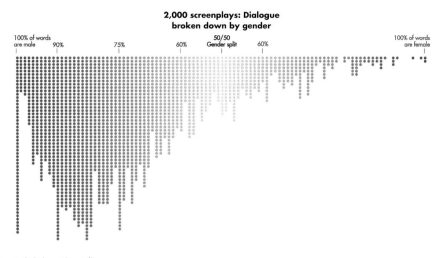

**Figure 0.3** (continued)

people. Indeed, a central goal of this book is to show how governments and corporations have long employed data and statistics as management techniques to preserve an unequal status quo. Working with data from a feminist perspective requires knowing and acknowledging this history. To frame the trouble with data in another way: it's not a coincidence that the institution that employed Christine Darden and enabled her professional rise is the same that wielded the results of her data analysis to assert the technological superiority of the United States over its communist adversaries and to plant an American flag on the moon. But this flawed history does not mean ceding control of the future to the powers of the past. Data are part of the problem, to be sure. But they are also part of the solution. Another central goal of this book is to show how the power of data can be wielded back.

To guide us in this work, we have developed seven core principles. Individually and together, these principles emerge from the foundation of intersectional feminist thought. Each of the following chapters is structured around a single principle. The seven principles of data feminism are as follows:

1. **Examine power.** Data feminism begins by analyzing how power operates in the world.

2. **Challenge power.** Data feminism commits to challenging unequal power structures and working toward justice.

3. **Elevate emotion and embodiment.** Data feminism teaches us to value multiple forms of knowledge, including the knowledge that comes from people as living, feeling bodies in the world.

4. **Rethink binaries and hierarchies.** Data feminism requires us to challenge the gender binary, along with other systems of counting and classification that perpetuate oppression.

5. **Embrace pluralism.** Data feminism insists that the most complete knowledge comes from synthesizing multiple perspectives, with priority given to local, Indigenous, and experiential ways of knowing.

6. **Consider context.** Data feminism asserts that data are not neutral or objective. They are the products of unequal social relations, and this context is essential for conducting accurate, ethical analysis.

7. **Make labor visible.** The work of data science, like all work in the world, is the work of many hands. Data feminism makes this labor visible so that it can be recognized and valued.

Each of the following chapters takes up one of these principles, drawing upon examples from the field of data science, expansively defined, to show how that principle can be put into action. Along the way, we introduce key feminist concepts like the matrix of domination (Patricia Hill Collins; see chapter 1), situated knowledge (Donna Haraway; see chapter 3), and emotional labor (Arlie Hochschild; see chapter 8), as well as some of our own ideas about what data feminism looks like in theory and practice. To this end, we introduce you to people at the cutting edge of data and justice. These include engineers and software developers, activists and community organizers, data journalists, artists, and scholars. This range of people, and the range of projects they have helped to create, is our way of answering the question: What makes a project feminist? As will become clear, a project may be feminist in *content*, in that it challenges power by choice of subject matter; in *form*, in that it challenges power by shifting the aesthetic and/or sensory registers of data communication; and/or in *process*, in that it challenges power by building participatory, inclusive processes of knowledge production. What unites this broad scope of data-based work is a commitment to action and a desire to remake the world.

Our overarching goal is to take a stand against the status quo—against a world that benefits us, two white college professors, at the expense of others. To work toward this goal, we have chosen to feature the voices of those who speak from the margins, whether because of their gender, sexuality, race, ability, class, geographic location, or any combination of those (and other) subject positions. We have done so, moreover, because of our belief that those with direct experience of inequality know better than we do about what actions to take next. For this reason, we have attempted to prioritize the work of people in closer proximity to issues of inequality over those who study inequality from a distance. In this book, we pay particular attention to inequalities at the intersection of gender and race. This reflects our location in the United States, where the most entrenched issues of inequality have racism at their source. Our values statement, included as an appendix to this book, discusses the rationale for these authorial choices in more detail.

Any book involves making choices about whose voices and whose work to include and whose voices and work to omit. We ask that those who find their perspectives insufficiently addressed or their work insufficiently acknowledged view these gaps as additional openings for conversation. Our sincere hope is to contribute in a small way to a much larger conversation, one that began long before we embarked upon this writing process and that will continue long after these pages are through.

This book is intended to provide concrete steps to action for data scientists seeking to learn how feminism can help them work toward justice, and for feminists seeking to learn how their own work can carry over to the growing field of data science. It is also addressed to professionals in all fields in which data-driven decisions are being made, as well as to communities that want to resist or mobilize the data that surrounds them. It is written for everyone who seeks to better understand the charts and statistics that they encounter in their day-to-day lives, and for everyone who seeks to communicate the significance of such charts and statistics to others.

Our claim, once again, is that data feminism is for everyone. It's for people of all genders. It's by people of all genders. And most importantly: it's about much more than gender. Data feminism is about power, about who has it and who doesn't, and about how those differentials of power can be challenged and changed using data. We invite you, the readers of this book, to join us on this journey toward justice and toward remaking our data-driven world.

# 1  The Power Chapter

**Principle: Examine Power**

*Data feminism begins by analyzing how power operates in the world.*

When tennis star Serena Williams disappeared from Instagram in early September 2017, her six million followers assumed they knew what had happened. Several months earlier, in March of that year, Williams had accidentally announced her pregnancy to the world via a bathing suit selfie and a caption that was hard to misinterpret: "20 weeks." Now, they thought, her baby had finally arrived.

But then they waited, and waited some more. Two weeks later, Williams finally reappeared, announcing the birth of her daughter and inviting her followers to watch a video that welcomed Alexis Olympia Ohanian Jr. to the world.[1] The video was a montage of baby bump pics interspersed with clips of a pregnant Williams playing tennis and having cute conversations with her husband, Reddit cofounder Alexis Ohanian, and then, finally, the shot that her fans had been waiting for: the first clip of baby Olympia. Williams was narrating: "So we're leaving the hospital," she explains. "It's been a long time. We had a lot of complications. But look who we got!" The scene fades to white, and the video ends with a set of stats: Olympia's date of birth, birth weight, and number of grand slam titles: 1. (Williams, as it turned out, was already eight weeks pregnant when she won the Australian Open earlier that year.)

Williams's Instagram followers were, for the most part, enchanted. But soon, the enthusiastic congratulations were superseded by a very different conversation. A number of her followers—many of them Black women like Williams herself—fixated on the comment she'd made as she was heading home from the hospital with her baby girl. Those "complications" that Williams experienced—other women had had them too. In Williams's case, the complications had been life-threatening, and her self-advocacy in the hospital played a major role in her survival.

On Williams's Instagram feed, dozens of women began posting their own experiences of childbirth gone horribly wrong. A few months later, Williams returned to social media—Facebook, this time—to continue the conversation (figure 1.1). Citing a 2017 statement from the US Centers for Disease Control and Prevention (CDC), Williams wrote that "Black women are over 3 times more likely than white women to die from pregnancy- or childbirth-related causes."[2]

These disparities were already well-known to Black-women-led reproductive justice groups like SisterSong, the Black Mamas Matter Alliance, and Raising Our Sisters Everywhere (ROSE), some of whom had been working on the maternal health crisis for decades. Williams helped to shine a national spotlight on them. The mainstream media also recently had begun to pay more attention to the crisis as well. A few months earlier, Nina Martin of the investigative journalism outfit ProPublica, working with Renee Montagne of NPR, had reported on the same phenomenon.[3] "Nothing Protects Black Women from Dying in Pregnancy and Childbirth," the headline read. In addition to the study cited by Williams, Martin and Montagne cited a second study from 2016, which showed that neither education nor income level—the

Serena Williams ✓
January 15, 2018 · Facebook Creator · 🌐

I didn't expect that sharing our family's story of Olympia's birth and all of complications after giving birth would start such an outpouring of discussion from women — especially black women — who have faced similar complications and women whose problems go unaddressed.

These aren't just stories: according to the CDC, (Center for Disease Control) black women are over 3 times more likely than White women to die from pregnancy- or childbirth-related causes. We have a lot of work to do as a nation and I hope my story can inspire a conversation that gets us to close this gap.

Let me be clear: EVERY mother, regardless of race, or background deserves to have a healthy pregnancy and childbirth. I personally want all women of all colors to have the best experience they can have. My personal experience was not great but it was MY experience and I'm happy it happened to me. It made me stronger and it made me appreciate women -- both women with and without kids -- even more. We are powerful!!!

I want to thank all of you who have opened up through online comments and other platforms to tell your story. I encourage you to continue to tell those stories. This helps. We can help others. Our voices are our power.

👍😮❤ 27K                                  1.6K Comments  1.4K Shares  840K Views

↪ Share

**Figure 1.1**
A Facebook post by Serena Williams responding to her Instagram followers who had shared their stories of pregnancy and childbirth-related complications with her. Image from Serena Williams, January 15, 2018.

factors usually invoked when attempting to account for healthcare outcomes that diverge along racial lines—impacted the fates of Black women giving birth.[4] On the contrary, the data showed that Black women with college degrees suffered more severe complications of pregnancy and childbirth than white women without high school diplomas.

So what were these complications, more precisely? And how many women had actually died as a result? Nobody was counting. A 2014 United Nations report, coauthored by SisterSong, described the state of data collection on maternal mortality in the United States as "particularly weak."[5] The situation hadn't improved in 2017, when ProPublica began its reporting. In 2018, *USA Today* investigated these racial disparities, and found what was an even more fundamental problem: there was still no national system for tracking complications sustained in pregnancy and childbirth, even though similar systems had long been in place for tracking any number of other health issues, such as teen pregnancy, hip replacements, or heart attacks.[6] They also found that there was still no reporting mechanism for ensuring that hospitals follow national safety standards, as is required for both hip surgery and cardiac care. "Our maternal data is embarrassing," stated Stacie Geller, a professor of obstetrics and gynecology at the University of Illinois, when asked for comment. The chief of the CDC's Maternal and Infant Health branch, William Callaghan, makes the significance of this "embarrassing" data more clear: "What we choose to measure is a statement of what we value in health," he explains.[7] We might edit his statement to add that it's a measure of *who* we value in health, too.[8]

Why did it take the near-death of an international sports superstar for the media to begin paying attention to an issue that less famous Black women had been experiencing and organizing around for decades? Why did it take reporting by the predominantly white mainstream press for US cities and states to begin collecting data on the issue?[9] Why are those data still not viewed as big enough, statistically significant enough, or of high enough quality for those cities and states, and other public institutions, to justify taking action? And why didn't those institutions just #believeblackwomen in the first place?[10]

The answers to these questions are directly connected to larger issues of power and privilege. Williams recognized as much when asked by *Glamour* magazine about the fact that she had to demand that her medical team perform additional tests in order to diagnose her own postnatal complications—and because she was Serena Williams, twenty-three-time grand slam champion, they complied.[11] "If I wasn't who I am, it could have been me," she told *Glamour*, referring to the fact that the privilege she experienced as a tennis star intersected with the oppression she experienced as a Black woman, enabling her to avoid becoming a statistic herself. As Williams asserted, "that's not fair."[12]

Needless to say, Williams is right. It's absolutely not fair. So how do we mitigate this unfairness? We begin by examining systems of power and how they intersect—like how the influences of racism, sexism, and celebrity came together first to send Williams into a medical crisis and then, thankfully, to keep her alive. The complexity of these intersections is the reason that *examine power* is the first principle of data feminism, and the focus of this chapter. Examining power means naming and explaining the forces of oppression that are so baked into our daily lives—and into our datasets, our databases, and our algorithms—that we often don't even see them. Seeing oppression is especially hard for those of us who occupy positions of privilege. But once we identify these forces and begin to understand how they exert their potent force, then many of the additional principles of data feminism—like challenging power (chapter 2), embracing emotion (chapter 3), and making labor visible (chapter 7)—become easier to undertake.

## Power and the Matrix of Domination

But first, what do we mean by *power*? We use the term *power* to describe the current configuration of structural privilege and structural oppression, in which some groups experience unearned advantages—because various systems have been designed by people like them and work for people like them—and other groups that experience systematic disadvantages—because those same systems were not designed by them or with people like them in mind. These mechanisms are complicated, and there are "few pure victims and oppressors," notes influential sociologist Patricia Hill Collins. In her landmark text, *Black Feminist Thought*, first published in 1990, Collins proposes the concept of the *matrix of domination* to explain how systems of power are configured and experienced.[13] It consists of four domains: the structural, the disciplinary, the hegemonic, and the interpersonal. Her emphasis is on the intersection of gender and race, but she makes clear that other dimensions of identity (sexuality, geography, ability, etc.) also result in unjust oppression, or unearned privilege, that become apparent across the same four domains.

The *structural domain* is the arena of laws and policies, along with schools and institutions that implement them. This domain organizes and codifies oppression. Take, for example, the history of voting rights in the United States. The US Constitution did not originally specify who was authorized to vote, so various states had different policies that reflected their local politics. Most had to do with owning property, which,

**Table 1.1**

The four domains of the matrix of domination

| Structural domain | Disciplinary domain |
|---|---|
| Organizes oppression: laws and policies. | Administers and manages oppression. Implements and enforces laws and policies. |
| **Hegemonic domain** | **Interpersonal domain** |
| Circulates oppressive ideas: culture and media. | Individual experiences of oppression. |

Chart based on concepts introduced by Patricia Hill Collins in *Black Feminist Thought: Knowledge, Consciousness, and the Politics of Empowerment.*

conveniently, only men could do. But with the passage of the Fourteenth Amendment in 1868, which granted the rights of US citizenship to those who had been enslaved, the nature of those rights—including voting—were required to be spelled out at the national level for the first time. More specifically, voting was defined as a right reserved for "male citizens." This is a clear instance of codified oppression in the structural domain.

It would take until the passage of the Nineteenth Amendment in 1920 for most (but not all) women to be granted the right to vote.[14] Even still, many state voting laws continued to include literacy tests, residency requirements, and other ways to indirectly exclude people who were not property-owning white men. These restrictions persist today, in the form of practices like dropping names from voter rolls, requiring photo IDs, and limits to early voting—the burdens of which are felt disproportionately by low-income people, people of color, and others who lack the time or resources to jump through these additional bureaucratic hoops.[15] This is the *disciplinary domain* that Collins names: the domain that administers and manages oppression through bureaucracy and hierarchy, rather than through laws that explicitly encode inequality on the basis of someone's identity.[16]

Neither of these domains would be possible without *the hegemonic domain*, which deals with the realm of culture, media, and ideas. Discriminatory policies and practices in voting can only be enacted in a world that already circulates oppressive ideas about, for example, who counts as a citizen in the first place. Consider an antisuffragist pamphlet from the 1910s that proclaims, "You do not need a ballot to clean out your sink spout."[17] Pamphlets like these, designed to be literally passed from hand to hand, reinforced preexisting societal views about the place of women in society. Today, we have animated GIFs instead of paper pamphlets, but the hegemonic function is the same: to consolidate ideas about who is entitled to exercise power and who is not.

The final part of the matrix of domination is the *interpersonal domain*, which influ-ences the everyday experience of individuals in the world. How would you feel if you were a woman who read that pamphlet, for example? Would it have more or less of an impact if a male family member gave it to you? Or, for a more recent example, how would you feel if you took time off from your hourly job to go cast your vote, only to discover when you got there that your name had been purged from the official voting roll or that there was a line so long that it would require that you miss half a day's pay, or stand for hours in the cold, or ... the list could go on. These are examples of how it *feels* to know that systems of power are not on your side and, at times, are actively seek-ing to take away the small amount of power that you do possess.[18]

The matrix of domination works to uphold the undue privilege of *dominant* groups while unfairly oppressing *minoritized* groups. What does this mean? Beginning in this chapter and continuing throughout the book, we use the term *minoritized* to describe groups of people who are positioned in opposition to a more powerful social group. While the term *minority* describes a social group that is comprised of fewer people, *minoritized* indicates that a social group is actively devalued and oppressed by a domi-nant group, one that holds more economic, social, and political power. With respect to gender, for example, men constitute the dominant group, while all other genders constitute minoritized groups. This remains true even as women actually constitute a majority of the world population. *Sexism* is the term that names this form of oppres-sion. In relation to race, white people constitute the dominant group (racism); in rela-tion to class, wealthy and educated people constitute the dominant group (classism); and so on.[19]

Using the concept of the matrix of domination and the distinction between domi-nant and minoritized groups, we can begin to examine how power unfolds in and around data. This often means asking uncomfortable questions: who is doing the work of data science (and who is not)? Whose goals are prioritized in data science (and whose are not)? And who benefits from data science (and who is either overlooked or actively harmed)?[20] These questions are uncomfortable because they unmask the inconvenient truth that there are groups of people who are disproportionately benefitting from data science, and there are groups of people who are disproportionately harmed. Asking these *who questions* allows us, as data scientists ourselves, to start to see how privilege is baked into our data practices and our data products.[21]

## Data Science by Whom?

It is important to acknowledge the elephant in the server room: the demograph-ics of data science (and related occupations like software engineering and artificial intelligence research) do not represent the population as a whole. According to the most recent data from the US Bureau of Labor Statistics, released in 2018, only 26 percent of those in "computer and mathematical occupations" are women.[22] And across all of those women, only 12 percent are Black or Latinx women, even though Black and Latinx women make up 22.5 percent of the US population.[23] A report by the research group AI Now about the diversity crisis in artificial intelligence notes that women comprise only 15 percent of AI research staff at Facebook and 10 percent at Google.[24] These numbers are probably not a surprise. The more surprising thing is that those numbers are getting worse, not better. According to a research report published by the American Association of University Women in 2015, women computer science graduates in the United States peaked in the mid-1980s at 37 percent, and we have seen a steady decline in the years since then to 26 percent today (figure 1.2).[25] As "data analysts" (low-status number crunchers) have become rebranded as "data scientists" (high status

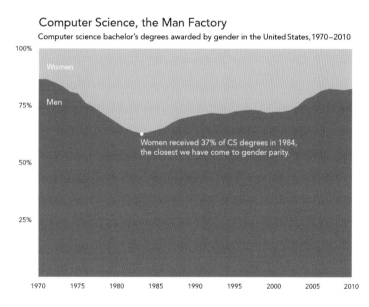

**Figure 1.2**
Computer science has always been dominated by men and the situation is worsening (even while many other scientific and technical fields have made significant strides toward gender parity). Women awarded bachelor's degrees in computer science in the United States peaked in the mid-1980s at 37 percent, and we have seen a steady increase in the ratio of men to women in the years since then. This particular report treated gender as a binary, so there was no data about nonbinary people. Graphic by Catherine D'Ignazio. Data from the National Center for Education Statistics.

researchers), women are being pushed out in order to make room for more highly valued and more highly compensated men.[26]

There are not disparities only along gender lines in the higher education pipeline. The same report noted specific underrepresentation for Native American women, multiracial women, white women, and all Black and Latinx people. So is it really a surprise that each day brings a new example of data science being used to disempower and oppress minoritized groups? In 2018, it was revealed that Amazon had been developing an algorithm to screen its first-round job applicants. But because the model had been trained on the resumes of prior applicants, who were predominantly male, it developed an even stronger preference for male applicants. It downgraded resumes with the word *women* and graduates of women's colleges. Ultimately, Amazon had to cancel the project.[27] This example reinforces the work of Safiya Umoja Noble, whose book, *Algorithms of Oppression*, has shown how both gender and racial biases are encoded into some of the most pervasive data-driven systems—including Google search, which boasts over five billion unique web searches per day. Noble describes how, as recently as 2016, comparable searches for "three Black teenagers" and "three white teenagers" turned up wildly different representations of those teens. The former returned mugshots, while the latter returned wholesome stock photography.[28]

The problems of gender and racial bias in our information systems are complex, but some of their key causes are plain as day: the data that shape them, and the models designed to put those data to use, are created by small groups of people and then scaled up to users around the globe. But those small groups are not at all representative of the globe as a whole, nor even of a single city in the United States. When data teams are primarily composed of people from dominant groups, those perspectives come to exert outsized influence on the decisions being made—to the exclusion of other identities and perspectives. This is not usually intentional; it comes from the ignorance of being on top. We describe this deficiency as a *privilege hazard*.

How does this come to pass? Let's take a minute to imagine what life is like for someone who epitomizes the dominant group in data science: a straight, white, cisgender man with formal technical credentials who lives in the United States. When he looks for a home or applies for a credit card, people are eager for his business. People smile when he holds his girlfriend's hand in public. His body doesn't change due to childbirth or breastfeeding, so he does not need to think about workplace accommodations. He presents his social security number in jobs as a formality, but it never hinders his application from being processed or brings him unwanted attention. The ease with which he traverses the world is invisible to him because it has been designed for people just like him. He does not think about how life might be different for everyone else. In fact, it is difficult for him to imagine that at all.

This is the *privilege hazard*: the phenomenon that makes those who occupy the most privileged positions among us—those with good educations, respected credentials, and professional accolades—so poorly equipped to recognize instances of oppression in the world.[29] They lack what Anita Gurumurthy, executive director of IT for Change, has called "the empiricism of lived experience."[30] And this lack of lived experience—this evidence of how things truly *are*—profoundly limits their ability to foresee and prevent harm, to identify existing problems in the world, and to imagine possible solutions.

The privilege hazard occurs at the level of the individual—in the interpersonal domain of the matrix of domination—but it is much more harmful in aggregate because it reaches the hegemonic, disciplinary and structural domains as well. So it matters deeply that data science and artificial intelligence are dominated by elite white men because it means there is a collective privilege hazard so great that it would be a profound surprise if they could actually identify instances of bias prior to unleashing them onto the world. Social scientist Kate Crawford has advanced the idea that the biggest threat from artificial intelligence systems is not that they will become smarter than humans, but rather that they will hard-code sexism, racism, and other forms of discrimination into the digital infrastructure of our societies.[31]

What's more, the same cis het white men responsible for designing those systems lack the ability to detect harms and biases in their systems once they've been released into the world.[32] In the case of the "three teenagers" Google searches, for example, it was a young Black teenager that pointed out the problem and a Black scholar who wrote about the problem. The burden consistently falls upon those more intimately familiar with the privilege hazard—in data science as in life—to call out the creators of those systems for their limitations.

For example, Joy Buolamwini, a Ghanaian-American graduate student at MIT, was working on a class project using facial-analysis software.[33] But there was a problem—the software couldn't "see" Buolamwini's dark-skinned face (where "seeing" means that it detected a face in the image, like when a phone camera draws a square around a person's face in the frame). It had no problem seeing her lighter-skinned collaborators. She tried drawing a face on her hand and putting it in front of the camera; it detected that. Finally, Buolamwini put on a white mask, essentially going in "whiteface" (figure 1.3).[34] The system detected the mask's facial features perfectly.

Digging deeper into the code and benchmarking data behind these systems, Buolamwini discovered that the dataset on which many of facial-recognition algorithms are tested contains 78 percent male faces and 84 percent white faces. When she did

**Figure 1.3**
Joy Buolamwini found that she had to put on a white mask for the facial detection program to "see" her face. Buolamwini is now founder of the Algorithmic Justice League. Courtesy of Joy Buolamwini.

an intersectional breakdown of another test dataset—looking at gender and skin type together—only 4 percent of the faces in that dataset were women and dark-skinned. In their evaluation of three commercial systems, Buolamwini and computer scientist Timnit Gebru showed that darker-skinned women were up to forty-four times more likely to be misclassified than lighter-skinned males.[35] It's no wonder that the software failed to detect Buolamwini's face: both the training data and the benchmarking data relegate women of color to a tiny fraction of the overall dataset.[36]

This is the privilege hazard in action—that no coder, tester, or user of the software had previously identified such a problem or even thought to look. Buolamwini's work has been widely covered by the national media (by the *New York Times*, by CNN, by the *Economist*, by *Bloomberg BusinessWeek*, and others) in articles that typically contain a hint of shock.[37] This is a testament to the social, political, and technical importance of the work, as well as to how those in positions of power—not just in the field of data science, but in the mainstream media, in elected government, and at the heads of corporations—are so often surprised to learn that their "intelligent technologies" are not so intelligent after all. (They need to read data journalist Meredith Broussard's book *Artificial Unintelligence*).[38] For another example, think back to the introduction of this book, where we quoted Shetterly as reporting that Christine Darden's white male manager was "shocked at the disparity" between the promotion rates of men and women. We can speculate that Darden herself wasn't shocked, just as Buolamwini and Gebru likely were not entirely shocked at the outcome of their study either. When sexism, racism, and other forms of oppression are publicly unmasked, it is almost never surprising to those who experience them.

For people in positions of power and privilege, issues of race and gender and class and ability—to name only a few—are OPP: other people's problems. Author and antiracist educator Robin DiAngelo describes instances like the "shock" of Darden's boss or the surprise in the media coverage of Buolamwini's various projects as a symptom of the "racial innocence" of white people.[39] In other words, those who occupy positions of privilege in society are able to remain innocent of that privilege. Race becomes something that only people of color have. Gender becomes something that only women and nonbinary people have. Sexual orientation becomes something that all people *except* heterosexual people have. And so on. A personal anecdote might help illustrate this point. When we published the first draft of this book online, Catherine told a colleague about it. His earnestly enthusiastic response was, "Oh great! I'll show it to my female graduate students!" To which Catherine rejoined, "You might want to show it to your other students, too."

If things were different—if the 79 percent of engineers at Google who are male were specifically trained in structural oppression before building their data systems (as social workers are before they undertake social work)—then their overrepresentation might be very slightly less of a problem.[40] But in the meantime, the onus falls on the individuals who already feel the adverse effects of those systems of power to prove, over and over again, that racism and sexism exist—in datasets, in data systems, and in data science, as in everywhere else.

Buolamwini and Gebru identified how pale and male faces were overrepresented in facial detection training data. Could we just fix this problem by diversifying the data set? One solution to the problem would appear to be straightforward: create a more representative set of training and benchmarking data for facial detection models. In fact, tech companies are starting to do exactly this. In January 2019, IBM released a database of one million faces called Diversity in Faces (DiF).[41] In another example, journalist Amy Hawkins details how CloudWalk, a startup in China in need of more images of faces of people of African descent, signed a deal with the Zimbabwean government for it to provide the images the company was lacking.[42] In return for sharing its data, Zimbabwe will receive a national facial database and "smart" surveillance infrastructure that it can install in airports, railways, and bus stations.

It might sound like an even exchange, but Zimbabwe has a dismal record on human rights. Making things worse, CloudWalk provides facial recognition technologies to the Chinese police—a conflict of interest so great that the global nonprofit Human Rights Watch voiced its concern about the deal.[43] Face harvesting is happening in the US as well. Researchers Os Keyes, Nikki Stevens and Jacqueline Wernimont have shown how immigrants, abused children, and dead people are some of the groups whose faces have been used to train software—without their consent.[44] So is a diverse database of faces really a good idea? Voicing his concerns in response to the announcement of Buolamwini and Gebru's 2018 study on Twitter, an Indigenous Marine veteran shot back, "I hope facial recognition software has a problem identifying my face too. That'd come in handy when the police come rolling around with their facial recognition truck at peaceful demonstrations of dissent, cataloging all dissenters for 'safety and security.'"[45]

Better detection of faces of color cannot be characterized as an unqualified good. More often than not, it is enlisted in the service of increased oppression, greater surveillance, and targeted violence. Buolamwini understands these potential harms and has developed an approach that works across all four domains of the matrix of domination to address the underlying issues of power that are playing out in facial analysis technology. Buolamwini and Gebru first quantified the disparities in the dataset—a technical audit, which falls in the disciplinary domain of the matrix of domination. Then, Buolamwini went on to launch the Algorithmic Justice League, an organization that works to highlight and intervene in instances of algorithmic bias. On behalf of the AJL, Buolamwini has produced viral poetry projects and given TED talks—taking action in the hegemonic domain, the realm of culture and ideas. She has advised on legislation and professional standards for the field of computer vision and called for a moratorium on facial analysis in policing on national media and in Congress.[46] These are actions operating in the structural domain of the matrix of domination—the realm of law and policy. Throughout these efforts, the AJL works with students and researchers to help guide and shape their own work—the interpersonal domain. Taken together, Buolamwini's various initiatives demonstrate how any "solution" to bias in algorithms and datasets must tackle more than technical limitations. In addition, they present a compelling model for the data scientist as public intellectual—who, yes, works on technical audits and fixes, but also works on cultural, legal, and political efforts too.

While equitable representation—in datasets and data science workforces—is important, it remains window dressing if we don't also transform the institutions that produce and reproduce those biased outcomes in the first place. As doctoral health student

Arrianna Planey, quoting Robert M. Young, states, "A racist society will give you a racist science."[47] We cannot filter out the downstream effects of sexism and racism without also addressing their root cause.

## Data Science for Whom?

One of the downstream effects of the privilege hazard—the risks incurred when people from dominant groups create most of our data products—is not only that datasets are biased or unrepresentative, but that they never get collected at all. Mimi Onuoha—an artist, designer, and educator—has long been asking *who questions* about data science. Her project, *The Library of Missing Datasets* (figure 1.4), is a list of datasets that one might expect to already exist in the world, because they help to address pressing social issues, but that in reality have never been created. The project exists as a website and as an art object. The latter consists of a file cabinet filled with folders labeled

**Figure 1.4**
*The Library of Missing Datasets*, by Mimi Onuoha (2016) is a list of datasets that are not collected because of bias, lack of social and political will, and structural disregard. Courtesy of Mimi Onuoha. Photo by Brandon Schulman.

with phrases like: "People excluded from public housing because of criminal records," "Mobility for older adults with physical disabilities or cognitive impairments," and "Total number of local and state police departments using stingray phone trackers (IMSI-catchers)." Visitors can tab through the folders and remove any particular folder of interest, only to reveal that it is empty. They all are. The datasets that should be there are "missing."

By compiling a list of the datasets that are missing from our "otherwise data-saturated" world, Onuoha explains, "we find cultural and colloquial hints of what is deemed important" and what is not. "Spots that we've left blank reveal our hidden social biases and indifferences," she continues. And by calling attention to these datasets as "missing," she also calls attention to how the matrix of domination encodes these "social biases and indifferences" across all levels of society.[48] Along similar lines, foundations like Data2X and books like *Invisible Women* have advanced the idea of a systematic "gender data gap" due to the fact that the majority of research data in scientific studies is based around men's bodies. The downstream effects of the gender data gap range from annoying—cell phones slightly too large for women's hands, for example—to fatal. Until recently, crash test dummies were designed in the size and shape of men, an oversight that meant that women had a 47 percent higher chance of car injury than men.[49]

The *who question* in this case is: Who benefits from data science and who is overlooked? Examining those gaps can sometimes mean calling out missing datasets, as Onuoha does; characterizing them, as *Invisible Women* does; and advocating for filling them, as Data2X does. At other times, it can mean collecting the missing data yourself. Lacking comprehensive data about women who die in childbirth, for example, ProPublica decided to resort to crowdsourcing to learn the names of the estimated seven hundred to nine hundred US women who died in 2016.[50] As of 2019, they've identified only 140. Or, for another example: in 1998, youth living in Roxbury—a neighborhood known as "the heart of Black culture in Boston"[51]—were sick and tired of inhaling polluted air. They led a march demanding clean air and better data collection, which led to the creation of the AirBeat community monitoring project.[52]

Scholars have proposed various names for these instances of ground-up data collection, including *counterdata* or *agonistic data* collection, *data activism*, *statactivism*, and *citizen science* (when in the service of environmental justice).[53] Whatever it's called, it's been going on for a long time. In 1895, civil rights activist and pioneering data journalist Ida B. Wells assembled a set of statistics on the epidemic of lynching that was sweeping the United States.[54] She accompanied her data with a meticulous exposé of the fraudulent claims made by white people—typically, that a rape, theft, or assault of some kind had occurred (which it hadn't in most cases) and that lynching was a justified response. Today, an organization named after Wells—the Ida B. Wells Society for Investigative Reporting—continues her mission by training up a new generation of journalists of color in the skills of data collection and analysis.[55]

A counterdata initiative in the spirit of Wells is taking place just south of the US border, in Mexico, where a single woman is compiling a comprehensive dataset on femicides—gender-related killings of women and girls.[56] María Salguero, who also goes by the name Princesa, has logged more than five thousand cases of femicide since 2016.[57] Her work provides the most accessible information on the subject for journalists, activists, and victims' families seeking justice.

The issue of femicide in Mexico rose to global visibility in the mid-2000s with widespread media coverage about the deaths of poor and working-class women in Ciudad Juárez. A border town, Juárez is the site of more than three hundred *maquiladoras*: factories that employ women to assemble goods and electronics, often for low wages and in substandard working conditions. Between 1993 and 2005, nearly four hundred of these women were murdered, with around a third of those murders exhibiting signs of exceptional brutality or sexual violence. Convictions were made in only three of those deaths. In response, a number of activist groups like Ni Una Más (Not One More) and Nuestras Hijas de Regreso a Casa (Our Daughters Back Home) were formed, largely motivated by mothers demanding justice for their daughters, often at great personal risk to themselves.[58]

These groups succeeded in gaining the attention of the Mexican government, which established a Special Commission on Femicide. But despite the commission and the fourteen volumes of information about femicide that it produced, and despite a 2009 ruling against the Mexican state by the Inter-American Human Rights Court, and despite a United Nations Symposium on Femicide in 2012, and despite the fact that sixteen Latin American countries have now passed laws defining femicide—despite all of this, deaths in Juárez have continued to rise.[59] In 2009 a report pointed out that one of the reasons that the issue had yet to be sufficiently addressed was the lack of data.[60] Needless to say, the problem remains.

How might we explain the missing data around femicides in relation to the four domains of power that constitute Collins's matrix of domination? As is true in so many cases of data collected (or not) about women and other minoritized groups, the collection environment is compromised by imbalances of power.

The most grave and urgent manifestation of the matrix of domination is within the interpersonal domain, in which cis and trans women become the victims of violence and murder at the hands of men. Although law and policy (the structural domain) have recognized the crime of femicide, no specific policies have been implemented to ensure adequate information collection, either by federal agencies or local authorities. Thus the disciplinary domain, in which law and policy are enacted, is characterized by a deferral of responsibility, a failure to investigate, and victim blaming. This persists in a somewhat recursive fashion because there are no consequences imposed within the structural domain. For example, the Special Commission's definition of femicide as a "crime of the state" speaks volumes to how the government of Mexico is deeply complicit through inattention and indifference.[61]

Of course, this inaction would not have been tolerated without the assistance of the hegemonic domain—the realm of media and culture—which presents men as strong and women as subservient, men as public and women as private, trans people as deviating from "essential" norms, and nonbinary people as nonexistent altogether. Indeed, government agencies have used their public platforms to blame victims. Following the femicide of twenty-two-year-old Mexican student Lesvy Osorio in 2017, researcher Maria Rodriguez-Dominguez documented how the Public Prosecutor's Office of Mexico City shared on social media that the victim was an alcoholic and drug user who had been living out of wedlock with her boyfriend.[62] This led to justified public backlash, and to the hashtag #SiMeMatan (If they kill me), which prompted sarcastic tweets such as "#SiMeMatan it's because I liked to go out at night and drink a lot of beer."[63]

It is into this data collection environment, characterized by extremely asymmetrical power relations, that María Salguero has inserted her femicides map. Salguero manually plots a pin on the map for every femicide that she collects through media reports or through crowdsourced contributions (figure 1.5a). One of her goals is to "show that these victims [each] had a name and that they had a life," and so Salguero logs as many details as she can about each death. These include name, age, relationship with the perpetrator, mode and place of death, and whether the victim was transgender, as well as the full content of the news report that served as the source. Figure 1.5b shows a detailed view for a single report from an unidentified transfemicide, including the date, time, location, and media article about the killing. It can take Salguero three to four hours a day to do this unpaid work. She takes occasional breaks to preserve her mental health, and she typically has a backlog of a month's worth of femicides to add to the map.

Although media reportage and crowdsourcing are imperfect ways of collecting data, this particular map, created and maintained by a single person, fills a vacuum created by her national government. The map has been used to help find missing women, and Salguero herself has testified before Mexico's Congress about the scope of the problem.

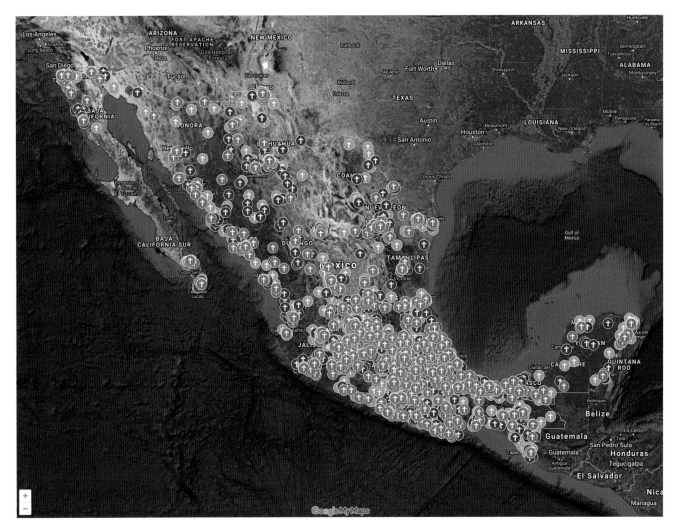

**Figure 1.5**
María Salguero's map of femicides in Mexico (2016–present) can be found at https://feminicidiosmx
.crowdmap.com/. (a) Map extent showing the whole country. (b) A detailed view of Ciudad Juárez
with a focus on a single report of an anonymous transfemicide. Salguero crowdsources points on
the map based on reports in the press and reports from citizens to her. Courtesy of María Salguero.

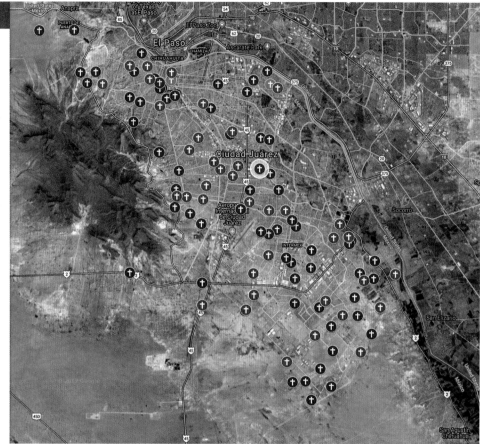

**Figure 1.5** (continued)

Salguero is not affiliated with an activist group, but she makes her data available to activist groups for their efforts. Parents of victims have called her to give their thanks for making their daughters visible, and Salguero affirms this function as well: "This map seeks to make visible the sites where they are killing us, to find patterns, to bolster arguments about the problem, to georeference aid, to promote prevention and try to avoid femicides."

It is important to make clear that the example of missing data about femicides in Mexico is not an isolated case, either in terms of subject matter or geographic location. The phenomenon of missing data is a regular and expected outcome in all societies characterized by unequal power relations, in which a gendered, racialized order is maintained through willful disregard, deferral of responsibility, and organized neglect for data and statistics about those minoritized bodies who do not hold power. So too are examples of individuals and communities using strategies like Salguero's to fill in the gaps left by these missing datasets—in the United States as around the world.[64] If "quantification is representation," as data journalist Jonathan Stray asserts, then this offers one way to hold those in power accountable. Collecting counterdata demonstrates how data science can be enlisted on behalf of individuals and communities that need more power on their side.[65]

### Data Science with Whose Interests and Goals?

Far too often, the problem is not that data about minoritized groups are missing but the reverse: the databases and data systems of powerful institutions are built on the excessive surveillance of minoritized groups. This results in women, people of color, and poor people, among others, being overrepresented in the data that these systems are premised upon. In *Automating Inequality*, for example, Virginia Eubanks tells the story of the Allegheny County Office of Children, Youth, and Families in western Pennsylvania, which employs an algorithmic model to predict the risk of child abuse in any particular home.[66] The goal of the model is to remove children from potentially abusive households before it happens; this would appear to be a very worthy goal. As Eubanks shows, however, inequities result. For wealthier parents, who can more easily access private health care and mental health services, there is simply not that much data to pull into the model. For poor parents, who more often rely on public resources, the system scoops up records from child welfare services, drug and alcohol treatment programs, mental health services, Medicaid histories, and more. Because there are far more data about poor parents, they are oversampled in the model, and so their children are overtargeted as being at risk for child abuse—a risk that results in children being removed from their families and homes. Eubanks argues that the model "confuse[s] parenting while poor with poor parenting."

This model, like many, was designed under two flawed assumptions: (1) that more data is always better and (2) that the data are a neutral input. In practice, however, the reality is quite different. The higher proportion of poor parents in the database, with more complete data profiles, the more likely the model will be to find fault with poor parents. And data are never neutral; they are always the biased output of unequal social, historical, and economic conditions: this is the matrix of domination once again.[67] Governments can and do use biased data to marshal the power of the matrix of domination in ways that amplify its effects on the least powerful in society. In this case, the model becomes a way to administer and manage classism in the disciplinary domain—with the consequence that poor parents' attempts to access resources and improve their lives, when compiled as data, become the same data that remove their children from their care.

So this raises our next *who question*: Whose goals are prioritized in data science (and whose are not)? In this case, the state of Pennsylvania prioritized its bureaucratic goal of efficiency, which is an oft-cited reason for coming up with a technical solution to a social and political dilemma. Viewed from the perspective of the state, there were simply not enough employees to handle all of the potential child abuse cases, so it needed a mechanism for efficiently deploying limited staff—or so the reasoning goes. This is what Eubanks has described as a *scarcity bias*: the idea that there are not enough resources for everyone so we should think small and allow technology to fill the gaps. Such thinking, and the technological "solutions" that result, often meet the goals of their creators—in this case, the Allegheny County Office of Children, Youth, and Families—but not the goals of the children and families that it purports to serve.

Corporations also place their own goals ahead of those of the people their products purport to serve, supported by their outsize wealth and the power that comes with it. For example, in 2012, the *New York Times* published an explosive article by Charles Duhigg, "How Companies Learn Your Secrets,"[68] which soon became the stuff of legend in data and privacy circles. Duhigg describes how Andrew Pole, a data scientist working at Target, was approached by men from the marketing department who asked, "If we wanted to figure out if a customer is pregnant, even if she didn't want us to know, can you do that?"[69] He proceeded to synthesize customers' purchasing histories with the timeline of those purchases to give each customer a so-called pregnancy prediction score (figure 1.6).[70] Evidently, pregnancy is the second major life event, after leaving for college, that determines whether a casual shopper will become a customer for life.

Target turned around and put Pole's pregnancy detection model into action in an automated system that sent discount coupons to possibly pregnant customers. Win-win—or so the company thought, until a Minneapolis teenager's dad saw the coupons for baby clothes that she was getting in the mail and marched into his local Target to read the manager the riot act. Why was his daughter getting coupons for pregnant women when she was only a teen?!

It turned out that the young woman was indeed pregnant. Pole's model informed Target before the teenager informed her family. By analyzing the purchase dates of approximately twenty-five common products, such as unscented lotion and large bags of cotton balls, the model found a set of purchase patterns that were highly correlated

**Figure 1.6**
Screenshot from a video of statistician Andrew Pole's presentation at Predictive Analytics World about Target's pregnancy detection model in October 2010, titled "How Target Gets the Most out of Its Guest Data to Improve Marketing ROI." He discusses the model at 47:50. Image by Andrew Pole for Predictive Analytics World.

with pregnancy status and expected due date. But the win-win quickly became a lose-lose, as Target lost the trust of its customers in a PR disaster and the Minneapolis teenager lost far worse: her control over information related to her own body and her health.

This story has been told many times: first by Pole, the statistician; then by Duhigg, the *New York Times* journalist; then by many other commentators on personal privacy and corporate overreach. But it is not only a story about privacy: it is also a story about gender injustice—about how corporations approach data relating to women's bodies and lives, and about how corporations approach data relating to minoritized populations more generally. Whose goals are prioritized in this case? The corporation's, of course. For Target, the primary motivation was maximizing profit, and quarterly financial reports to the board are the measurement of success. Whose goals are *not* prioritized? The teenager's and those of every other pregnant woman out there.

How did we get to the point where data science is used almost exclusively in the service of profit (for a few), surveillance (of the minoritized), and efficiency (amidst scarcity)? It's worth stepping back to make an observation about the organization of the data economy: data are expensive and resource-intensive, so only already powerful institutions—corporations, governments, and elite research universities—have the means to work with them at scale. These resource requirements result in data science that serves the primary goals of the institutions themselves. We can think of these goals as the *three Ss*: science (universities), surveillance (governments), and selling (corporations). This is not a normative judgment (e.g., "all science is bad") but rather an observation about the organization of resources. If science, surveillance, and selling are the main goals that data are serving, because that's who has the money, then what other goals and purposes are going underserved?

Let's take "the cloud" as an example. As server farms have taken the place of paper archives, storing data has come to require large physical spaces. A project by the Center for Land Use Interpretation (CLUI) makes this last point plain (figure 1.7). In 2014, CLUI set out to map and photograph data centers around the United States, often in those seemingly empty in-between areas we now call *exurbs*. In so doing, it called attention to "a new kind of physical information architecture" sprawling across the United States: "windowless boxes, often with distinct design features such as an appliqué of surface graphics or a functional brutalism, surrounded by cooling systems." The environmental impacts of the cloud—in the form of electricity and air conditioning—are enormous. A 2017 Greenpeace report estimated that the global IT sector, which is largely US-based, accounted for around 7 percent of the world's energy use. This is more than some of largest countries in the world, including Russia, Brazil, and Japan.[71] Unless that energy comes from renewable sources (which the Greenpeace report shows that it does not), the cloud has a significant accelerating impact on global climate change.

So the cloud is not light and it is not airy. And the cloud is not cheap. The cost of constructing Facebook's newest data center in Los Lunas, New Mexico, is expected to reach $1 billion.[72] The electrical cost of that center alone is estimated at $31 million per year.[73] These numbers return us to the question about financial resources: Who has the money to invest in centers like these? Only powerful corporations like Facebook and Target, along with wealthy governments and elite universities, have the resources to collect, store, maintain, analyze, and mobilize the largest amounts of data. Next, who is in charge of these well-resourced institutions? Disproportionately men, even more disproportionately white men, and even more than that, disproportionately rich white men. Want the data on that? Google's Board of Directors is comprised of 82 percent white men. Facebook's board is 78 percent male and 89 percent white. The 2018 US Congress was 79 percent male—actually a better percentage than in previous years—and with a median net worth of five times more than the average American household.[74] These are the people who experience the most privilege within the matrix

**Figure 1.7**
Photographs from *Networked Nation: The Landscape of the Internet in America*, an exhibition by the Center for Land Use Interpretation staged in 2013. The photos show four data centers located in North Bergen, NJ; Dalles, OR; Ashburn, VA; and Lockport, NY (counterclockwise from top right). They show how the "cloud" is housed in remote locations and office parks around the country. Images by the Center for Land Use Interpretation.

**Figure 1.7** (continued)

of domination, and they are also the people who benefit the most from the current status quo.[75]

In the past decade or so, many of these men at the top have described data as "the new oil."[76] It's a metaphor that resonates uncannily well—even more than they likely intended. The idea of data as some sort of untapped natural resource clearly points to the potential of data for power and profit once they are processed and refined, but it also helps highlight the exploitative dimensions of extracting data from their source— people—as well as their ecological cost. Just as the original oil barons were able to use their riches to wield outsized power in the world (think of John D. Rockefeller, J. Paul Getty, or, more recently, the Koch brothers), so too do the Targets of the world use their corporate gain to consolidate control over their customers. But unlike crude oil, which is extracted from the earth and then sold to people, data are both extracted from people and sold back to them—in the form of coupons like the one the Minneapolis teen received in the mail, or far worse.[77]

This extractive system creates a profound asymmetry between who is collecting, storing, and analyzing data, and whose data are collected, stored, and analyzed.[78] The goals that drive this process are those of the corporations, governments, and well-resourced universities that are dominated by elite white men. And those goals are neither neutral nor democratic—in the sense of having undergone any kind of participatory, public process. On the contrary, focusing on those *three Ss*—science, surveillance, and selling—to the exclusion of other possible objectives results in significant oversights with life-altering consequences. Consider the Target example as the flip side of the missing data on maternal health outcomes. Put crudely, there is no profit to be made collecting data on the women who are dying in childbirth, but there is significant profit in knowing whether women are pregnant.

 How might we prioritize different goals and different people in data science? How might data scientists undertake a feminist analysis of power in order to tackle bias at its source? Kimberly Seals Allers, a birth justice advocate and author, is on a mission to do exactly that in relation to maternal and infant care in the United States. She followed the Serena Williams story with great interest and watched as Congress passed the Preventing Maternal Deaths Act of 2018. This bill funded the creation of maternal health review committees in every state and, for the first time, uniform and comprehensive data collection at the federal level. But even as more data have begun to be collected about maternal mortality, Seals Allers has remained frustrated by the public conversation: "The statistics that are rightfully creating awareness around the Black maternal mortality crisis are also contributing to this gloom and doom deficit narrative. White

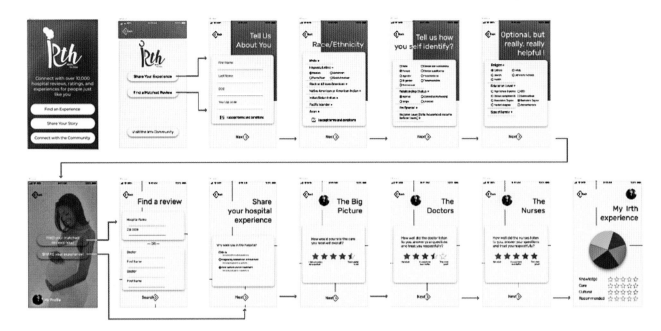

**Figure 1.8**
Irth is a mobile app and web platform focused on removing bias from birth (including prenatal, birth, and postpartum health care). Users post intersectional reviews of the care they received from individual nurses and doctors, as well as whole practices and hospitals. When parents to be are searching for providers, they can consult Irth to see what kind of care people like them received in the hands of specific caregivers. Wireframes from Irth's first prototype are shown here. Images by Kimberly Seals Allers and the Irth team, 2019.

people are like, 'how can we save Black women?' And that's not the solution that we need the data to produce."[79]

Seals Allers—and her fifteen-year-old son, Michael—are working on their own data-driven contribution to the maternal and infant health conversation: a platform and app called Irth—from *birth*, but with the *b* for *bias* removed (figure 1.8). One of the major contributing factors to poor birth outcomes, as well as maternal and infant mortality, is biased care. Hospitals, clinics, and caregivers routinely disregard Black women's expressions of pain and wishes for treatment.[80] As we saw, Serena Williams's own story almost ended in this way, despite the fact that she is an international tennis star. To combat this, Irth operates like an intersectional Yelp for birth experiences. Users post ratings and reviews of their prenatal, postpartum, and birth experiences at specific hospitals and in the hands of specific caregivers. Their reviews include important details like their race, religion, sexuality, and gender identity, as well as whether they felt that those identities were respected in the care that they received. The app also has a taxonomy of bias and asks users to tick boxes to indicate whether and how they may have experienced different types of bias. Irth allows parents who are seeking care to search for a review from someone like them—from a racial, ethnic, socioeconomic, and/or gender perspective—to see how they experienced a certain doctor or hospital.

Seals Allers's vision is that Irth will be both a public information platform, for individuals to find better care, and an accountability tool, to hold hospitals and providers responsible for systemic bias. Ultimately, she would like to present aggregated stories and data analyses from the platform to hospital networks to push for change grounded in women's and parents' lived experiences. "We keep telling the story of maternal mortality from the grave," she says. "We have to start preventing those deaths by sharing the stories of people who actually lived."[81]

Irth illustrates the fact that "doing good with data" requires being deeply attuned to the things that fall outside the dataset—and in particular to how datasets, and the data science they enable, too often reflect the structures of power of the world they draw from. In a world defined by unequal power relations, which shape both social norms and laws about how data are used and how data science is applied, it remains imperative to consider who gets to do the "good" and who, conversely, gets someone else's "good" done to them.

## Examine Power

Data feminism begins by examining how power operates in the world today. This consists of asking *who questions* about data science: Who does the work (and who is pushed out)? Who benefits (and who is neglected or harmed)? Whose priorities get turned into products (and whose are overlooked)? These questions are relevant at the level of individuals and organizations, and are absolutely essential at the level of society. The current answer to most of these questions is "people from dominant groups," which has resulted in a *privilege hazard* so acute that it explains the near-daily revelations about another sexist or racist data product or algorithm. The *matrix of domination* helps us to understand how the privilege hazard—the result of unequal distributions of power—plays out in different domains. Ultimately, the goal of examining power is not only to understand it, but also to be able to challenge and change it. In the next chapter, we explore several approaches for challenging power with data science.

# 2   Collect, Analyze, Imagine, Teach

**Principle: Challenge Power**

*Data feminism commits to challenging unequal power structures and working toward justice.*

In 1971, the Detroit Geographic Expedition and Institute (DGEI) released a provocative map, *Where Commuters Run Over Black Children on the Pointes-Downtown Track*. The map (figure 2.1) uses sharp black dots to illustrate the places in the community where the children were killed. On one single street corner, there were six Black children killed by white drivers over the course of six months. On the map, the dots blot out that entire block.

The people who lived along the deadly route had long recognized the magnitude of the problem, as well as its profound impact on the lives of their friends and neighbors. But gathering data in support of this truth turned out to be a major challenge. No one was keeping detailed records of these deaths, nor was anyone making even more basic information about what had happened publicly available. "We couldn't get that information," explains Gwendolyn Warren, the Detroit-based organizer who headed the unlikely collaboration: an alliance between Black young adults from the surrounding neighborhoods and a group led by white male academic geographers from nearby universities.[1] Through the collaboration, the youth learned cutting-edge mapping techniques and, guided by Warren, leveraged their local knowledge in order to produce a series of comprehensive reports, covering topics such as the social and economic inequities among neighborhood children and proposals for new, more racially equitable school district boundaries.

Compare the DGEI map with another map of Detroit made thirty years earlier, *Residential Security Map* (figure 2.2). Both maps use straightforward cartographic techniques: an aerial view, legends and keys, and shading. But the similarities end there. The maps differ in terms of visual style, of course. But more profound is how they diverge in terms of the worldviews of their makers and the communities they seek to support. The latter map was made by the Detroit Board of Commerce, which consisted

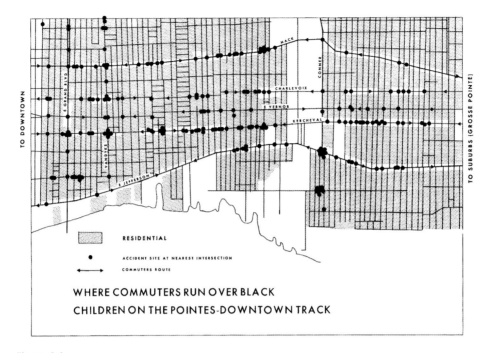

**Figure 2.1**

*Where Commuters Run Over Black Children on the Pointes-Downtown Track* (1971) is one image from a report, "Field Notes No. 3: The Geography of Children" which documented the racial inequities of Detroit children. The map was created by Gwendolyn Warren, the administrative director of the Detroit Geographic Expedition and Institute (DGEI), in a collaboration between Black young adults in Detroit and white academic geographers that lasted from 1968–1971. The group worked together to map aspects of the urban environment related to children and education. Warren also worked to set up a free school at which young adults could take college classes in geography for credit. Courtesy of Gwendolyn Warren and the Detroit Geographical Expedition and Institute.

of only white men, in collaboration with the Federal Home Loan Bank Board, which consisted mostly of white men. Far from emancipatory, this map was one of the earliest instances of the practice of *redlining*, a term used to describe how banks rated the risk of granting loans to potential homeowners on the basis of neighborhood demographics (specifically race and ethnicity), rather than individual creditworthiness.

Redlining gets its name because the practice first involved drawing literal red lines on a map. (Sometimes the areas were shaded red instead, as in the map in figure 2.2.) All of Detroit's Black neighborhoods fall into red areas on this map because housing discrimination and other forms of structural oppression predated the practice.[2] But denying home loans to the people who lived in these neighborhoods reinforced those

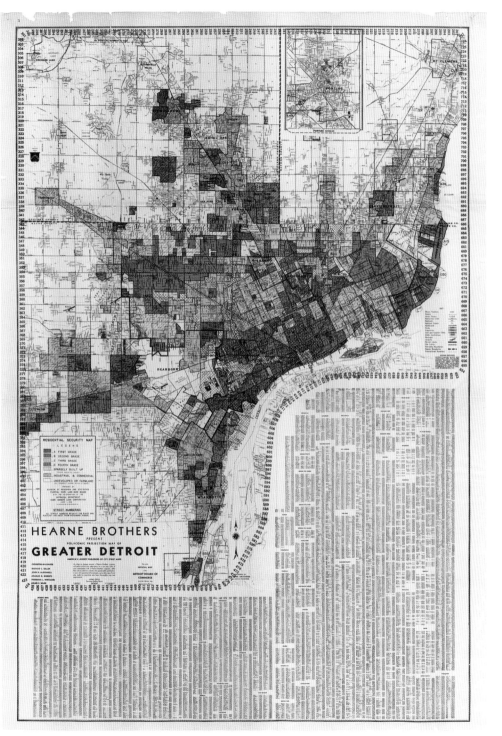

**Figure 2.2**

*Residential Security Map*, a redlining map of Detroit published in 1939. Created as a collaboration between the (all white and male) Detroit Chamber of Commerce and the (majority white and male) Federal Home Loan Bank Board, the red colors signify neighborhoods that these institutions deemed red neighborhoods, at "high risk" for bank loans. Courtesy of Robert K. Nelson, LaDale Winling, Richard Marciano, Nathan Connolly, et al., *Mapping Inequality: Redlining in New Deal America.*

existing inequalities and, as decades of research have shown, were directly responsible for making them worse.[3]

Early twentieth-century redlining maps had an aura very similar to the "big data" approaches of today. These high-tech, scalable "solutions" were deployed across the nation, and they were one method among many that worked to ensure that wealth remained attached to the racial category of whiteness.[4] At the same time that these maps were being made, the insurance industry, for example, was implementing simi-lar data-driven methods for granting (or denying) policies to customers based on their demographics. Zoning laws that were explicitly based on race had already been declared unconstitutional; but within neighborhoods, so-called covenants were nearly as exclusionary and completely legal.[5] This is a phenomenon that political philosopher Cedric Robinson famously termed *racial capitalism*, and it continues into the present in the form of algorithmically generated credit scores that are consistently biased and in the consolidation of "the 1 percent" through the tax code, to give only two examples of many.[6] What's more, the benefits of whiteness accrue: "Whiteness retains its value as a 'consolation prize,'" civil rights scholar Cheryl Harris explains. "It does not mean that all whites will win, but simply that they will not lose."[7]

Who makes maps and who gets mapped? The redlining map is one that secures the power of its makers: the white men on the Detroit Board of Commerce, their families, and their communities. This particular redlining map is even called *Residential Security Map*. But the title reflects more than a desire to secure property values. Rather, it reveals a broader desire to protect and preserve home ownership as a method of accumulating wealth, and therefore status and power, that was available to white people only. In far too many cases, data-driven "solutions" are still deployed in similar ways: in support of the interests of the people and institutions in positions of power, whose worldviews and value systems differ vastly from those of the communities whose data the systems rely upon.[8]

The DGEI map, by contrast, challenges this unequal distribution of data and power. It does so in three key ways. First, in the face of missing data, DGEI compiled its own counterdata. Warren describes how she developed relationships with "political people in order to use them as a means of getting information from the police department in order to find out exactly what time, where, how and who killed [each] child."[9] Second, the DGEI map plotted the data they collected with the deliberate aim of quantify-ing structural oppression. They intentionally and explicitly focused on the problems of "death, hunger, pain, sorrow and frustration in children," as they explain in the report.[10] Finally, the DGEI map was made by young Black people who lived in the community, under the leadership of a Black woman who was an organizer in the com-munity, with support provided by the academic geographers.[11] The identities of these makers matter, their proximity to the subject matter matters, the terms of their collaboration matter, and the leadership of the project matters.[12]

For these reasons, the DGEI provides a model of the second principle of data feminism: *challenge power.* Challenging power requires mobilizing data science to push back against existing and unequal power structures and to work toward more just and equitable futures. As we will discuss in this chapter, the goal of challenging power is closely linked to the act of examining power, the first principle of data feminism. In fact, the first step of challenging power is to examine that power. But the next step—and the reason we have chosen to dedicate two principles to the topic of power—is to take action against an unjust status quo.

Taking action can itself take many forms, and in this chapter we offer four starting points: (1) *Collect:* Compiling counterdata—in the face of missing data or institutional neglect—offers a powerful starting point as we see in the example of the DGEI, or in María Salguero's femicide maps discussed in chapter 1. (2) *Analyze:* Challenging power often requires demonstrating inequitable outcomes across groups, and new computational methods are being developed to audit opaque algorithms and hold institutions accountable. (3) *Imagine:* We cannot *only* focus on inequitable outcomes, because then we will never get to the root cause of injustice. In order to truly dismantle power, we have to imagine our end point not as "fairness," but as co-liberation. (4) *Teach:* The identities of data scientists matter, so how might we engage and empower newcomers to the field in order to shift the demographics and cultivate the next generation of data feminists?

## Analyze and Expose Oppression

One can make a direct comparison between yesterday's redlining maps and today's risk assessment algorithms. The latter are used in many cities in the United States today to inform judgments about the length of a particular prison sentence, the amount of bail that should be set, and even whether bail should be set in the first place. The "risk" in their name has to do with the likelihood of a person detained by the police committing a future crime. Risk assessment algorithms produce scores that influence whether a person is sent to jail or set free, effectively altering the course of their life.

But risk assessment algorithms, like redlining maps, are neither neutral nor objective. In 2016, Julia Angwin led a team at ProPublica to investigate one of the most widely used risk assessment algorithms in the United States, created by the company Northpointe (now Equivant).[13] Her team found that white defendants are more often mislabeled as low risk than Black defendants and, conversely, that Black defendants are mislabeled as high risk more often than white defendants.[14] Digging further into the process, the journalists uncovered a 137-question worksheet that each detainee is required to fill out (figure 2.3). The detainee's answers feed into the software, in which they are compared with other data to determine that person's risk score. Although the questionnaire does not ask directly about race, it asks questions that, given the structural inequalities embedded in US culture, serve as proxies for race. These include questions like whether you were raised by a single mother, whether you have ever been suspended from school, or whether you have friends or family that have been arrested. In the United States, each of those questions is linked to a set of larger social, cultural, and political—and, more often than not, racial—realities. For instance, it has been demonstrated that 67 percent of Black kids grow up in single-parent households,

**The next few questions are about the family or caretakers that mainly raised you when growing up.**

31. Which of the following best describes who principally raised you?
    ☐ Both Natural Parents
    ☐ Natural Mother Only
    ☐ Natural Father Only
    ☐ Relative(s)
    ☐ Adoptive Parent(s)
    ☐ Foster Parent(s)
    ☑ Other arrangement

32. If you lived with both parents and they later separated, how old were you at the time?
    ☑ Less than 5 ☐ 5 to 10 ☐ 11 to 14 ☐ 15 or older ☐ Does Not Apply

33. Was your father (or father figure who principally raised you) ever arrested, that you know of?
    ☑ No ☐ Yes

34. Was your mother (or mother figure who principally raised you) ever arrested, that you know of?
    ☑ No ☐ Yes

35. Were your brothers or sisters ever arrested, that you know of?
    ☐ No ☑ Yes

36. Was your wife/husband/partner ever arrested, that you know of?
    ☑ No ☐ Yes

37. Did a parent or parent figure who raised you ever have a drug or alcohol problem?
    ☑ No ☐ Yes

38. Was one of your parents (or parent figure who raised you) ever sent to jail or prison?
    ☑ No ☐ Yes

**Figure 2.3**
Equivant's risk assessment algorithm is called Correctional Offender Management Profiling for Alternative Sanctions (COMPAS) and is derived from a defendant's answers to a 137-question survey about their upbringing, personality, family, and friends, including many questions that can be considered proxies for race, such as whether they were raised by a single mother. Note that evidence of family criminality would not be admissible evidence in a court case for a crime committed by an individual, but here it is used as a factor in making important decisions about a person's freedom. Courtesy of Julia Angwin, Jeff Larson, Surya Mattu, and Lauren Kirchner for ProPublica, 2016.

whereas only 25 percent of white kids do.[15] Similarly, studies have shown that Black kids are punished more harshly than are white kids for the same minor infractions, starting as early as preschool.[16] So, though the algorithm's creators claim that they do not consider race, race is embedded into the data they are choosing to employ. What's more, they are using that information to further disadvantage Black people, whether because of an erroneous belief in the objectivity of their data, or because they remain unmoved by the evidence of how racism is operating through their technology.

Sociologist Ruha Benjamin has a term for these situations: the *New Jim Code*—where software code and a false sense of objectivity come together to contain and control the lives of Black people, and of other people of color.[17] In this regard, the redlining map and the Equivant risk assessment algorithm share some additional similarities. Both use aggregated data about *social groups* to make decisions about *individuals*: Should we grant a loan to this person? What's the risk that this person will reoffend? Furthermore, both use past data to predict future behavior—and to constrain it. In both cases, the past data in question (like segregated housing patterns or single parentage) are *products* of structurally unequal conditions. These unequal conditions are true across large social groups, and yet the technology uses those data as *predictive elements* that will influence one person's future. Surya Mattu, a former ProPublica reporter who worked on the story, makes this point directly: "Equivant didn't account for the fact that African Americans are more likely to be arrested by the police regardless of whether they committed a crime or not. The system makes an assumption that if you have been arrested you are probably at higher risk."[18] This is one of the challenges of using data about people as an input into a system: the data are never "raw." Data are always the product of unequal social relations—relations affected by centuries of history. As computer scientist Ben Green states, "Although most people talk about machine learning's ability to predict the future, what it really does is predict the past."[19] Effectively such "predictive" software reinforces existing demographic divisions, amplifying the social inequities that have limited certain groups for generations. The danger of the New Jim Code is that these findings are actively promoted as objective, and they track individuals and groups through their lives and limit their future potential.

But machine learning algorithms don't just predict the past; they also reflect current social inequities. A less well-known finding from the ProPublica investigation of Equivant, for example, is that it also surfaced significantly different treatment of women by the algorithm. Due to a range of factors, women tend to recidivate—to commit new crimes—less than men do. That means the risk scale for women "is such that somebody with a high risk score that's a woman is generally about the level of a medium risk score for a man. So, it's actually really shocking that judges are looking at these and thinking that high risk means the same thing for a man and a woman when it doesn't," explains lead reporter Julia Angwin.[20]

Angwin decided to focus the story on race in part because of the prior work of criminologists such as Kristy Holtfreter, which had already highlighted some of these gender differentials.[21] But there was another factor at play in her editorial decision: workplace sexism faced by women reporters like Angwin herself. Angwin explains how she had always been wary of working on stories about women and gender because she wanted to avoid becoming pigeonholed as a reporter who *only* worked on stories about women and gender. But, she explains, "one of the things I woke up to during the #MeToo movement was how many decisions like that I had made over the years"—an internalized form of oppression that had discouraged her from covering those important issues. In early 2018, when we conducted this interview, Angwin was hiring for her own data journalism startup, the Markup, founded with a goal of using data-driven methods to investigate the differential harms and benefits of new technologies on society. She was encouraged to see how many job candidates of all genders were pitching stories on issues relating to gender inequality. "In the era of data and AI, the challenge is that accountability is hard to prove and hard to trace," she explains. "The challenge for journalism is to try to make as concrete as possible those linkages when we can so we can show the world what the harms are."

Angwin is pointing out a tricky issue that is unlikely to go away. The field of journalism has long prided itself on "speaking truth to power." But today, the location of that power has shifted from people and corporations to the datasets and models that they create and employ. These datasets and models require new methods of interrogation, particularly when they—like Equivant's—are proprietary. How does one report on a *black box*, as these harmful algorithms are sometimes described?[22] Much like the situation encountered by Gwendolyn Warren when she looked into the data on the Detroit children's deaths, or like María Salguero when she started logging femicides in Mexico, ProPublica found no existing studies that examined whether the risk scores were racially biased, or existing datasets they could use to point them to answers. To write the risk assessment story, ProPublica had to assemble a dataset of their own. The researchers looked at ten thousand criminal defendants from a single county in Florida and compared their recidivism risk scores with people who actually reoffended in a two-year period. After doing some initial exploratory analysis, they created their own regression model that considered race, age, criminal history, future recidivism, charge degree, and gender. They found that age, race, and gender were the strongest predictors of who received a high risk score—with Black defendants 77 percent more likely than white ones to receive a higher violent recidivism score. Their analysis also included creating models to test the overall accuracy of the COMPAS model over time and an investigation of errors to see if there were racial differences in the distribution of false positives and false negatives. As it turns out, there were: the system was more likely to predict that white people would not commit additional crimes if released, when they actually did recidivate.[23]

Angwin and her coauthors used data science to challenge data science. By collecting missing data and reverse-engineering the algorithm that was judging each defendant's risk, they were able to prove systemic racial bias. This analysis method is called *auditing algorithms* and it is being increasingly used in journalism and in academic research in order show how the harms and benefits of automated systems are differentially distributed. Computational journalism researcher Nicholas Diakopoulos has proposed that work like this become formalized into an algorithm accountability beat, which would help to make the practice more widespread.[24] He and computer scientist Sorelle Friedler have asserted that algorithms need to be held "publicly accountable" for their consequences, and the press is one place where this accounting can take place.[25] By providing proof of how racism and sexism, among other oppressions, create unequal outcomes across social groups, analyzing data is a powerful strategy for challenging power and working toward justice.

**The Pitfalls of Proof**

Let's pause here for a feminist *who question*, as we introduced in chapter 1. Who is it, exactly, that needs to be shown the harms of such differentials of power? And what kind of proof do they require to believe that oppression is real? Women who experience instances of sexism, as Angwin did in her workplace, already know the harms of that oppressive behavior. The young adults whom Gwendolyn Warren worked with in Detroit already knew intimately that the white commuters were killing their Black neighbors and friends. They had no need to prove to their own communities that structural racism was a factor in these deaths. Rather, their goal in partnering with the DGEI was to prove the structural nature of the problem to those in positions of power. Those dominant groups and institutions were the ones that, by privileging their own social, political, and economic interests, bore much of the responsibility for the problem; and they also, because of the phenomenon we have described as a privilege hazard, were unlikely to see that such problem existed in the first place. The theory of change that motivates these efforts to use data as evidence, or "proof," is that by being made aware of the extent of the problem, those in power will be prompted to take action.

These kinds of data-driven revelations can certainly be compelling. When the analysis appears in a high-profile newspaper or blog or TV show (in other words: a place white enough and male enough to be considered mainstream), it can indeed prompt people in power to act. The ProPublica story on risk assessment algorithms, for example, prompted a New York City council member to propose an algorithmic accountability bill. Enacted in 2018, the bill became the first legal measure to tackle algorithmic discrimination in the United States and led to the creation of a task force focused on "equity and fairness" in city algorithms.[26] Should the city implement some of the task force's recommendations, it would influence the work of software vendors, as well as legislation in other cities. This path of influence—from community problem to gathering proof to informed reporting to policy change—represents the best aspirations of speaking truth to power.[27]

While analyzing and exposing oppression in order to hold institutions accountable can be extremely useful, its efficacy comes with two caveats. Proof can just as easily become part of an endless loop if not accompanied by other tools of community engagement, political organizing, and protest. Any data-based evidence can be minimized because it is not "big" enough, not "clean" enough, or not "newsworthy" enough to justify a meaningful response from institutions that have a vested interest in maintaining the status quo.[28] As we saw in chapter 1, María Salguero's data on femicides was augmented by government commissions, reports from international agencies, and rulings of international courts. But none of those data-gathering efforts have been enough to prompt comprehensive action.

Another feminist *who question*: On whom is the burden of proof is placed? In 2015, communications researcher Candice Lanius wrote a widely shared blog post, "Fact Check: Your Demand for Statistical Proof is Racist," in which she summarizes the ample research on how those in positions of power accept anecdotal evidence from those like themselves, but demand endless statistics from minoritized groups.[29] In those cases, she argues convincingly, more data will never be enough.

Proof can also unwittingly compound the harmful narratives—whether sexist or racist or ableist or otherwise oppressive—that are already circulating in the culture, inadvertently contributing to what are known as *deficit narratives*. These narratives reduce a group or culture to its "problems," rather than portraying it with the strengths, creativity, and agency that people from those cultures possess. For example, in their book *Indigenous Statistics*, Maggie Walter and Chris Anderson describe how statistics used by settler colonial groups to describe Indigenous populations have mainly functioned as "documentation of difference, deficit, and dysfunction."[30] This can occur even when the creators have good intentions—for example, as Kimberly Seals Allers notes (see chapter 1), a great deal of the media reporting on Black maternal mortality data falls into the deficit narrative category. It portrays Black women as victims and fails to amplify the efforts of the Black women who have been working on the issue for decades.

This goes for gender data as well. "What little data we collect about women tends to be either about their experience of violence or reproductive health," explains Nina Rabinovitch Blecker, who directs communications for Data2X, a nonprofit aimed at improving the quality of data related to gender in a global context.[31] The current data encourage additional deficit narratives—in which women are relentlessly and reductively portrayed as victims of violent crimes like murder, rape, or intimate partner violence. These narratives imply that the subjects of the data have no agency and need "saving" from governments, international institutions, or concerned citizens. As one step to counteract that, Blecker chose to publish an example from Uruguay that didn't focus on violence, but rather on quantifying women's unseen contributions to the economy.[32]

So, though collecting counterdata and analyzing data to provide proof of oppression remain worthy goals, it is equally important to remain aware of how the subjects of oppression are portrayed. Working with communities directly, which we talk more about in chapter 5, is the surest remedy to these harms. Indigenous researcher Maggie Walter explains that ownership of the process is key in order to stop the propagation of deficit narratives: "We [Indigenous people] must have real power in how statistics about us are done—where, when and how."[33] Key too is a sustained attention to the ways in which communities themselves are already addressing the issues. These actions are often more creative, more effective, and more culturally grounded than the actions that any outside organization would take.

## Envision Equity, Imagine Co-liberation

As the examples discussed thus far in this book clearly demonstrate, one of the most dangerous outcomes of the tools of data and data science being consolidated in the hands of dominant groups is that these groups are able to obscure their politics and their goals behind their technologies. Benjamin, whose book *Race after Technology: Abolitionist Tools for the New Jim Code* (mentioned earlier), describes this phenomenon as the "imagined objectivity of data and technology" because data-driven systems like redlining and risk assessment algorithms are not really objective at all.[34] Her concept of *imagined objectivity* emphasizes the role that cultural assumptions and personal preconceptions play in upholding this false belief: one imagines (wrongly) that datasets and algorithms are less partial and less discriminatory than people and thus more "objective."[35] But as we discuss in chapter 1, these data products seem objective only because the perspectives of those who produce them—elite, white men and the institutions they control—pass for the default. Assumptions about objectivity are becoming a major focus in data science and related fields as algorithm after algorithm is revealed to be sexist, racist, or otherwise flawed. What can the people who design these computational systems do to avoid these pitfalls? And what can everyone else do to help them and hold them accountable?

The quest for answers to these questions has prompted the development of a new area of research known as *data ethics*. It represents a growing interdisciplinary effort—both critical and computational—to ensure that the ethical issues brought about by our increasing reliance on data-driven systems are identified and addressed. Thus far, the major trend has been to emphasize the issue of "bias," and the values of "fairness, accountability, and transparency" in mitigating its effects.[36] This is a promising development, especially for technical fields that have not historically foregrounded ethical issues, and as funding mechanisms for research on data and ethics proliferate.[37] However, as Benjamin's concept of imagined objectivity helps to show, addressing bias in a dataset is a tiny technological Band-Aid for a much larger problem. Even the values mentioned here, which seek to address instances of bias in data-driven systems, are themselves non-neutral, as they locate the source of the bias in individual people and specific design decisions. So how might we develop a practice that results in data-driven systems that challenge power at its source?

The following chart (table 2.1) introduces an alternate set of orienting concepts for the field: these are the six ideals that we believe should guide data ethics work. These

**Table 2.1**

From data ethics to data justice

| Concepts That Secure Power<br>Because they locate the source of the problem in individuals or technical systems | Concepts That Challenge Power<br>Because they acknowledge structural power differentials and work toward dismantling them |
|---|---|
| Ethics | Justice |
| Bias | Oppression |
| Fairness | Equity |
| Accountability | Co-liberation |
| Transparency | Reflexivity |
| Understanding algorithms | Understanding history, culture, and context |

concepts all have legacies in intersectional feminist activism, collective organizing, and critical thought, and they are unabashedly explicit in how they work toward justice.

In the left-hand column, we list some of the major concepts that are currently circulating in conversations about the uses of data and algorithms in public (and private) life. These are a step forward, but they do not go far enough. On the right-hand side, we list adjacent concepts that emerge from a grounding in intersectional feminist activism and critical thought. The gap between these two columns represents a fundamental difference in view of why injustice arises and how it operates in the world. The concepts on the left are based on the assumption that injustice arises as a result of flawed individuals or small groups ("bad apples," "racist cops," "brogrammers") or flawed technical systems ("the algorithm/dataset did it"). Although flawed individuals and flawed systems certainly exist, they are not the root cause of the problems that occur again and again in data and algorithms.

What is the root cause? If you've read chapter 1, you know the answer: the matrix of domination, the matrix of domination, and the matrix of domination. The concepts on the left may do good work, but they ultimately keep the roots of the problem in place. In other words, they maintain the current structure of power, even if they don't intend to, because they let the matrix of domination off the hook. They direct data scientists' attention toward seeking technological fixes. Sometimes those fixes are necessary and important. But as technology scholars Julia Powles and Helen Nissenbaum assert, "Bias is real, but it's also a captivating diversion."[38] There is a more fundamental problem that must also be addressed: we do not all arrive in the present with equal power or privilege. Hundreds of years of history and politics and culture have brought us to the present moment. This is a reality of our lives as well as our data. A broader focus on *data justice*, rather than *data ethics* alone, can help to ensure that past inequities are not distilled into black-boxed algorithms that, like the redlining maps of the twentieth century, determine the course of people's lives in the twenty-first.

In proposing this chart, we are not suggesting that ethics have no place in data science, that bias in datasets should not be addressed, or that issues of transparency should go ignored.[39] Rather, the main point is that the concepts on the left are inadequate *on their own* to account for the root causes of structural oppression. By not taking root causes into account, they limit the range of responses possible to challenge power and work toward justice. In contrast, the concepts on the right start from the basic feminist belief that oppression is real, historic, ongoing, and worth dismantling.

Media theorist and designer Sasha Costanza-Chock proposes a restorative approach to data justice.[40] Drawing from theories of restorative justice—meaning that decisions should be made in ways that recognize and rectify any harms of the past—Costanza-Chock asserts that any notion of algorithmic *fairness* must also acknowl-edge the systematic nature of the *unfairness* that has long been perpetrated by certain groups on others. They give the example of college admissions—a topic that always seems to be in the news, not the least because it's a major mechanism of protecting privilege.[41] A restorative approach to college admissions would entail making decisions about who gets admitted in the present on the basis of who was historically not admit-ted in the past—like women, who were excluded from MIT, where Constanza-Chock teaches, for decades. According to this model, a "fair" present-day entering class that accounts for history might be composed of 90 percent women and people of color.[42]

Does this approach make fairness political? Emphatically yes, because all systems are political. In fact, the appeal to avoid politics is a very familiar way for those in power to attempt to hold onto it.[43] The ability to sidestep politics is a privilege in itself—held only by those whose existence does not challenge the status quo. If you are a Black woman or a Muslim man or a transgender service member and you live in the United States today, your being in the world is political, whether or not you want it to be.[44] So rather than design algorithms that purport to be "color-blind" (since color-blindness is of course a myth), Costanza-Chock explains that we should be designing algorithms that are *just*.[45] This means shifting from the ahistorical notion of fairness to a model of *equity*.

Equity is justice of a specific flavor, and it is different than equality. Equality is mea-sured from a starting point in the present: $t = 0$, where $t$ equals time and 0 indicates that no time has elapsed since now. Based on this formula, the principle of equality would hold that resources and/or punishments should be doled out according to what is hap-pening in the present moment—the time when $t = 0$. But this formula for equal treat-ment means that those who are ahead in the present can go further, achieve more, and stay on top, whereas those who start out behind can never catch up. Kiddada Green, executive director of the Black Mothers Breastfeeding Association, makes the case that in a country where Black babies are dying at twice the rate of white babies, equality is actually systematically unfair: "There is a level of political correctness in America that causes some people to believe that equality is the way to go. Even when equality is unfair, some say that it's the right thing to do."[46] Working toward a world in which everyone is treated equitably, not equally, means taking into account these present power differentials and distributing (or redistributing) resources accordingly. Equity is much harder to model computationally than equality—as it needs to take time, history, and differential power into account—but it is not impossible.[47]

This difficulty also underscores the point that bias (in individuals, in datasets, in statistical models, or in algorithms) is not a strong enough concept in which to anchor ideas about equity and justice. In writing about the creation of New York's Welfare Management System in the early 1970s, for example, Virginia Eubanks describes: "These early big data systems were built on a specific understanding of what constitutes discrimination: personal bias."[48] The solution at the time was to remove the humans from the loop, and it remains so today: without potentially bad—in this case, racist—apples, there would be less discrimination. But this line of thinking illustrates what whiteness studies scholar Robin DiAngelo would call the *new racism*: the belief that racism is due to individual bad actors, rather than structures or systems.[49] In relation to welfare management, Eubanks emphasizes that this often meant replacing social workers, who were often women of color, and who had empathy and flexibility and listening skills, with an automated system that applies a set of rigid criteria, no matter what the circumstances.

While bias remains a serious problem, it should not be viewed as something that can be fixed after the fact. Instead, we must look to understand and design systems that address the source of the bias: structural oppression. In truth, oppression is itself an outcome, one that results from the matrix of domination. In this model, majoritized bodies are granted undeserved advantages and minoritized bodies must survive undeserved hardships. Starting from the assumption that oppression is the problem, not bias, leads to fundamentally different decisions about what to work on, who to work with, and when to stand up and say that a problem cannot and should not be solved by data and technology.[50] Why should we settle for retroactive audits of potentially flawed systems if we can design with a goal of co-liberation from the start?[51] And here, *co-liberation* doesn't mean "free the data," but rather "free the people." The people in question are not only those with less privilege, but also those with more privilege: data scientists, designers, researchers, and educators—in other words, those like ourselves—who play a role in upholding oppressive systems.

The key to co-liberation is that it requires a commitment to and a belief in mutual benefit, from members of both dominant groups and minoritized groups; that's the *co* in the term. Too often, acts of data service performed by tech companies are framed as charity work (we discuss the limits of "data for good" in chapter 5). The frame of co-liberation equalizes this exchange as a form of relationship building and demographic healing. There is a famous saying credited to aboriginal activists in Queensland, Australia, from the 1970s: "If you have come here to help me, you are wasting your time. But if you have come because your liberation is bound up with mine, then let us work together."[52]

What does this mean? As poet and community organizer Tawana Petty explains in relation to efforts around antiracism in the United States: "We need whites to firmly believe that their liberation, their humanity, is also dependent upon the destruction of racism and the dismantling of white supremacy."[53] The same goes for gender: men are often not prompted to think about how unequal gender relations seep into the institutions they dominate, resulting in harm for everyone.

This goal of co-liberation motivates the Our Data Bodies (ODB) project. Led by a group of five women, including Gangadharan and Petty, who sit at the intersection of academia and organizing work, this project is a community-centered initiative focused on data collection efforts that disproportionately impact minoritized people. Working with community organizations in three US cities, the ODB project has led participatory research initiatives and educational workshops, culminating in the recently released *Digital Defense Playbook*, a set of activities, tools, and tip sheets intended to be used by and for marginalized communities to understand how data-driven technologies impact their lives.[54]

*Digital Defense Playbook* was born out of many years of relationship-building and research, as well as a deliberate shift. The group explains in the playbook's introduction, "We wanted to shift who gets to define problems around data collection, data privacy, and data security—from elites to impacted communities; shine a light on how communities have been confronting data-driven problems as well as how they wish to confront these problems; and forge an analysis of data and data-driven technologies from and with allied struggles."[55] In so doing, the ODB project demonstrates how co-liberation requires not only transparency of methods but also *reflexivity*: the ability to reflect on and take responsibility for one's own position within the multiple, intersecting dimensions of the matrix of domination. Along the way, the scholars and organizers involved in the project decided to shift their research agenda, which had begun as a general project about data profiling and resistance, to surveillance, in response to the problems voiced by the communities themselves.[56]

Even within big tech itself, there is evidence of an increasing sense of reflexivity among employees for their role in creating harmful data systems. Employees have pushed back against Google's work with the Department of Defense (DoD) on Project Maven, which uses AI to improve drone strike accuracy; Microsoft's decision to take $480 million from the Department of Defense to develop military applications of its augmented reality headset HoloLens; and Amazon's contract with US Immigration and Customs Enforcement (ICE) to develop its Rekognition platform for use in targeting individuals for detention and deportation at US borders.[57] This pushback has led to the cancelling of the Google and Microsoft projects, as well as political consciousness raising across the sector, which we discuss further in the book's conclusion.[58]

Designing datasets and data systems that dismantle oppression and work toward justice, equity, and co-liberation requires new tools in our collective toolbox. We have some good starting points; building more understandable algorithms is a laudable, worthy research goal. And yet what we need to explain and account for are not only the inner workings of machine learning, but also the history, culture, and context that lead to discriminatory outputs in the first place. For example, it is not an isolated incident that facial analysis software couldn't "see" Joy Buolamwini's face, as we discussed in chapter 1. It is not an isolated incident that the "Lena" image used to test most image-processing algorithms was the centerfold from the November 1972 issue of *Playboy*, cropped demurely at the shoulders.[59] It is not an isolated incident that the women who worked on the ENIAC computer were not invited to the fiftieth anniversary celebration in 1995. It is not an isolated incident that Christine Darden was not promoted as quickly as her male coworkers. None of these are isolated incidents: they are connected data points and eminently measurable and predictable outcomes of the matrix of domination. But you can only detect the pattern if you know the history, culture, and context that surrounds it.

Data people, generally speaking, have choices—choices in who they work for, which projects they work on, and what values they reject.[60] Starting from the assumption that oppression is the problem, equity is the path, and co-liberation is the desired goal leads to fundamentally different projects that challenge power at their source. It also leads to different metrics of success. These extend beyond the efficiency of a database under load, the precision of a classification algorithm, or the size of a user base one year after launch. The success of a project designed with co-liberation in mind would also depend on how much trust was built between institutions and communities, how effectively those with power and resources shared their power and resources, how much learning happened in both directions, how much the people and organizations were transformed in the process, and how much inspiration for future work, together, was co-conspired. These metrics are a little more squishy than the numbers and rankings that we tend to believe are our only option, but utterly and entirely measurable nonetheless.

## Teach Data Like an Intersectional Feminist

When Gwendolyn Warren and the DGEI researchers collected their data about hit and runs on Black children or scoured Detroit playgrounds to weigh and measure the broken glass they found, they were not only doing this work to make a data-driven case for change. The "institute" part of the Detroit Geographic Expedition and Institute described the educational wing of the organization that ran classes in data collection, mapping, and cartography. It came about at Warren's insistence that the academic geographers give something back to the community whose knowledge they were drawing upon for their research. She recognized that while a single map or project could make a focused intervention, education would enable her community to come away with a longer-term strategy for challenging power. As it turned out, the institutional affiliations of the academic geographers enabled them to offer free, for-credit college courses, which they taught in the community for community members.

In her emphasis on education, Warren recognized its enduring role as a mechanism of both empowerment and transformation. This belief is not new; as American educational reformer Horace Mann stated famously in 1848, "Education, then, beyond all other divides of human origin, is a great equalizer of conditions of men—the balance wheel of the social machinery." But here is the thing—it really matters how we do that equalizing and who we imagine that equalizing to serve. For his part, Mann was literal about the "men": education was to be an equalizer of men, but only certain men (read: white, Anglo, Christian) and explicitly not women.[61] Warren, on the other hand, recognized that access to education—and to data science education in particular—would have to be expanded in order for it to achieve its equalizing force.

Unfortunately, Warren's transformative vision has still yet to enter the data science classroom. As was true in Mann's era, men still lead. Women faculty comprise less than a third of computer science and statistics faculty. More than 80 percent of artificial intelligence professors are men.[62] This gender imbalance, and the narrowness of vision that results, is compounded by the fact that data science is often framed as an *abstract and technical* pursuit. Steps like cleaning and wrangling data are presented as solely technical conundrums; there is less discussion of the social context, ethics, values, or politics of data.[63] This perpetuates the myth that data science about astrophysics is the same as data science about criminal justice is the same as data science about carbon emissions. This limits the transformative work that can be done. Finally, because the goal of learning data science is modeled as *individual mastery of technical concepts and skills*, communities are not engaged and conversations are restricted. Instead, teachers impart technical knowledge via lectures, and students complete assignments and quizzes individually. We might call this model of teaching "the Horace Mann Factory Model of Data Science," because it represents the exclusionary view that Mann himself advanced. But let's just call it the *Man Factory* for short.

The Man Factory is really good at producing men, mainly elite white men like the ones who already lead the classes. It's not as good at producing women data scientists, or nonbinary data scientists, or data scientists of color. For years, researchers and advocacy organizations have recognized that there are problems with this "pipeline" for technical fields; yet this research is framed around questions like "Why are there so few women computer scientists?" and "Why are women leaving computing?"[64] Note that these questions imply that it is the women who have the problem, inadvertently perpetuating a deficit narrative. Feminist scholars who are studying the issue are, not surprisingly, asking very different questions, like "How can the men running the Man Factory share their power?" and "How can we structurally transform STEM education together?"[65]

One person currently modeling an answer to these questions is Laurie Rubel, the math educator behind the Local Lotto project. If you were on the city streets of Brooklyn or the Bronx in the past five years, you may have inadvertently crossed paths with one of her data science classes. You probably didn't realize it because the classes looked nothing like a traditional classroom (figure 2.4). Teenagers from the neighborhood wandered around in small groups. They were outfitted with tablets, pen and paper, cameras, and maps. They periodically took pictures on the street, walked into bodegas, chatted with passersby in Spanish or English, and entered information on their tablets.

Rubel is a leader in an area called mathematics for spatial justice, which aims to show how mathematical concepts can be taught in ways that relate to justice concerns arising from students' everyday lives, and to do so in dialogue with people in their neighborhoods and communities. The goal of Local Lotto was to develop a place-specific way of teaching concepts related to data and statistics grounded in considerations of equity.[66] Specifically, Rubel and the other organizers of Local Lotto wanted young learners to come up with a data-driven answer to the question: "Is the lottery good or bad for your neighborhood?"

In New York, as in other US states that operate lotteries, lottery ticket sales go back into the state budget—sometimes, but not always, to fund educational programs.[67] But lottery tickets are not purchased equally across all income brackets or all neighborhoods. Low-wage workers buy more tickets than their higher-earning counterparts. What's more, the revenue from ticket purchases is not allocated back to those workers or the places they live. Because of this, scholars have argued that the lottery system is a form of regressive taxation—essentially a "poverty tax"—whereby low-income neighborhoods are "taxed" more because they play more, but do not receive a proportional share of the profit.[68]

The Local Lotto curriculum was designed to expose high school learners to this instance of social inequality. They begin by talking about the lottery and the idea of probability by playing chance-based games. Then they consider jackpot games like the Sweet Millions lottery, advertised by New York State as "your best chance from the New York Lottery to win a million for just a buck." The best chance to win one million, however, turns out to be about one in four million; an entire class session is

**Figure 2.4**
A data science classroom. The Local Lotto project (2012–2015) taught local high school students statistics and data analysis rooted in neighborhood and justice concerns. Courtesy of the City Digits Project Team, including Brooklyn College, the Civic Data Design Lab at MIT and the Center for Urban Pedagogy. This work was supported by the National Science Foundation under Grant No. DRL-1222430.

devoted to a discussion about other instances of "four million" that more closely relate to the learners' lives.[69] The learners then leave the classroom with the goal of collecting data about how other people experience the lottery, which takes them back into their neighborhoods. They map stores that sell lottery tickets. They record interviews with shopkeepers and ticket buyers on their tablets and then geolocate them on their maps. They take pictures of lottery advertising. Afterward, the learners analyze their results and present them to the class. They examine choropleth maps of income levels, they make ratio tables, and they correlate state spending of lottery profits with median family income. (No surprise: there is no correlation.) Finally, they create a data-driven argument: an opinion piece supported with evidence from their statistical and spatial analyses, as well as their fieldwork (figures 2.5 and 2.6).

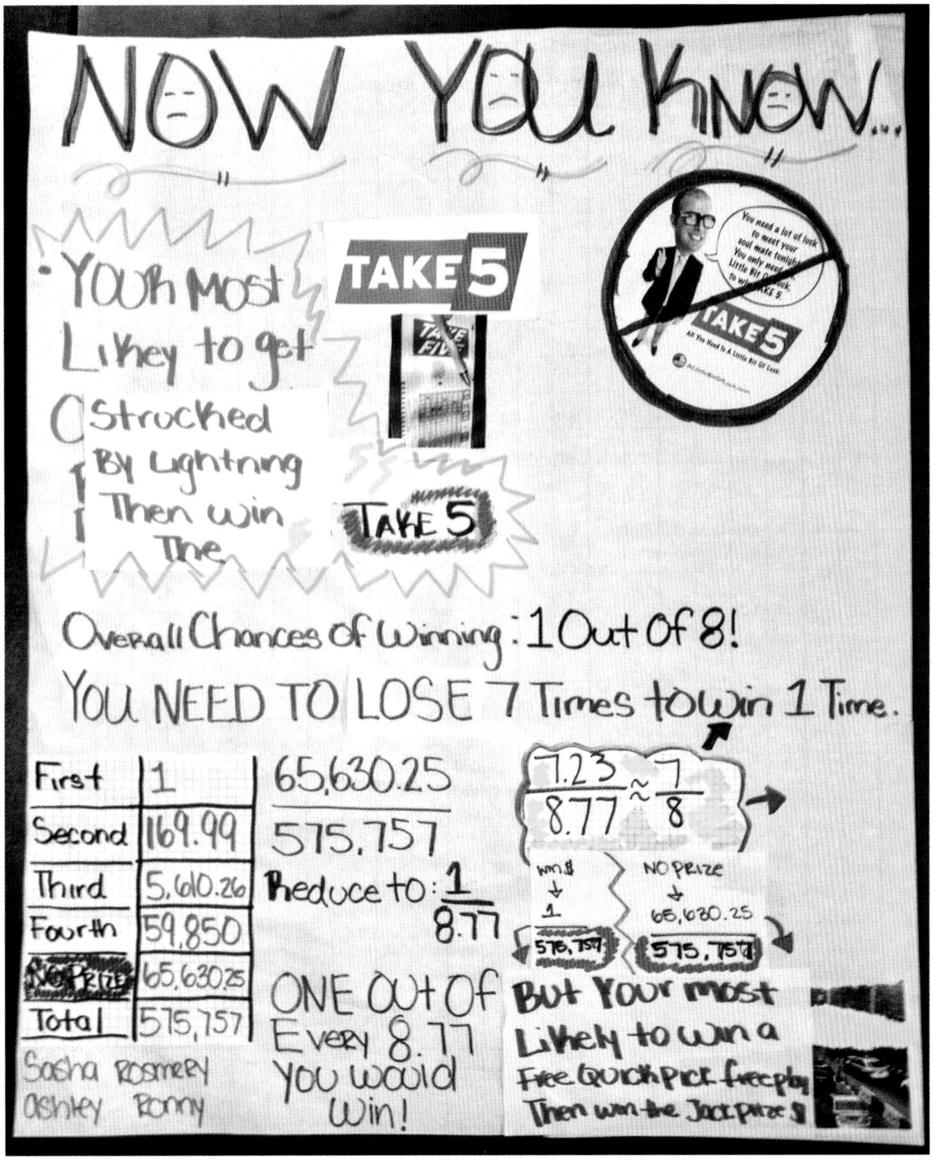

**Figure 2.5**
"Now You Know." A student's infographic poster responding to the New York Lottery advertising message, "Hey, you never know … ," that explores the probability of winning anything—ranging from a million dollars to another ticket. Courtesy of the City Digits Project Team. This work was supported by the National Science Foundation under Grant No. DRL-1222430.

**Figure 2.6**
Final collaborative opinion piece asserting that the lottery is not good for the students' neighborhood and presenting evidence they collected to back up their case. See the multimedia slideshow at http://citydigits.mit.edu/locallotto#tours-tab. Courtesy of Emmanuela, Angel, Robert, and Janeva. This work was supported by the National Science Foundation under Grant No. DRL-1222430.

By formal measures, the Local Lotto approach worked: before one school's implementation of Local Lotto, only two of forty-seven learners were able to determine the correct number of possible combinations in a lottery example. Later, almost half (twenty-one of forty-seven) were successfully able to calculate the number of combinations. But perhaps more importantly, the Local Lotto approach made math and statistics relevant to the students' lives. One student shared that what he learned was "something new that could help me in my local environment, in my house actually," and that after the course, he tried to convince his mother to spend less money on the lottery by "showing her my math book and all the work." Spanish-speaking women in the class who didn't often participate in classroom discussion became essential translators during the participatory mapping module. Several students went on to teach other teachers about the curriculum, both locally and nationally.[70]

What's different about the Local Lotto approach to teaching data analysis and statistical concepts compared to the Man Factory? How is Local Lotto challenging power both inside and outside the classroom? First, it was woman-led: the project was conceived by three women leaders representing three institutions.[71] Just as with the DGEI map and school, led by Gwendolyn Warren, the identities of the creators matter. Second, rather than modeling data science as abstract and technical, Local Lotto modeled a data science that was grounded in solving ethical questions around social inequality that had relevance for learners' everyday lives: Is the lottery good or bad for your neighborhood? The project valued lived experience: the learners came in as "domain experts" in their neighborhoods. And it valued both qualitative data and quantitative data: the learners spoke with neighborhood residents and connected their beliefs, attitudes, and concerns to probability calculations. Learners used community members' voices as evidence in their final projects. Third, rather than valorizing individual mastery of technical skills as the gold standard, learners worked together during every phase of the project. They used methods from art and design (like the creation of infographics and digital slideshows) to practice communicating with data.

Even as we celebrate these intentional pedagogical choices, the Local Lotto project still had its shortcomings, as the organizers noted in a 2016 paper for *Cognition and Instruction*.[72] Many of these stemmed from a basic fact: the teachers and course designers of the project were white and Asian, whereas the youth in the classes were predominantly Latinx and Black. This led to several issues. For instance, the curriculum designers had intended to focus primarily on income inequality, but they discovered that "the students consistently surfaced race." Because race and ethnicity were not part of the teaching material, the teachers felt that they did not have the experience or background to discuss them explicitly and deflected those conversations. As they write in the paper, "Youth, and in this case youth of color, have different understandings about racial boundaries; theirs are differently nuanced and scaled than affluent, white, or adult perspectives." The organizers are now taking steps to explicitly integrate discussions about race into the curriculum, as well as to include race, ethnicity, and age data in the course projects.[73]

The course designers also encountered "limited but recurring instances of resistance from students" to the project's central focus on income inequality. They attribute this resistance to the fact that the course was developed and taught by outsiders and could be seen as passing judgment on the people in their neighborhoods: that because they were not from the community, the teachers were perpetuating a deficit narrative about low-income people. This is both a sophisticated and very fair pushback from the young learners. Most people, regardless of their wealth or level of education, know they are not going to win the lottery, after all. There is an element of imaginative fantasy in purchasing a ticket. The campaign slogan, "Hey, you never know …" appeals as much to this fantasy as it does to the reality of the odds, and this fantasy has value too. In reflecting on the unintended sense of judgment experienced by the students, the course designers determined that, in the next iteration of the course, they would work to connect students with people in the communities themselves who are actively working to address issues of income inequality.

In both its successes and its failures, as well as its commitment to iteration and trying again, Local Lotto encapsulates what it means to challenge power and privilege and work toward justice. Justice is a journey. The discomfort that comes along with this journey is par for the course. There is no such thing as mastery of feminism because those who hold positions of privilege—like those in data science, like the Local Lotto course designers, and like us, the authors of this book—are constantly learning how to be better allies and accomplices across difference. In this process, what becomes most important is to "stay with the trouble," as feminist philosopher Donna Haraway would say.[74] Staying with the trouble means persisting in your work, especially when it becomes uncomfortable, unclear, or outright upsetting. One of the biggest strengths of the Local Lotto project is the courage of its creators to publicly, transparently, and reflexively interrogate themselves and their process, to detail their stumbling blocks, and to describe their commitments to doing better in the future.

**Challenge Power**

After examining power, the next step is to challenge it—map by map, audit by audit, community by community, and classroom by classroom. Collecting counterdata to quantify and visualize structural oppression, as Gwendolyn Warren and the DGEI did with their map, helps those who occupy positions of power understand the scope, scale, and character of the problems from which they are otherwise far removed. Analyzing biased algorithms, as Julia Angwin and ProPublica did, can show the real, material harms of automated systems, as well as build a base of evidence for political or institutional change. At the same time, it is important to remember that minoritized individuals and groups should not have to repeatedly prove that their experiences of oppression are real. And data alone do not always lead to change—especially when that change also requires dominant groups to share their resources and their power.

Those of us who use data in our work must alter some of our most basic assumptions and imagine new starting points. Shifting the frame from concepts that secure power, like fairness and accountability, to those that challenge power, like equity and co-liberation, can help to ensure that data scientists, designers, and researchers take oppression and inequality as their grounding assumption for creating computational products and systems. We must learn from—and design with—the communities we seek to support. A commitment to data justice begins with an acknowledgment of the fact that oppression is real, historic, ongoing, and worth dismantling. This commitment is one that we must teach the next generation of data scientists and data citizens, in communities and in classrooms, if we want to broaden our path toward justice.

# 3   On Rational, Scientific, Objective Viewpoints from Mythical, Imaginary, Impossible Standpoints

**Principle: Elevate Emotion and Embodiment**

*Data feminism teaches us to value multiple forms of knowledge, including the knowledge that comes from people as living, feeling bodies in the world.*

In 2012, twenty kindergarten children and six adults were shot and killed at an elementary school in Sandy Hook, Connecticut. In the wake of this unconscionable tragedy, and of the additional acts of gun violence that followed, the design firm Periscopic began a new project: to visualize all the gun deaths that took place in the United States over the course of a calendar year. Although there is no shortage of prior work on the subject in the form of bar charts or line graphs, Periscopic, a company with the tagline "Do good with data," took a different approach.

When you load the project's webpage, you first see a single orange line that arcs up from the x-axis on the left-hand side of the screen. Then, the color abruptly changes to white. A small dot drops down, and you see the phrase, "Alexander Lipkins, killed at 29" (figure 3.1a). The line continues to arc up across the screen and then down, coming back to rest on the x-axis, where a second phrase appears: "Could have lived to be 93." Then, a second line appears—the arc of another life. The animation speeds up and the arcs multiply. A counter at the top right displays how many years of life have been "stolen" from these victims of gun violence. After several excruciating minutes, the visualization completes its count for the year: 11,419 people killed, totaling 502,025 stolen years (figure 3.1b).

The visualization uses demographic data and rigorous statistical methods to arrive at these numbers, as is explained in the methods section on the site. But what makes Periscopic's visualization so very different from a more conventional bar chart of similar information, such as "The Era of 'Active Shooters'" from the *Washington Post* (figure 3.2)? The projects share the proposition that gun deaths present a serious threat. But unlike the *Washington Post* bar chart, Periscopic's work is framed around an emotion:

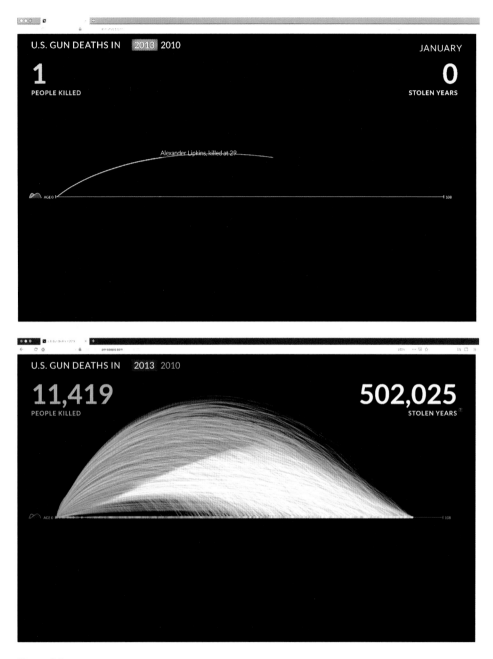

**Figure 3.1**
An animated visualization of the "stolen years" of people killed by guns in the United States in 2013. The first image (a) shows the beginning state of the animation and the second image (b) shows the end state. Images by Periscopic.

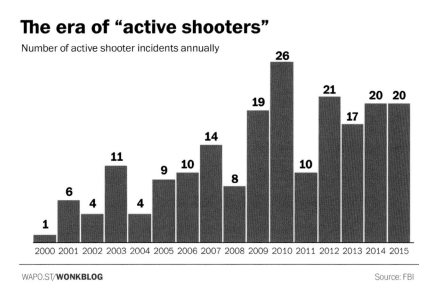

**Figure 3.2**
A bar chart of the number of "active shooter" incidents in the United States between 2000 and 2015. Images by Christopher Ingraham for the *Washington Post*.

loss. People are dying; their remaining time on earth has been stolen from them. These people have names and ages. They have parents and partners and children who suffer from that loss as well.

This message was clearly received, as was the project overall. It was featured in *Wired* magazine, and even won an Information is Beautiful award. But it also caused some stewing on the part of the visualization community. Alberto Cairo, the author of the visualization book *The Truthful Art*, expressed his concerns about the use of emotion and persuasion in the project: "Is it clear to a general audience that what they see is the work of professionals who actively shape data to support a cause, and not the product of automated processes?"[1] At root for Cairo was the question of how detached and "neutral" a visualization should be. He wondered, Should a visualization be designed to evoke emotion?

The received wisdom in technical communication circles is, emphatically, "No." In the recent book *A Unified Theory of Information Design*, authors Nicole Amare and Alan Manning state: "The plain style normally recommended for technical visuals is directed toward a deliberately neutral emotional field, a blank page in effect, upon which viewers are more free to choose their own response to the information."[2] Here, plainness is equated with the absence of design and thus greater freedom on the part of the viewer to interpret the results for themselves. Things like colors and icons work only to stir up emotions and cloud the viewer's rational mind.

They're not the first ones to posit this belief. In the field of data communication, any kind of ornament has long been viewed as suspect. Why? As historian of science Theodore Porter puts it, "Quantification is a technology of distance."[3] And distance, he explains, is closely related to objectivity because it puts literal space between people and the knowledge they produce. This desire for separation is what underlies the nineteenth-century statistician Karl Pearson's exhortation, echoed in Cairo's comments about the Periscopic visualization, for people to set aside their "own feelings and emotions" when performing statistical work.[4] The more plain, the more neutral; the more neutral, the more objective; and the more objective, the more true—or so this line of reasoning goes. At a data visualization master class in 2013, workshop leaders from the *Guardian* newspaper held up spreadsheet data—spreadsheet data!—as an ideal for the communication of quantitative data, calling it: "Clarity without persuasion."[5]

But persuasion is everywhere, even in spreadsheets, and—as feminist philosopher Donna Haraway would likely argue—especially in spreadsheets. In the 1980s, Haraway was among the first to connect the seeming neutrality and objectivity of data and their visual display to the ideas about distance that we've just discussed. She described data visualization, in particular, as "the god trick of seeing everything from nowhere." The view from nowhere—from a distance, from up above, like a god—may be data visualization's most signature feature. It's also the most ethically complicated to navigate for the ways in which it masks the people, the methods, the questions, and the messiness that lies behind clean lines and geometric shapes. Haraway calls it a trick because it makes the viewer *believe* that they can see everything, all at once, from an imaginary and impossible standpoint. But it's also a trick because what appears to be everything, and what appears to be neutral, is always what she terms a *partial perspective*. And in most cases of seemingly "neutral" visualizations, this perspective is the one of the dominant, default group. Think back to the presumption of whiteness as default that we discussed in the introduction, or—for an example of an actual visualization—to the redlining map discussed in chapter 2. This is a good example of the god trick at work.[6]

The god trick and its underlying assumptions about neutrality and truth are baked into today's best practices for data visualization. This is largely due to the influence of one man: the renowned statistical graphics expert Edward Tufte. Back in the 1980s, Tufte invented a metric for measuring the amount of superfluous information included in a chart. He called it the *data-ink ratio*.[7] In his view, a visualization designer should strive to use ink to display data alone. Any ink devoted to something other than the data themselves—such as background color, iconography, or embellishment—is a suspect and intruder to the graphic. Visual minimalism, according to this logic, appeals to reason first. As police officer Joe Friday says to every woman character on the American TV series *Dragnet*, "Just the facts, ma'am." Decorative elements, on the other hand, are associated with messy feelings—or, worse, represent stealthy (and, according to Tufte, unscientific) attempts at emotional persuasion. Data visualization has even been named as "the unempathetic art" by designer Mushon Zer-Aviv because of its emphatic rejection of emotion.[8]

The logic that sets up this false binary between emotion and reason is gendered, of course, because the belief that women are more emotional than men (and, by contrast, that men are more reasoned than women) is one of the most persistent stereotypes across many Western cultures. Indeed, psychologists have called it a *master stereotype* and puzzled over how it endures even when certain emotions—even extreme ones, like anger and pride—are simultaneously associated with men.[9] A central focus of feminist scholarship has been to challenge false binaries like this one between reason and emotion and to point out how they establish hierarchies as well. (We discuss this more in chapter 4.) For now, the important thing to note is how false binaries work to benefit a single one of Haraway's partial perspectives: that of the group already at the top—elite white men.

How can we let go of this binary logic? Two additional questions help challenge this reductive way of thinking and the oppressive hierarchies that it supports. First, is visual minimalism really more neutral? And second, how might activating emotion—leveraging, rather than resisting, emotion in data visualization—help us learn, remember, and communicate with data? Exploring these questions helps get us closer to the third principle of data feminism: *embrace emotion and embodiment*.

## Visualization as Rhetoric

Information visualization has diverse origins. Its history is often traced from the explosion of European men mapping their colonial conquests in the late fifteenth and early sixteenth centuries, through the development of new visual typologies like the timeline and the bar chart in the seventeenth and eighteenth centuries, to the adoption of those forms by powerful nations as they amassed increasing amounts of data on the populations they sought to control. But feminist scholars are increasingly challenging this simple narrative of progress, as well as its cast of characters, which is predominantly white and male. Whitney Battle-Baptiste and Britt Rusert recently published a new edition of the visualization work of W. E. B. Du Bois, the renowned Black sociologist and civil rights activist, who created his "data portraits" of African American life for the 1900 Paris Exposition. Laura Bliss, in a blog post that went viral, called attention to the "narrative maps" of Shanawdithit, a member of the Beothuk (Newfoundland) tribe, which she created around 1829 at the urging of a visiting anthropologist. And Lauren, one of the authors of this book, created a website that reanimates the historical charts of Elizabeth Palmer Peabody, the nineteenth-century editor and educator, who used visualization in her teaching (figure 3.3).[10]

Each of these early visualization designers understood how their images could function rhetorically. But in more recent history, many of data visualization's theorists and practitioners have come from technical disciplines aligned with engineering and computer science and have not been trained in that most fundamental of Western communication theories. In his ancient Greek treatise, Aristotle defines *rhetoric* as "the faculty of observing in any given case the available means of persuasion."[11] But rhetoric isn't only found in political speeches made by men dressed in tunics with wreaths on their heads.[12] Any communicating object that reflects choices about the selection and representation of reality is a rhetorical object. Whether or not it is rhetorical (it always is) has nothing to do with whether or not it is *true* (it may or may not be).

The question of rhetoric matters because "a rhetorical dimension is present in every design," says visualization researcher Jessica Hullman.[13] This includes visualizations that do not deliberately intend to persuade people of a certain message. It *especially and definitively* includes those so-called neutral visualizations that do not appear to have an editorial hand. In fact, those might even be the most perniciously persuasive visualizations of all!

Editorial choices become most apparent when compared with alternative choices. For example, in his book *The Curious Journalist's Guide to Data*, journalist Jonathan Stray discusses a data story from the *New York Times* about the September 2012 jobs report.[14] The *New York Times* created two graphics from the report: one framed from the perspective of Democrats (the party in power at the time; figure 3.4a) and one framed from the perspective of Republicans (figure 3.4b).

Either of these graphics, considered in isolation, appears to be neutral and factual. The data are presented with standard methods (line chart and area chart respectively) and conventional positionings (time on the x-axis, rates expressed as percentages on the y-axis, title placed above the graphic). There is a high data-ink ratio in both cases and very little in the way of ornamentation. But the graphics have significant editorial differences. The Democrats' graphic emphasizes that unemployment is decreasing—in its title, the addition of the thick blue arrow pointing downward, and the annotation "Friday's drop was larger than expected." Whereas the Republicans' graphic highlights

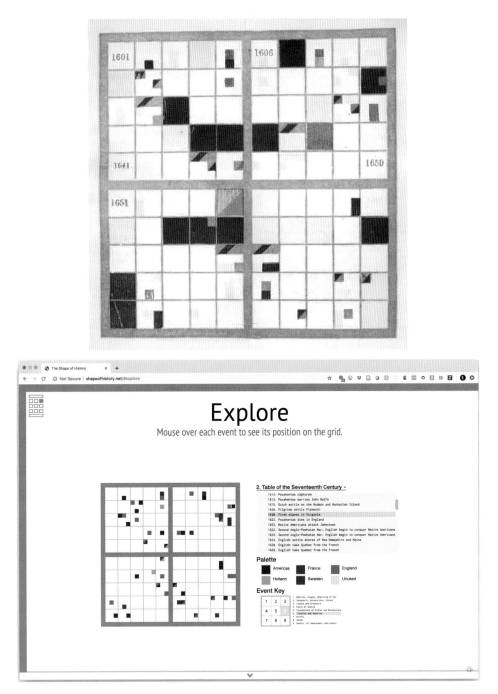

**Figure 3.3**

(a) Elizabeth Palmer Peabody's chart of "Significant Events of the 17th Century United States" (1865).
(b) Peabody's chart recreated in digital form by Lauren's Digital Humanities Lab (2017). (c) A rendering
of Peabody's chart reimagined as an interactive quilt by Lauren's Digital Humanities Lab (2019). Images
by (a) Elizabeth Palmer Peabody, *A Chronological History of the United States* (1856), (b) the Georgia Tech
Digital Humanities Lab, and (c) Courtney Allen for the Georgia Tech Digital Humanities Lab.

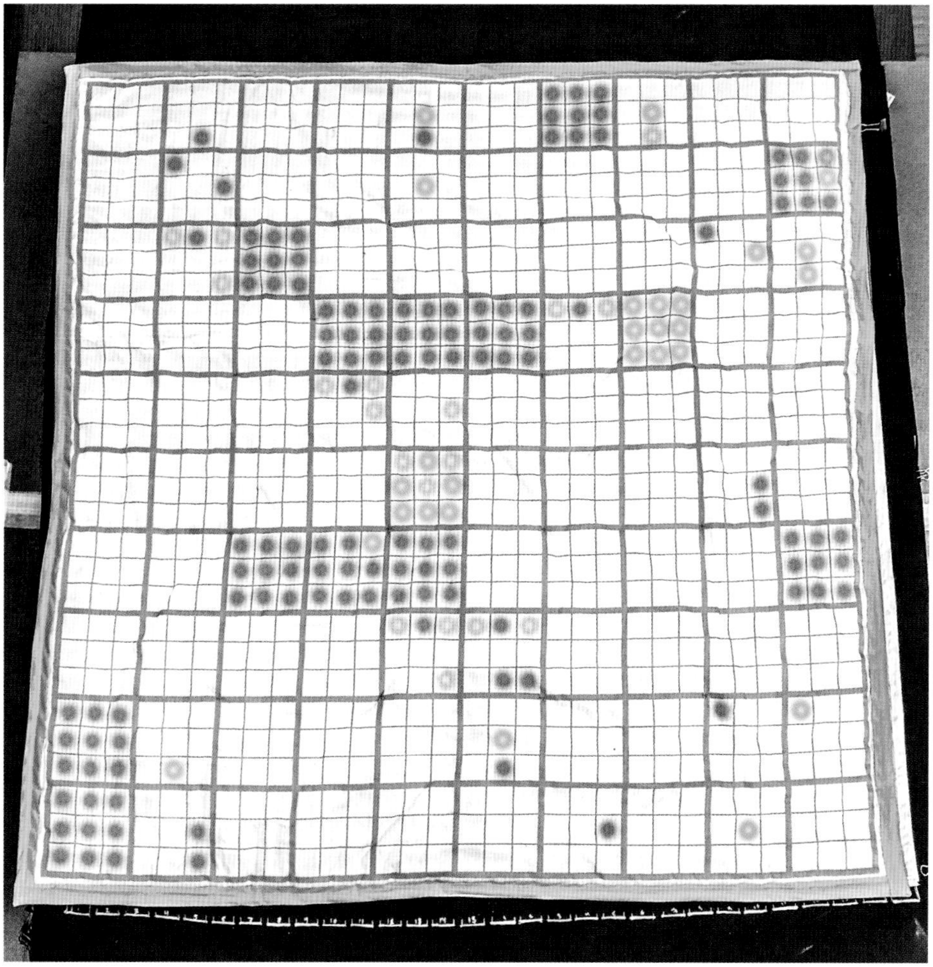

**Figure 3.3** (continued)

the fact that unemployment has been steadily high for the past three years—through the use of the "8 percent unemployment" reference line, the choice to use an area chart instead of a line, and, of course, the title of the graphic.[15] So neither graphic is neutral, but both graphics are factual. As Jonathan Stray says, "The constraints of truth leave a very wide space for interpretation."[16] When visualizing data, the only certifiable fact is that it's impossible to avoid interpretation (unless you simply republish the September jobs report as your visualization, but then it wouldn't be a visualization).

Fields very close to visualization, like cartography, have long seen their work as ideological. But discussions of rhetoric, editorial choices, and power have been far less

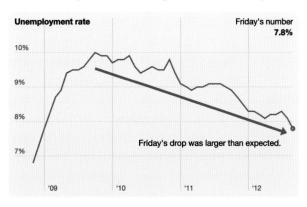

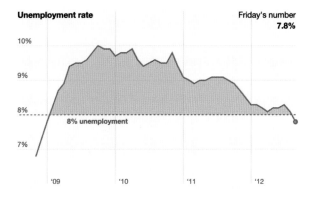

**Figure 3.4**

A data visualization of the September 2012 jobs report from the perspective of Democrats (a) and Republicans (b). The *New York Times* data team shows how simple editorial changes lead to large differences in framing and interpretation. As data journalist Jonathan Stray remarks on these graphics, "The constraints of truth leave a very wide space for interpretation." Images by Mike Bostock, Shan Carter, Amanda Cox, and Kevin Quealy, for the *New York Times*, as cited in *The Curious Journalist's Guide to Data* by Jonathan Stray.

frequent in the field of data visualization. In 2011, Hullman and coauthor Nicholas Diakopoulos wrote an influential paper reasserting the importance of rhetoric for the data visualization community.[17] Their main argument was that visualizing data involves editorial choices: some things are necessarily highlighted, while others are necessarily obscured. When designers make these choices, they carry along with them *framing effects*, which is to say they have an impact on how people interpret the graphics and what they take away from them.

For example, it is standard practice to cite the source of one's data. This functions on a practical level—so that readers may go out and download the data themselves. But this choice also functions as what Hullman and Diakopoulos call *provenance rhetoric* designed to signal the transparency and trustworthiness of the presentation source to end users. Establishing trust between the designers and their audience in turn increases the likelihood that viewers will believe what they see.

Other aspects of data visualization also work to displace viewers' attention from editorial choices to reinforce a graphic's perceived neutrality and "truthiness." After doing a sociological analysis, Helen Kennedy and coauthors determined that four conventions of data visualization reinforce people's perceptions of its factual basis: (1) two-dimensional viewpoints, (2) clean layouts, (3) geometric shapes and lines, and (4) the inclusion of data sources at the bottom.[18] These conventions contribute to the perception of data visualization as objective, scientific, and neutral. Both unemployment graphics from the *New York Times* employ these conventions: the image space is two-dimensional and abstract; the layout is "clean," meaning minimal and lacking embellishment beyond what is necessary to communicate the data; the lines representing employment rates vary smoothly and faithfully against a geometrically gridded background; and the source of the data is noted at the bottom. Either the Democrat or the Republican graphic would have been entirely plausible as a *New York Times* visualization, and very few of us would have thought to question the graphic's framing of the data.

So if plain, "unemotional" visualizations are not neutral, but are actually extremely persuasive, then what does this mean for the concept of neutrality in general? Scientists and journalists are just some of the people who get nervous and defensive when questions about neutrality and objectivity come up. Auditors and accountants get nervous, too. They often assume that the only alternative to objectivity is a retreat into complete relativism and a world in which alternative facts reign and everyone gets a gold medal for having an opinion. But there are other options.

Rather than valorizing the neutrality ideal and trying to expunge all human traces from a data product because of their bias, feminist philosophers have proposed a goal of more complete knowledge. Donna Haraway's idea of the god trick comes from a larger argument about the importance of developing *feminist objectivity*. It's not just data visualization but all forms of knowledge that are *situated*, she explains, meaning that they are produced by specific people in specific circumstances—cultural, historical, and geographic.[19] Feminist objectivity is a tool that can account for the situated nature of knowledge and can bring together multiple—what she terms *partial*—perspectives. Sandra Harding, who developed her ideas alongside Haraway, proposes a concept of *strong objectivity*. This form of objectivity works toward more inclusive knowledge production by centering the perspectives—or *standpoints*—of groups that are otherwise excluded from knowledge-making processes.[20] This has come to be known as *standpoint theory*. To supplement these ideas, Linda Alcoff has introduced the idea of *positionality*, a concept that emphasizes how individuals come to knowledge-making processes from multiple positions, each determined by culture and context.[21] All of these ideas offer alternatives to the quest for a universal objectivity—which is, of course, an unattainable goal.

The belief that universal objectivity should be our goal is harmful because it's always only partially put into practice. This flawed belief is what provoked renowned cardiologist Dr. Nieca Goldberg to title her book *Women Are Not Small Men* because she found that heart disease in women unfolds in a fundamentally different way than in men.[22] The vast majority of scientific studies—not just of heart disease, but of most medical conditions—are conducted on men, with women viewed as varying from this "norm" only by their smaller size.[23] The key to fixing this problem is to acknowledge that all science, and indeed all work in the world, is undertaken by individuals. Each person occupies a particular perspective, as Haraway might say; a particular standpoint, as Harding might say; or a particular set of positionalities, as Alcott might say. And all would agree that research by only men and about only men cannot be universalized to make knowledge claims about all other people in the world.

Disclosing your subject position(s) is an important feminist strategy for being transparent about the limits of your—or anyone's—knowledge claims. Thus, for example, we (the authors) included statements about our own positionalities in the introduction in order to disclose the gender, race/ethnicity, class, ability, education, and other subject positions that informed the writing of this book. Rather than viewing these positionalities as threats or as influences that might have biased our work, we embraced them as offering a set of valuable perspectives that could *frame* our work. This is an approach that we would like to see others embrace as well. Each person's intersecting subject positions are unique, and when applied to data science, they can generate creative and wholly new research questions.

**Data Visceralization**

This embrace of multiple perspectives and positionalities helps to rebalance the hierarchy of reason over emotion in data visualization.[24] How? Since the early 2000s, there
has been an explosion of research about *affect*—the term that academics use to refer to
emotions and other subjective feelings—from fields as diverse as neuroscience, geography, and philosophy. (We discuss affect further in chapter 7.) This work challenges
the thinking that casts emotion out as irrational and illegitimate, even as it undeniably
influences the social, political, and scientific processes of the world. Evelyn Fox Keller,
a physicist turned philosopher, famously employed the Nobel Prize–winning research
of geneticist Barbara McClintock to show how even the most profound of scientific
discoveries are generated from a combination of experiment and insight, reason and
emotion.[25]

Once we embrace the idea of leveraging emotion in data visualization, we can truly
appreciate what sets Periscopic's "US Gun Deaths" graphic apart from the *Washington
Post* graphic or from any number of other gun death charts that have appeared in
newspapers and policy documents. The *Washington Post* graphic, for example, represents death counts as blue ticks on a generic bar chart. If we didn't read the caption, we
wouldn't know whether we were counting gun deaths in the United States or haystacks
in Kansas or exports from Malaysia or any other statistic. In contrast, the Periscopic
visualization leads with loss, grief, and mourning—primarily through its rhetorical
emphasis on counting "stolen years." This draws the attention of viewers to "what
could have been." The counting is reinforced by the visual language for representing
the "stolen years" as grey lines, appropriate for numbers that are rigorously determined
but not technically facts because they come from a statistical model. The visualization
also uses animation and pacing to help us first appreciate the scale of one life, and
then compound that scale 11,419-fold. The magnitude of the loss, especially when
viewed in aggregate and over time, makes a statement of profound truth revealed to us
through our own emotions. It is important to note that emotion and visual minimalism are not incompatible here; the Periscopic visualization shows us how emotion can
be leveraged alongside visual minimalism for maximal effect.

Skilled data artists and designers know these things already, or at least intuit them.
Like the Periscopic team, others are pushing the boundaries of what affective and
embodied data visualization could look like. In 2010, Kelly Dobson founded the Data
Visceralization research group at the Rhode Island School of Design (RISD) Digital +
Media graduate program. The goal for this group was not to visualize data but to *vis-
ceralize* it. Visual things are for the eyes, but visceralizations are representations of data

that the whole body can experience, emotionally as well as physically—data that "we see, hear, feel, breathe and even ingest," writes media theorist Luke Stark.[26]

The reasons for visceralizing data have to do with more than simply creative experimentation. First, humans are not two eyeballs attached by stalks to a brain computer. We are embodied, multisensory beings with cultures and memories and appetites.[27] Second, people with visual disabilities need a way to access the data encoded in charts and dashboards as well. According to the World Health Organization, 253 million people globally live with some form of visual impairment, on the spectrum from limited vision to complete blindness.[28] For reasons of accessibility, Aimi Hamraie, the director of the Mapping Access project at Vanderbilt University, advocates for a form of data visceralization, although not in those exact terms: "Rather than relying entirely on visual representations of data," they explain, "digital-accessibility apps could expand access by incorporating 'deep mapping,' or collecting and surfacing information in multiple sensory formats."[29]

At the moment, however, examples of objects and events that make use of multiple sensory formats are more likely to be found in the context of research labs and galleries and museums. For example, in *A Sort of Joy (Thousands of Exhausted Things)*, the theater troupe Elevator Repair Service joined forces with the data visualization firm the Office of Creative Research to script a live performance based on metadata about the artworks held by New York's Museum of Modern Art (MoMA).[30] With 123,951 works in its collection, MoMA's metadata consists of the names of artists, the titles of artworks, their media formats, and their time periods. But how does an artwork make it into the museum collection to begin with? Major art museums and their collection policies have long been the focus of feminist critique because the question of whose work gets collected translates into the question of whose work is counted in the annals of history.[31] As you might guess, this history has mostly consisted of a parade of white male European "masters," as the Guerrilla Girls project pictured in figure 3.5 reminds us.

In 1989, the Guerrilla Girls, an anonymous collective of women artists, published an infographic: *Do Women Have to Be Naked to Get into the Met. Museum?* The graphic makes a data-driven argument by comparing the gender statistics of artists collected by another New York museum, the Metropolitan Museum of Art (the Met) to the gender statistics of the subjects and models in the artworks. It was designed to be displayed on a billboard, but it was rejected by the sign company because it "wasn't clear enough."[32] If you ask us, it's pretty clear: the Met readily collects paintings in which women are the (naked) subjects but it collects very few artworks created by women artists themselves.

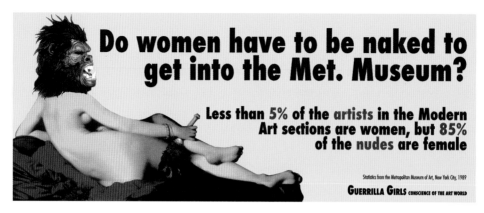

**Figure 3.5**
*Do Women Have to Be Naked to Get into the Met. Museum?* An infographic (of a sort) created by the Guerrilla Girls in 1989, intended to be displayed on a bus billboard. Courtesy of the Guerrilla Girls.

After being thwarted by the sign company, the Guerrilla Girls then paid for the infographic to be printed on posters displayed throughout the New York City bus system, until the Metropolitan Transportation Authority (MTA) cancelled the contract, stating that the figure seemed to have more than a fan in her hand. It is *definitely* more than a fan, but this deliberate understatement reveals the MTA's discomfort with this provocative, activist image.[33]

*A Sort of Joy* deploys wholly different tactics to similar ends. The performance starts with a group of white men standing in a circle in the center of the room. They face out toward the audience, which surrounds them. The men are dressed like stereotypical museum visitors: collared shirts, slacks, and so on. Each wears headphones and holds an iPad on which the names of artists in the collection scroll by. "John," the men say together. We see the iPads scrolling through all the names of the artists in the MoMA collection whose first name is John: John Baldessari, John Cage, John Lennon, John Waters, and so on. Three female performers, also wearing headphones and carrying iPads with scrolling names, pace around the circle of men. "Robert," the men say together, and the names scroll through the Roberts alphabetically. The women are silent and keep walking. "David," the men say together. It soon becomes apparent that the artists are sorted by first name, and then ordered by which first name has the most works in the collection. Thus, the Johns and Roberts and Davids come first, because they have the most works in the collection. But Marys have fewer works, and Mohameds and Marías are barely in the register. Several minutes later, after the men say "Michael,"

"James," "George," "Hans," "Thomas," "Walter," "Edward," "Yan," "Joseph," "Martin," "Mark," "José," "Louis," "Frank," "Otto," "Max," "Steven," "Jack," "Henry," "Henri," "Alfred," "Alexander," "Carl," "Andre," "Harry," "Roger," and "Pierre," "Mary" finally gets her due, spoken by the female performers, the first sound they've made.

For audience members, the experience is slightly confusing at first. Why are the men in a circle? Why do they randomly speak someone's name? And why are those women walking around so intently? But "Mary" becomes a kind of aha moment, highlighting the highly gendered nature of the collection—exactly the same kind of experience of insight that data visualization is so good at producing, according to researcher Martin Wattenberg.[34] From that point on, audience members start to listen differently, eagerly awaiting the next female name. It takes more than three minutes for "Mary" to be spoken, and the next female name, "Joan," doesn't come for a full minute longer. "Barbara" follows immediately after that, and then the men return to reading: "Werner," "Tony," "Marcel," "Jonathan."

From a data analysis perspective, *A Sort of Joy* consists of simple operations: counting and grouping. A bar chart or a tree map of first names could easily have represented the same results. But presenting the dataset as a time-based experience makes the audience wait and listen and experience. It also runs counter to the mantra in information visualization expressed by researcher Ben Shneiderman in the mid-1990s: "Overview first, zoom and filter, then details-on-demand."[35] In this data performance, we do not see the overview first. We hear and see and experience each datapoint one at a time and only slowly construct a sense of the whole. The different gender expressions, body movements, and verbal tones of the performers draw our collective attention to the issue of gender in the MoMA collection. We start to anticipate when the next woman's name will arise. We *feel* the gender differential, rather than *see* it.

This feeling is affect. It comprises the emotions that arise when experiencing the performance, as well as the physiological reactions to the sounds and movements made by the performers, as well as the desires and drives that result—even if that drive is to walk into another room because the performance is disconcerting or just plain long.

Data visceralizations that leverage affect aren't limited to major art institutions. Catherine and artist Andi Sutton led walking tours of the future coastline of Boston based on sea level rise.[36] Interactive artist Mikhail Mansion made a leaning, bobbing chair that animatronically shifts based on real-time shifts in river currents.[37] Nonprofit organizations in Tanzania staged a design competition for data-driven clothing that incorporated statistics about gender inequality and closed the project with a fashion runway show.[38] Artist Teri Rueb stages "sound encounters" between the geologic layers of a landscape and the human body that is affected by them.[39] Simon Elvins drew a giant paper map of pirate radio stations in London that you can actually listen to.[40] A robot designed by Annina Rust decorates real pies with pie charts about gender equality, and then visitors eat them.[41]

These projects may seem to be speaking to another part of brain (or belly) than your standard histograms or network maps, but there is something to be learned from the opportunities opened up by visceralizing data. Deliberately embracing emotions like wonder, confusion, humor, and solidarity enables a valuable form of data maximalism, one that allows for multisensory entry points, greater accessibility, and a range of learning types.

**Visceralizing Uncertainty**

Scientific researchers are now proving by experiment what designers and artists have known through practice: activating emotion, leveraging embodiment, and creating novel presentation forms help people grasp and learn more from data-driven arguments, as well as remember them more fully.[42] As it turns out, visceralizing data may help designers solve one particularly pernicious problem in the visualization community: how to represent uncertainty in a medium that's become rhetorically synonymous with the truth. To this end, designers have created a huge array of charts and techniques for quantifying and representing uncertainty. These include box plots, violin plots (figure 3.6), gradient plots, and confidence intervals.[43] Unfortunately, however, people are terrible at recognizing uncertainty in data visualizations, even when they're explicitly told that something is uncertain. This remains true even for some researchers who use data themselves![44]

For example, let's consider the Total Electoral Votes graphic displayed as part of the *New York Times* live online coverage of the 2016 presidential election (figure 3.7). The blue and red lines represent the *New York Times*'s best guess at the outcomes over the course of election night and into the following day. The gradient areas show the degree of uncertainty that surrounded those guesses, with the darker inner area showing electoral vote outcomes that came up 25 percent to 75 percent of the time, and the lighter outer areas showing outcomes that came up 75 percent to 95 percent and 5 percent to 25 percent of the time, respectively. If you look closely at the far left of the graphic, which represents election night (everything prior to the 12:00 a.m. axis label), the outcome of Trump winning and Clinton losing easily falls within the 5 to 25 percent likelihood range.

Although many election postmortems pronounced the 2016 election the Great Failure of Data and Statistics, because most simulations and other statistical models

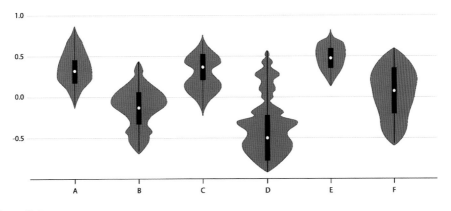

**Figure 3.6**

What is the best way to communicate uncertainty in a medium that looks so certain? Designers have created diverse chart forms to try to solve this problem. Depicted here are five violin plots; each shows the distribution of data along with their probability density (the purple part). You could also think of this form as a beautiful purple vagina, as the comic *xkcd* has observed; see https://www.xkcd.com/1967/. Images from the Data Visualisation Catalogue.

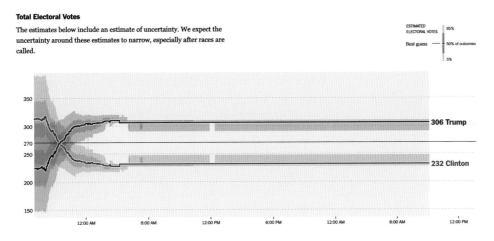

**Figure 3.7**

A 2016 chart from the *New York Times* that uses opacity—darker and lighter shades of blue and red—to indicate uncertainty. Images by Gregor Aisch, Nate Cohn, Amanda Cox, Josh Katz, Adam Pearce, and Kevin Quealy for the *New York Times*.

suggested that Clinton would win, most forecasts did include the possibility of a Trump victory. The underlying problem was not the failure of data but the difficulties of depicting uncertainty in visual form. People are just not sufficiently trained to recognize uncertainty in graphics such as this. Rather than interpreting the gradient bands as probabilities (e.g., Trump had a 20 percent chance of winning at 6 p.m.), people may interpret them as votes (e.g., Trump had 20 percent of the vote at 6 p.m.). Or they may ignore the gradient bands altogether and look only at the lines. Or they see Clinton on top and assume she is winning. This is called *heuristics* in psychology literature—using mental shortcuts to make judgments—and it happens all the time when people are asked to assess probabilities.[45] A large part of the problem is that visualization conventions reinforce these misjudgments. The graphics look so certain, even when they are trying their very hardest to visually illustrate *un*certainty!

Jessica Hullman, whose work on rhetoric we've already mentioned, offers one solution to this problem. Instead of creating fixed plots as in the *New York Times* example that represents uncertainty in aggregate or static form, Hullman advocates for rendering experiences of uncertainty.[46] In other words, *leverage emotion and affect* so that people experience uncertainty perceptually. Or, to invoke a common refrain from rhetorical training and design schools, "show, don't tell." Rather than *telling* people that they are looking at uncertainty while employing a certain-looking graphic style— which creates conditions ripe for those pesky heuristics to intervene—make them *feel* the uncertainty.

We can see a good example of showing uncertainty in action on the same *New York Times* live election coverage webpage. At the top of the page was a gauge (figure 3.8) that showed the *New York Times*'s real-time prediction of who was likely to win the race, with a gradient of categories that ranged from medium blue ("Very Likely" that Clinton would win) to medium red ("Very Likely" that Trump would win). But the needle did not stay in one place. It *jittered* between the twenty-fifth and seventy-fifth percentiles, showing the range of outcomes that the *New York Times* was then predicting, based on simulations using the most recent data. At the beginning of the day, the range of motion was fairly wide but still only showed the needle on the Hillary Clinton side. As the night went on, its range narrowed, and the center moved closer and closer to the red side of the gauge. By 9 p.m., the needle jittered just a little, and on the Trump side only.

A number of *New York Times* readers were aggressive in their dislike of the jitter, calling it "irresponsible" and "unethical" and "the most stressful thing I've ever looked at online and I've seen a lot of stressful shit."[47] In response, Gregor Aisch, one of the designers of the gauge, defended it, explaining that "we thought (and still think!) this

**Figure 3.8**
The controversial "jittering" election gauge featured in the *New York Times* coverage of the 2016 presidential election. Images by Gregor Aisch, Nate Cohn, Amanda Cox, Josh Katz, Adam Pearce, and Kevin Quealy for the *New York Times*.

movement actually helped demonstrate the uncertainty around our forecast, conveying the relative precision of our estimates."[48] So was this "unethical" design, or the sophisticated communication of uncertainty?

Building off of Hullman's work, we'd say that the answer is the latter. The jittering election gauge was actually exhibiting current best practices for communicating uncertainty. It gave people the perceptual, intuitive, *visceral, and emotional* experience of uncertainty to reinforce the quantitative depiction of uncertainty. The fact that it unsettled so much of the *New York Times* readership probably had less to do with the ethics of the visualization and more to do with the outcome of the election. So score one for emotion in the task of representing uncertainty.

## Don't Never Do a God Trick

So does this mean that all election graphics should jitter? Or that data visceralizations are categorically superior to visual graphics? Or that a data visualization framed around emotion is always the "better" choice for the design task at hand? Or that designers should never use the god trick to present a view from above?

The answer to these questions may surprise you: definitively no! If there is any single rule in design, it's that context is queen. A design choice made in one context or for one audience does not translate to other contexts or audiences. Simply stated: it is never a good idea to say "never" in design. We delve deeper into the importance of context when working with data in chapter 6.

Let's take the god trick as an example. Even though the god trick can do harm—for example, in the form of those racist, objective-looking redlining maps from chapter 2—there are also good reasons to use the god trick as a form of recuperation, contestation, or empowerment. As renowned data visualization designer Fernanda Viégas says, "The kind of overview that data visualization provides is one of the superpowers I treasure the most."[49]

It is this "superpower"—the aerial view from no body—that we see put into practice in the map *Coming Home to Indigenous Place Names in Canada* (figure 3.9a). Margaret Pearce, a cartographer and member of the Citizen Potawatomi Nation, spent fifteen months collecting Indigenous place names from First Nations, Métis, and Inuit peoples. The map depicts the land that is known in a contemporary Anglo-Western context as Canada, but without any of the common colonial orientation points, like the boundaries of the provinces or the locations of major cities like Ottawa, Montréal, and Nova Scotia. For example, you can see in figure 3.9b that places like Kinoomaagewaabi-kaang ("Teaching rocks") and Odawa ("Traders") and Kazabazua gajiibajiwan ("River runs under") fill the area usually described as Toronto. As the publisher's description states, "The names are ancient and recent, both in and outside of time, and they express and assert the Indigenous presence across the Canadian landscape in Indigenous languages."[50]

*Coming Home to Indigenous Place Names in Canada* leverages the authority of the god's-eye view to challenge the colonizer's view, to advocate for a "reseeing" of the land under terms of engagement that recognize Indigenous sovereignty and respect Indigenous homelands. The extent of geographic territory included and the sheer number of names asserts the Indigenous presence as major, originary, and ongoing. This is by design. Pearce intentionally created the map with the same paper size, fold, scale, and projection as the map published by Natural Resources Canada, the country's geographic authority.[51] By replicating these design features, *Coming Home* proposes an alternative yet equally authoritative conception of national identity.[52]

There is actually another twist on top of its intentionally authoritative view: the map does *not* reveal everything.[53] As the cartographer, Pearce proceeded with the Indigenous methodologies of respect, responsibility, and reciprocity.[54] What this meant in practice was caring for each name as Indigenous cultural property—securing the permissions for each name and respecting when communities did not want to share English translations of the name. She is definitive about the fact that the names are not data: "The place names aren't datasets. The place names are cultural property being shared with the map that come from people." Protecting that cultural property also meant protecting the exact geographic location of each site from being shared with outsiders. This is where the scale of the god trick has protective effects: because it is generalized, at

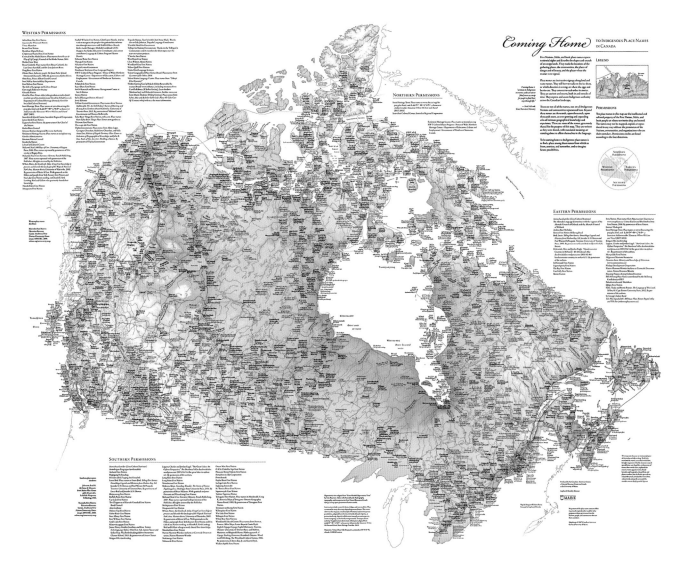

**Figure 3.9**

Overview and detail view of *Coming Home to Indigenous Place Names in Canada* (2017). (a) The map depicts First Nations, Métis, and Inuit place names collected from many tribes and nations across what is today more commonly called Canada. (b) The detail view depicts Indigenous names from the area around Toronto. Map by Margaret W. Pearce; map design copyright 2017 Canadian-American Center, University of Maine.

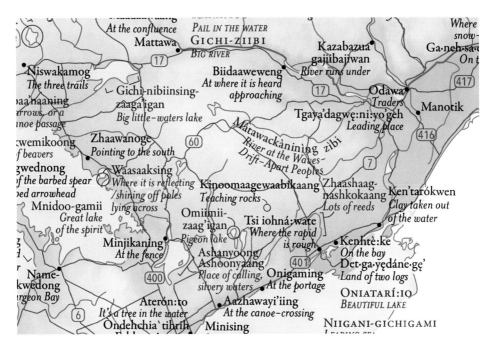

**Figure 3.9** (continued)

Place names in this detail image shared by permission of the following:

Alan Corbiere.

Hiio Delaronde and Jordan Engel, "Haudenosaunee Country in Mohawk," *The Decolonial Atlas*, decolonialatlas.wordpress.com/2015/02/04/haudenosaunee-country-in-mohawk-2/, by permission of the authors.

Charles Lippert and Jordan Engel, "The Great Lakes: An Ojibwe Perspective," *The Decolonial Atlas*, decolonialatlas.wordpress.com/2015/04/14/the-great-lakes-in-ojibwe-v2/, by permission of the authors.

Kitigan Zibi Anishinabeg.

Brian McInnes, *Sounding Thunder: The Stories of Francis Pegahmagabow* (East Lansing: Michigan State University Press, 2016), by permission of Brian McInnes, with gratitude to James Dumont and Wasauksing First Nation.

Woodland Cultural Centre, place names from Frances Froman, Alfred Keye, Lottie Keye, and Carrie Dyck, *English-Cayuga/Cayuga-English Dictionary* (Toronto, Ontario: University of Toronto Press); and Marianne Mithun and Reginald Henry, *Wadewayęstanih. A Cayuga Teaching Grammar* (Brantford, Ontario: Woodland Publishing, The Woodland Cultural Centre, 1984), by permission of Amos Key Jr. and Carrie Dyck.

1:5,000,000, it serves to communicate general location without pinpointing exact location. In this case, the god trick communicates Indigenous authority while preserving Indigenous autonomy.

Although the map has been published and the project is ostensibly finished, Pearce's commitment to and care for the Indigenous names continues. Each time the map is reproduced, such as in the detail image in figure 3.9b, Pearce writes to the communities to whom the names belong, explains the proposed context of the names, and requests permission for the names to be reproduced in that context. You can see these permissions as they were granted in the caption for the figure. Pearce proceeds with sensitivity to her own positionalities, as well as the depth of meaning of each particular place, its name, and the community. Place names are "relations," she says. "And they're not my relations, it's not my territory. ... They co-exist as relations that are incorporated into the community." Cartography, then, becomes not a straightforward representation of "what is" in some absolute sense. Rather, *Coming Home* is a map of relations, conversations, and shared investments across difference in the landscape.

## Elevate Emotion and Embodiment

The third principle of data feminism, and the theme of this chapter, is to *elevate emotion and embodiment*. As we have shown, these are crucial if often undervalued tools in the data communication toolbox. They help avoid inadvertently conveying the view from no body: the view from an imaginary and impossible standpoint that does not and cannot exist.

How has the whole picture, the overview, or the god trick come to be seen as rational and objective at all? How did the field of data visualization arrive at a set of conventions that prioritize rationality, devalue emotion, and completely ignore the nonseeing organs in the human body? Who is excluded when only vision is included?

Any knowledge community inevitably places certain things at the center and casts others out, in the same way that male bodies are almost always taken as the norm in scientific studies while female bodies are viewed as deviations, or that abled bodies are almost always taken as primary design cases while disabled bodies require a design retrofit. Feminist human-computer interaction (HCI) scholar Shaowen Bardzell asserts designers should look first to those at the margins: the people pushed to the margins in any particular design context demonstrate who and what the system is trying to exclude.[55] Subsequent work in HCI insists that designers then work to "demarginalize the 'margins' by recognizing intersections that exist, and engaging solidarity to navigate towards equity and inclusion."[56]

In the case of data visualization, what is excluded is emotion and affect, embodiment and expression, embellishment and decoration. These are the aspects of human experience associated with women, and thus devalued by the logic of our master stereotype. But Periscopic's gun violence visualization shows how visual minimalism can coexist with emotion for maximum impact. Works like *A Sort of Joy* demonstrate that data communication can be visceral—an experience for the whole body. And *Coming Home to Indigenous Place Names in Canada* establishes that the god trick itself can be used to simultaneously engender emotion and challenge injustice.

Rather than making universal rules and ratios (think: data-ink) that exclude some aspects of human experience in favor of others, our time is better spent working toward a more holistic and more inclusive ideal. All design fields, including visualization and data communication, are fields of possibility. Black feminist sociologist Patricia Hill Collins describes an ideal knowledge situation as one in which "neither ethics nor emotions are subordinated to reason."[57] Rebalancing emotion and reason opens up the data communication toolbox and allows us to focus on what truly matters in a design process: honoring context, architecting attention, and taking action to defy stereotypes and reimagine the world.[58]

# 4 "What Gets Counted Counts"

## Principle: Rethink Binaries and Hierarchies

*Data feminism requires us to challenge the gender binary, along with other systems of counting and classification that perpetuate oppression.*

"Sign in or create an account to continue." At a time in which every website seems to require its own user account, these words often elicit a groan—and the inevitability of yet another password that will soon be forgotten. But for people like Maria Munir, the British college student who famously came out as nonbinary to then president Barack Obama on live TV, the prospect of creating a new user account is more than mere annoyance.[1] Websites that require information about gender as part of their account registration process almost always only provide a binary choice: "male or female."[2] For Munir, those options are insufficient. They also take an emotional toll: "I wince as I'm forced to choose 'female' over 'male' every single time, because that's what my passport says, and ... being non-binary is still not legally recognised in the UK," Munir explains.[3]

For the millions of nonbinary people in the world—that is, people who are not *either* male *or* female, men *or* women—the seemingly simple request to "select gender" can be difficult to answer, if it can be answered at all.[4] Yet when creating an online user account, not to mention applying for a national passport, the choice between "male" or "female," and only "male" or "female," is almost always the only one.[5] These options (or the lack thereof) have consequences, as Munir clearly states: "If you refuse to register non-binary people like me with birth certificates, and exclude us in everything from creating bank accounts to signing up for mailing lists, you do not have the right to turn around and say that there are not enough of us to warrant change."[6]

"What gets counted counts," feminist geographer Joni Seager has asserted, and Munir is one person who understands that.[7] What is counted—like being a man or a woman—often becomes the basis for policymaking and resource allocation. By contrast, what is not counted—like being nonbinary—becomes invisible (although there are also good reasons for being invisible in some contexts, and we'll come back to those shortly). Seager's research focus is gender, the environment, and policy (see figure 4.1), and she points out that there is more global data on gender being collected than ever before. And yet, these data collection efforts often still leave many people out, including nonbinary people, lesbians, and older women. Even among those who are counted, they tend to be asked very narrow questions about their lives. "Women in poor countries seem to be asked about 6 times a day what kind of contraception they use," Seager quipped in a lecture at the Boston Public Library. "But they are not asked about whether they have access to abortion. They are not asked about what sports they like to play."[8]

The process of converting qualitative experience into data can be empowering, and even has the potential to be healing, as we address toward the end of this chapter. When thoughtfully collected, quantitative data can be empowering too. So many issues of structural inequality are problems of scale, and they can seem anecdotal until they are viewed as a whole. For instance, in 2014, when film professors Shelley Cobb and Linda Ruth Williams set out to count the women involved in the film industry in the United Kingdom, they encountered a woman screenwriter who had never before considered the fact that in the United Kingdom, women screenwriters are outnumbered by screenwriters of other genders at a rate of four to one.[9] She expressed surprise: "I didn't even know that because screenwriters never get to meet each other."[10]

A similar situation occurred in the example of ProPublica's reporting on maternal mortality in the United States, as discussed in chapter 1. The investigative team set out to count all the mothers who had died in childbirth or from complications shortly thereafter. They interviewed many families of women who had died while giving birth, but, like the screenwriter, few of the families were aware that the phenomenon extended beyond their own daughters and sisters, partners and friends. This lack of data, like the issue of maternal mortality itself, is another structural problem, and it serves as an example of why feminist sociologists like Ann Oakley have long advocated for the use of quantitative methods alongside qualitative ones. Without quantitative research, Oakley explains, "it is difficult to distinguish between personal experience and collective oppression."[11]

But before collective oppression can be identified through analyses like the one that ProPublica conducted, the data must exist in the first place. Which brings us back to Maria Munir and the importance of collecting data that reflects the population it purports to represent. On this issue, Facebook was ahead of the curve when, in 2014, it expanded the gender categories available to registered users from the standard two to over fifty choices, ranging from "Genderqueer" to "Neither"—a move that was widely praised by a range of LGBTQ+ advocacy groups (figure 4.2a).[12] One year later, when the

# Maternity and paternity leave
## Legal requirements and paid support

In most countries without government funding, employers are required to provide paid support

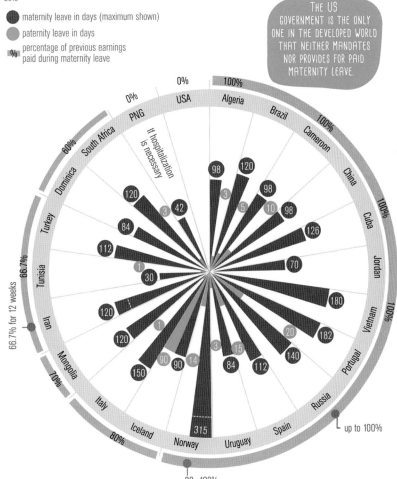

**Figure 4.1**
Maternity and paternity leave around the globe from *The Women's Atlas*, 5th edition (2018). Joni Seager and Annie Olson started working on the first women's atlas in 1980, when there was very little global data on women. The book is now in its fifth edition, but Seager highlights that there are still huge gender data gaps. Image courtesy of Joni Seager and Penguin Books.

company abandoned its select-from-options model altogether, replacing the "Gender" dropdown menu with a blank text field, the decision was touted as even more progressive (figure 4.2b).[13] Because Facebook users could input any word or phrase to indicate their gender, they were at last unconstrained by the assumptions imposed by any preset choice.[14]

But additional research by information studies scholar Rena Bivens has shown that below the surface, Facebook continues to resolve users' genders into a binary: either "male" or "female."[15] Evidently, this decision was made so that Facebook could allow its primary clients—advertisers—to more easily market to one gender or the other. Put another way, even if you can choose the gender that you show to your Facebook friends, you can't change the gender that Facebook provides to its paying customers (figure 4.3). And this discrepancy leads right back to the issues of power we've been discussing since the start of this book: it's corporations like Facebook, and not individuals like Maria Munir, who have the power to control the terms of data collection. This remains true even as it is people like Munir who have personally (and often painfully) run up against the limits of those classification systems—and who best know how they could be improved, remade, or in some cases, abolished altogether.

Feminists have spent a lot of time thinking about classification systems because the criteria by which people are divided into the categories of man and woman is exactly that: a classification system.[16] And while the gender binary is one of the most widespread classification systems in the world today, it is no less constructed than the Facebook advertising platform or, say, the Golden Gate Bridge. The Golden Gate Bridge is a physical structure; Facebook ads are a virtual structure; and the gender binary is a conceptual one. But all these structures were created by people: people living in a particular place, at a particular time, and who were influenced—as we all are—by the world around them.[17]

Many twentieth-century feminist scholars attempted to address the social construction of gender by treating gender as something separate from sex. But that distinction is increasingly breaking down. Both gender and sex are social constructs, as it turns out. Even sex, which today is sometimes still considered in biologically essential terms, has a distinct cultural history. It can be traced to a place (Europe) and a time (the Enlightenment) when new theories about democracy and what philosophers called "natural rights" began to emerge. Before then, there was a *hierarchy* of the sexes, with men on the top and women on the bottom. (Thanks, Aristotle![18]) But there wasn't exactly a *binary* distinction between those two (or any other) sexes. In fact, according to historian of sex and gender Thomas Laqueur, most people believed that women were just inferior men, with penises located inside instead of outside of their bodies and that— for reals!—could descend at any time in life.[19]

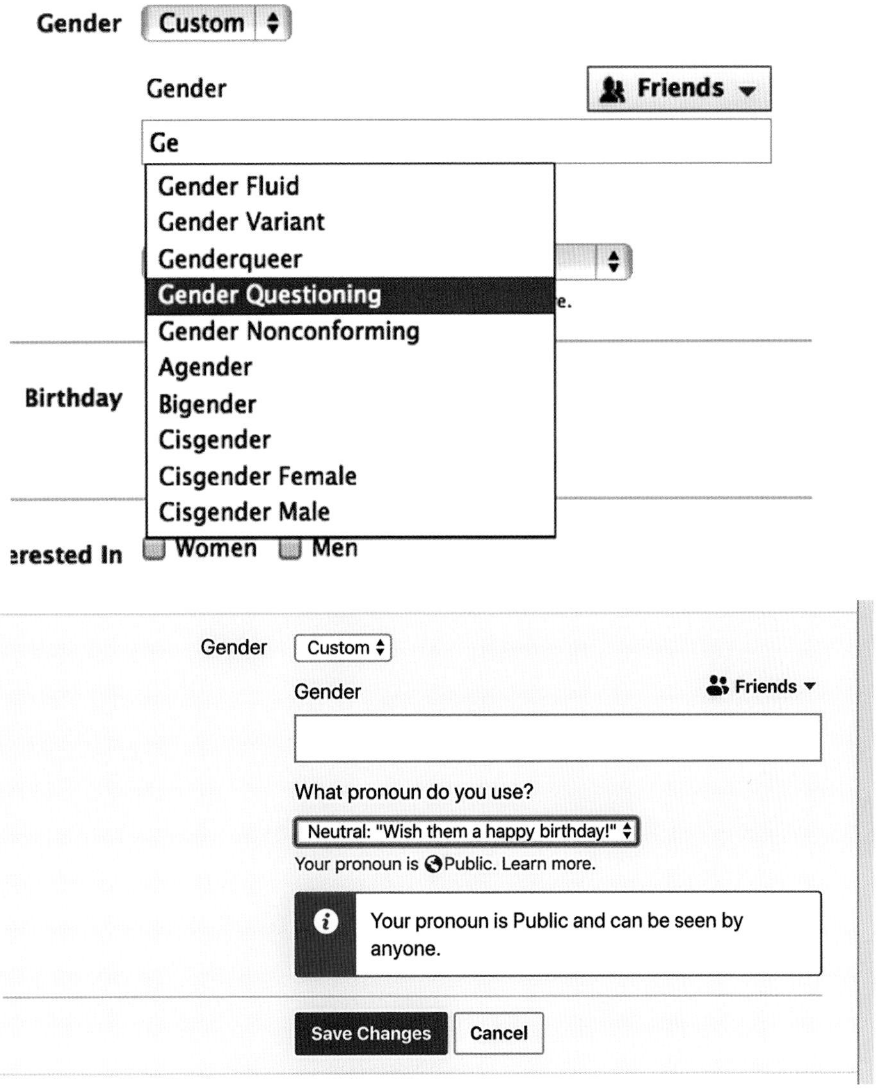

**Figure 4.2**
(a) Facebook's initial attempt to allow users to indicate additional genders, circa 2014. Image courtesy of *Slate*. (b) Facebook's updated gender field, circa 2018. Screenshot by Lauren F. Klein.

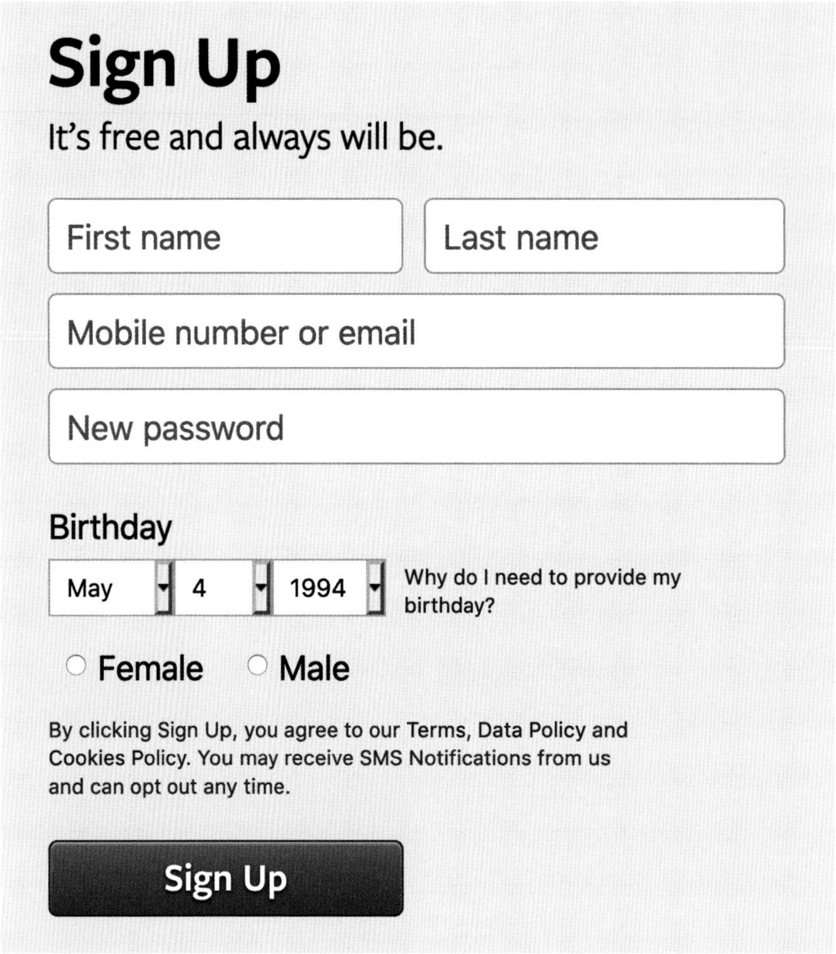

**Figure 4.3**
Detailed view of Facebook's new account creation page, circa 2018. Note that you still have to choose "Female" or "Male"—a binary choice—when you sign up. Screenshot by Lauren F. Klein.

For the idea of a sex binary to gain force, it would take figures like Thomas Jefferson declaring that all men were created equal, and entire countries like the United States to be founded on that principle. Once that happened, political leaders began to worry about what, exactly, they had declared: to whom did the principle of equality apply? All sorts of systems for classifying people have their roots in that era—not only sex but also, crucially, race.[20] Before the eighteenth century, Western societies understood race as a concept tied to religious affiliation, geographic origin, or some combination of both. Race had very little to do with skin color until the rise of the transatlantic slave trade, in the seventeenth century.[21] And even then, race was still a hazy concept. It would take the scientific racism of the mid-eighteenth century for race to begin to be defined by Western societies in terms of black and white.

Take Carl Linnaeus, for example, and the revolutionary classification system that he is credited with creating.[22] Linnaeus's system of binomial classification is the one that scientists still use to today to classify humans and all other living things. But Linnaeus's system didn't just include the category of *homo sapiens*, as it turns out. It also incorrectly—but as historians would tell you, unsurprisingly—included five subcategories of humans separated by race. (One of these five was set aside for mythological humans who didn't exist in real life, in case you're still ready to get behind his science.) But Linnaeus's classification system wasn't even the worst of the lot. Over the course of the eighteenth century, increasingly racist systems of classification began to emerge, along with pseudosciences like comparative anatomy and physiognomy. These allowed elite white men to provide a purportedly scientific basis for the differential treatment of people of color, women, disabled people, and gay people, among other groups. Although those fields have long since been discredited, their legacy is still visible in instances as far-ranging as the maternal health outcomes that we've already discussed, to the divergent rates of car insurance that are offered to Black vs. white drivers, as described in an investigation conducted by ProPublica and Consumer Reports.[23] What's more, as machine learning techniques are increasingly extended into new domains of human life, scientific racism is itself returning. Pointing to and debunking one machine learning technique that employs images of faces in an attempt to classify criminals, three prominent artificial intelligence researchers—Blaise Agüera y Arcas, Margaret Mitchell, and Alexander Todorov—have asserted that scientific racism has "entered a new era."[24]

A simple solution might be to say, "Fine, then. Let's just not classify anything or anyone!" But the flaw in that plan is that data must be classified in some way to be put to use. In fact, by the time that information becomes data, it's already been classified in some way. Data, after all, is information made *tractable*, to borrow a term from computer science. "What distinguishes data from other forms of information is that it can be processed by a computer, or by computer-like operations," as Lauren has written in an essay coauthored with information studies scholar Miriam Posner.[25] And to enable those operations, which range from counting to sorting and from modeling to visualizing, the data must be placed into some kind of category—if not always into a conceptual category like gender, then at the least into a computational category like *Boolean* (a type of data with only two values, like true or false), *integer* (a type of number with no decimal points, like 237 or –1), or *string* (a sequence of letters or words, like "this").

Classification systems are essential to any working infrastructure, as information theorists Geoffrey Bowker and Susan Leigh Star have argued in their influential book *Sorting Things Out*.[26] This is true not only for computational infrastructures and conceptual ones, but also for physical infrastructures like the checkout line at the grocery store. Think about how angry a shopper can get when they're stuck in the express line behind someone with more than the designated fifteen items or less. Or, closer to home, think of the system you use (or should use) to sort your clothes for the wash. It's not that we should reject these classification systems out of hand, or even that we could if we wanted to. (We're pretty sure that no one wants all their socks to turn pink.) It's just that once a system is in place, it becomes naturalized as "the way things are." This means we don't question how our classification systems are constructed, what values or judgments might be encoded into them, or why they were thought up in the first place. In fact—and this is another point made by Bowker and Star—we often forget to ask these questions until our systems become objects of contention, or completely break down.

Bowker and Star give the example of the public debates that took place in the 1990s around the categories of race employed on the US Federal Census. At issue was whether people should be able to choose multiple races on the census form. Multiracial people and their families were some of the main proponents of the option, who saw it as a way to recognize their multiple identities rather than forcing them to squeeze themselves into a single, inadequate box. Those opposed included the Congressional Black Caucus as well as some Black and Latinx civil rights groups that saw the option as potentially reducing their representative voice.[27] Ultimately, the 2000 census did allow people to choose multiple races, and millions of people took advantage of it. But the debates around that single category illustrate how classification gets complicated quickly, and with a range of personal and political stakes.[28]

Classification systems also carry significant material consequences, and the US Census provides an additional example of that. Census counts are used to draw voting districts, make policy decisions, and allocate billions of dollars in federal resources. The recent Republican-led proposal to introduce a question about citizenship status on the 2020 census represents an attempt to wield this power to very pointed political ends. Because undocumented immigrants know the risks, like deportation, that come with being counted, they are less likely to complete the census questionnaire. But because both political representation and federal funding are allocated according to the number and geographic areas of people counted in the census, undercounting undocumented immigrants (and the documented immigrants they often live with) means less voting power—and fewer resources—accorded to those groups. This is a clear example of what we term the *paradox of exposure*: the double bind that places those who stand to significantly gain from being counted in the most danger from that same counting (or classifying) act.

In each of these cases, as is true of any case of not fitting (or not wanting to fit) neatly into a box, it's important to ask whether it's the categories that are inadequate, or whether—and this is a key feminist move—it's the system of classification itself. Lurking under the surface of so many classification systems are false binaries and implied hierarchies, such as the artificial distinctions between men and women, reason and emotion, nature and culture, and body and world. Decades of feminist thinking have taught us to question why these distinctions have come about; what social, cultural, or political values they reflect; what hidden (or not so hidden) hierarchies they encode; and, crucially, whether they should exist in the first place.

## Questioning Classification Systems

Let's spend some time with an actual person who has started to question the classification systems that surround him: one Michael Hicks, an eight-year-old Cub Scout from New Jersey. Why is Mikey, as he's more commonly known, so concerned about classification? Well, Mikey shares his name with someone who has been placed on a terrorist watch list by the US federal government. As a result, Mikey has also been classified as a potential terrorist and is subjected to the highest level of airport security screening every time that he travels. "A terrorist can blow his underwear up and they don't catch him. But my 8-year-old can't walk through security without being frisked," his mother lamented to Lizette Alvarez, a reporter for the *New York Times* who covered the issue in 2010.[29]

Of course, in some ways, Mikey is lucky. He is white, so he does not run the risk of racial profiling—unlike, for example, the many Black women who receive TSA pat-downs due to their natural hair.[30] Moreover, Mikey's name sounds Anglo-European, so he does not need to worry about religious or ethnic profiling either—unlike, for another example, people named Muhammad who are disproportionately pulled over by the police due to their Muslim name.[31] But Mikey the Cub Scout still helps to expose the brokenness of some of the categories that structure the TSA's terrorist classification system; the combination of first and last name is simply insufficient to classify someone as a terrorist or not.

Or, consider another person with a history of bad experiences at the (literal) hands of the TSA. Sasha Costanza-Chock is nonbinary, like Maria Munir. They are also a design professor at MIT, so they have a lot of experience both living with and thinking through oppressive classification systems. In a 2018 essay, "Design Justice, A.I., and Escape from the Matrix of Domination," they give a concrete example of why design justice is needed in relation to data.[32] The essay describes how the seemingly simple system employed by the operators of those hands-in-the-air millimeter-wave airport security scanning machines is in fact quite complex—and also fundamentally flawed.

Few cisgender people are aware of the fact that before you step into a scanning machine, the TSA agent operating the machine looks you up and down, decides whether you are a man or a woman, and then pushes a button to select the corresponding gender on the scanner's touchscreen interface. That human decision loads the algorithmic profile for either male bodies or female ones, against which your body's measurements are compared. If your measurements diverge from the statistical norm of that gender's body—whether the discrepancy is because you're concealing a deadly weapon, because your body doesn't fit neatly into either of the two categories that the system has provided, or because the TSA agent simply made the wrong choice—you trigger a "risk alert." Then, in an act of what legal theorist Dean Spade terms *administrative violence*, you are subjected to the same full-body pat-down as a potential terrorist.[33] Here it's not that the scanning machines rely upon an insufficient number of categories, as in the case of Mikey the Cub Scout, or that they employ the wrong ones, as Mikey's mom would likely say. It's that the TSA scanners shouldn't rely on gender to classify air travelers to begin with. (And while we're going down that path, how about we imagine a future without a state agency that systematically pathologizes Black women and trans people and Cub Scouts in the first place?)

So when we say that what gets counted counts, it's folks like Sasha Costanza-Chock or Mikey Hicks or Maria Munir that we're thinking about. Because flawed classification systems—like the one that underlies the airport scanner's risk-detection algorithm or the one that determines which names end up on terrorist watch lists or simply (simply!) the gender binary—are not only significant problems in themselves, but also symptoms of a more global condition of inequality. The matrix of domination, which we introduced in chapter 1, describes how race, gender, and class (among other things) intersect to enhance opportunities for some people and constrain opportunities for others.[34] Under the matrix of domination, normative bodies pass through scanners, borders, and bathrooms with ease; these systems have been designed by people like them, for people like them, with an aim—sometimes explicit—of keeping people not like them out.[35]

As these examples help to show, the forces that operate through the matrix of domination are sneaky and diffuse. And they show up everywhere—even in pockets on pants. A recent journalistic investigation of the size of pockets in eighty pairs of men's and women's jeans confirmed what women (and men and nonbinary people who wear women's jeans) have been saying anecdotally for years: that their pants pockets just aren't big enough (figure 4.4).[36] More specifically, the pockets of jeans designed for women are 48 percent shorter and 6.5 percent narrower than the pockets of jeans designed for men. This size does matter! According to the same study, only 40 percent

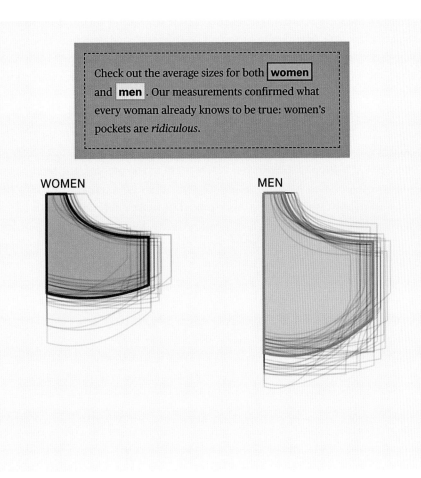

**Figure 4.4**
From "Someone Clever Once Said Women Were Not Allowed Pockets," a comparative study of pockets in women's and men's jeans by The Pudding (2018). Visualization by Jan Diehm and Amber Thomas for The Pudding.

of the front pockets of women's jeans can fit a smartphone, and less than half "can fit a wallet *specifically* designed to fit in front pockets." Hence the thriving market for women's handbags (to hold the aforementioned front-pocket wallet) and for replacement smartphone screens (for when your phone invariably falls out of your too-small pocket and cracks).

Now, the designers of any particular pair of women's jeans are almost certainly not thinking: "Let's oppress women by making their pockets too small." They are probably only thinking about what looks nice. But what looks nice has a history too. Before the seventeenth century, "pockets" were external sacks on strings that could be tied above or below other garments. But starting in the 1600s, men's clothing began to feature internal pockets. Meanwhile, women's clothing became increasingly close-cut. By the late eighteenth century, the women's pocket reached its breaking point, resulting in emergence of a new fashion item called a reticule, otherwise known as a purse. These tiny handbags were made out of cloth and, according to the Victoria and Albert Museum's helpful online history of pockets, could not hold very much.[37] And yet, as the museum curators point out, in an era in which most people shared all of their shelves and dressers, these reticules were one of the few places for women to store any items they wanted to keep to themselves. Fast forward to the present, and women (and people who wear women's fashion) must still carry their belongings outside of their clothes and on public display. They're also limited in their ability to use both of their hands at the same time. It's (mostly) a minor annoyance, but it's one way among many that the *patriarchy*—a term that describes the combination of legal frameworks, social structures, and cultural values that contribute to the continued male domination of society—inadvertently and invisibly reproduces itself. In this case, it's pants—perhaps even the ones you're wearing right now—that compound and consolidate the patriarchy's oppressive force.

In addition to pants pockets, one of the other things that upholds the patriarchy is, as it turns out, our ideas about gender itself. We've already asserted that *gender is a social construct*, but what does this phrase really mean? Queer theorist Judith Butler has long maintained that gender is best understood as a repeated performance, a set of categories that cohere by, for instance, wearing jeans with small pockets (or no pockets at all) or by participating in an activity that is similarly gender-coded, like child-rearing, or—importantly for Butler—having heterosexual sex.[38] These *performative acts*, as she terms them, repeated so many times that they become taken as fact, are what define the gender categories that we have today. Butler's idea of gender as performative moves away from an essentialist conception of the term: the idea that there is some innate or "essential" criteria that makes one, for instance, a woman or man. But these performances still reinforce the *categories* of gender, she reminds us, even if the actions and activities that determine them are not innate.

Gender is certainly complicated. This is one thing about which most contemporary scholars of gender largely agree. Conceptions of gender in health and clinical fields are also evolving as well. For example, the American Medical Association now calls gender a "spectrum" rather than a binary, and as of 2018 it issued a firm statement that "sex and gender are more complex than previously assumed."[39] But it's important to remember that there have always been more variations in gender identity and expression than most Anglo-Western societies have cared to acknowledge or to collectively remember. This is evidenced in the range of regional and vernacular terms, such as *kothi*, *hijra*, and *dhurani*, that are currently used to describe the genders of people across South Asia that fall outside the binary; we see it in the additional umbrella terms, such as *two-spirit*, that describe people in some North American Indigenous communities; and many more.[40] Not to mention that some people are gender-fluid, meaning their gender identity may shift from day to day, year to year, or situation to situation. And yet—at least in a US context—gender data is still almost always collected in the binary categories of "male" and "female" and visually represented by some form of binary division as well.[41] This remains true even as a 2018 Stanford study found that, when given the choice among seven points on a gender spectrum, more than two-thirds of the subjects polled placed themselves somewhere in the middle.[42]

As survey designers, and data scientists more generally, there would seem to be an obvious response to the Stanford report: collect gender data in more than binary categories, making sure to disaggregate the data—that is, compare the data by genders during the analysis phase. One recent alternative to the binary, developed by Public Health England in collaboration with LGBTQ+ organizations in the United Kingdom, is in evidence in figure 4.5. This two-item questionnaire was designed for use in routine national surveillance of HIV in England and Wales to determine self-identified gender and cis or trans status in a public health context. The designers offer three named genders, a catch-all fourth category, and an option for not disclosing gender identity. In a separate question, they ask about gender at birth, again giving an option for not disclosing. The survey design uses sensitive wording and inclusive terminology to allow trans and genderqueer populations to be counted. These questions are being considered for expanded use across other national health records and data collection systems in the United Kingdom.

Should all future gender data collection use this model? Not necessarily, and here's why: In a world in which quantification always leads to accurate representation, and accurate representation always leads to positive change, then always counting gender

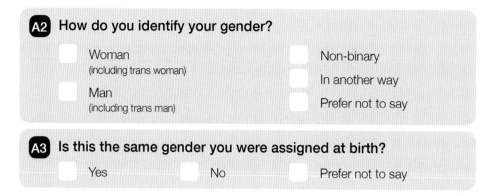

**Figure 4.5**
From the Positive Voices survey of people living with HIV in England and Wales developed by Public Health England in collaboration with several partner organizations. This represents current best practices for collecting nonbinary gender data in an Anglo-Western public health context, but it's still important to recognize that different decisions might be warranted depending on the context. Courtesy of Peter Kirwin, Public Health England, 2018.

identities outside the binary makes perfect sense. But being represented also means being made visible, and being made visible to the matrix of domination—which continuously develops laws, practices, and cultural norms to police the gender binary— poses significant risks to the health and safety of minoritized groups. Under the current administration in the United States, for example, transgender people are banned from serving in the military and, once identified as such, denied access to certain forms of healthcare.[43] This demonstrates some of the risks of having one's gender counted as something other than man or woman—risks that can occur in many contexts, depending on what data are being collected, by whom, and whether they are personally identifiable (or easily deanonymized). It's also important to recognize how trans and nonbinary people may possibly be identified even within otherwise large datasets simply because there are fewer of them relative to the larger population. This possibility poses additional risks, in the form of unwanted attention in the case of people who would prefer not to disclose their gender identity, or in the form of discrimination, violence, or even imprisonment, depending on the place they live.

As data scientists, what should we do amid these potential harms? Depending on the circumstances and the institution that is doing the collecting, the most ethical decision can vary. It might be to avoid collecting data on whether someone is cis or transgender, to make all gender data optional, to not collect gender data at all, or even to stick with binary gender categories. Social computation researcher Oliver Haimson has asserted that "in most non-health research, it's often not necessary to know participants' assigned gender at birth."[44] Heath Fogg Davis agrees: his book *Beyond Trans* argues that we don't need to classify people by sex on passports and licenses, for bathrooms or sports, among other things.[45] By contrast, J. Nathan Matias, Sarah Szalavitz, and Ethan Zuckerman chose to keep gender data in binary form for their application FollowBias, which detects gender from names, in order to avoid making a person's gender identity public against their wishes.[46]

The ethical complexity of whether to count gender, when to count gender, and how to count gender illuminates the complexity of acts of classification against the backdrop of structural oppression. Because when it comes to data collection, and the categories that structure it, there are power imbalances up and down, side to side, and everywhere in between. Because of these asymmetries, data scientists must proceed with awareness of context (discussed further in chapter 6) and an analysis of power in the collection environment (discussed further in chapter 1) to determine whose interests are being served by being counted, and who runs the risk of being harmed.

## Rethinking Binaries in Data Visualization

A feminist critique of counting, and of the binary classification systems that often structure those acts, is not limited to a focus on gender alone. A binary logic also pervades our thinking about race, for example, as feminist scholars Brittney Cooper and Margaret Rhee explain. Drawing from ideas about intersectionality, they call for "hacking" the Black/white binary that, on the one hand, helps to expose the racism experienced by Black people in the United States and, on the other, erases the other forms of racism experienced by Indigenous as well as Latinx, Asian American, and other minoritized groups. "Binary racial discourses elide our struggles for justice," they state plainly.[47] By challenging the binary thinking that erases the experiences of certain groups while elevating others, we can work toward more just and equitable data practices and consequently toward a more just and equitable future.

Sometimes, however, the goal of challenging binary thinking can be constrained by the realities of the field. Visualization designers, for example, do not typically have control over the collection practices of the data they are asked to visualize. They often inherit binary data that they then need to "hack" from within. What might this look like? We might point to the reporters on the Lifestyle Desk of the *Telegraph*, a British newspaper, who, in March 2018, were considering how to honor International Women's Day and were struck by the significant gender gap in the United Kingdom in terms of education, politics, business, and culture.[48] As journalists, they were working with multiple sources of data collected by other agencies, which all came in binary form. But they wanted to ensure that they didn't further reinforce any gender stereotypes. They paid particular attention to color. One line of designer logic would favor cultural convention for interpretability, like using pink for women and blue for men, but a feminist line would use color choices to hack those same conventions (figure 4.6).

Pink and blue is, after all, another hierarchy, and the goal of the *Telegraph* team members was to mitigate inequality, not reinforce it. So they took a different source for inspiration: the Votes for Women campaign of early twentieth-century England, in which purple was employed to represent freedom and dignity and green to represent

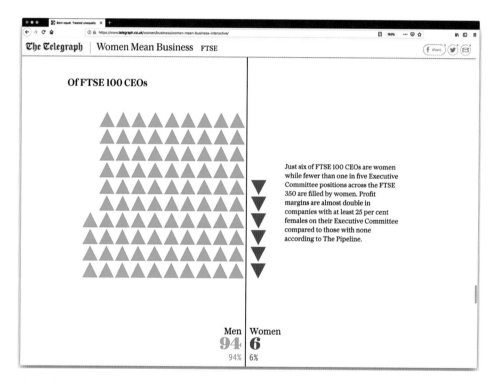

**Figure 4.6**
"Born Equal. Treated Unequally" was an interactive feature in the *Telegraph* in 2018 that examined the gender gap in the United Kingdom along a number of dimensions. Although the authors treated gender as a binary category, they used color to challenge stereotypically man/woman color coding. Feature by Claire Cohen, Patrick Scott, Ellie Kempster, Richard Moynihan, Oliver Edgington, Dario Verrengia, Fraser Lyness, George Ioakeimidis, and Jamie Johnson, for the *Telegraph*.

hope. When thinking about which of these colors to assign to each gender, they took a perceptual design principle as their guide: "Against white, purple registers with far greater contrast and so should attract more attention when putting alongside the green [sic], not by much but just enough to tip the scales. In a lot of the visualisations men largely outnumber women, so it was a fairly simple method of bringing them back into focus," Fraser Lyness, the *Telegraph*'s director of graphic journalism told visualization designer Lisa Charlotte Rost.[49] Here, one hierarchy—the hierarchy in which colors are perceived by the eye—was employed to challenge another one: the hierarchy of gender. When put into practice, this simple method had the result of communicating clearly without reinforcing stereotypes.

But the *Telegraph* journalists could have gone one step further to rethink binaries. They had an opportunity to communicate to the public that gender is not a binary by spelling that out—in the text of the story or in a caption under the graphics or by showing visually that there was no data for nonbinary people. Their colleagues at the *Guardian* recently adopted this latter strategy in their interactive piece "Does the New Congress Reflect You?" about the 2018 US midterm elections.[50] The piece presents three categories: cis male, cis female, and trans + nonbinary. When you click on "trans + nonbinary," as in figure 4.7, the interactive map displays all of the districts in grey, because "0 people in Congress are like you." The absence of data becomes an important takeaway, as meaningful as the data themselves.[51]

These examples have shown gender as a dimension of analysis, but how might we visually represent gender itself? This is a challenge of visualizing complexity of the highest degree, and Amanda Montañez, a designer for *Scientific American*, took this challenge head on (figure 4.8). She was tasked with creating an infographic to accompany an article on the evolving science of gender and sex—categories that she, like most people, viewed as distinct but related.[52] As she explains in a blog post on the *Scientific American* website, she first envisioned a simple spectrum, or perhaps two spectrums: one for sex and one for gender.[53] But she soon found confirmation of what we've been saying so far in this chapter: that few things in life can be truly reduced to binaries, and that insisting on binary categories of data collection—with respect to gender, to sex, to their relation, or to anything else—fails to acknowledge the value of what (or who) rests in between and outside.

We have already established that gender is more than binary; but it's less commonly acknowledged that sex is more than a binary too. As feminist biologist Anne Fausto-Sterling confirms, "There is no single biological measure that unassailably places each and every human into one of two categories—male or female."[54] Intersex people, who constitute an estimated 1.7 percent of the population, may have ovaries and a penis,

**Figure 4.7**
"Does the New Congress Reflect You?" is a 2018 interactive that appeared in the *Guardian*. Users select their own demographic characteristics to see how many people like them are in the 2018 Congress. Clicking on "trans + nonbinary" leads to a blank map showing zero people in Congress like you. Image by Sam Morris, Juweek Adolphe, and Erum Salam for the *Guardian*.

or "mosaic genetics" in which some of one's cells have XX chromosomes and some have XY.[55] It's also increasingly acknowledged that that sex, like gender, and sometimes together with gender, is multilayered and continuously unfolding throughout a person's life.

To begin to represent this complexity, Montañez had to begin by rejecting much of the data and research that she and her research assistant turned up, either on account of flawed categories or on account of flawed collection practices. She decided to focus on sex, and after an extensive design process, which included consulting with domain experts, Montañez and the design firm Pitch Interactive, which helped finalize the diagram, arrived at the result. *Beyond XX and XY* is a complex diagram, which employs a color spectrum to represent the sex spectrum, a vertical axis to represent change over time, and branching arrows to connect to text blocks that provide additional contextual information. The design offers a beautifully executed visual challenge to the scientifically incorrect idea that there are only two sexes, and even that the concepts of sex and gender are wholly distinct. Visualization is often thought of as a way to reduce complexity, but here it operates in the reverse—to push simple, oppressive ideas to be more complex, nuanced, and just.

**Refusing Data, Recovering Data**

Montañez's graphic made what was already counted count. In other words, she took what scientists and theorists knew to be true about the nature of sexual differentiation and made that knowledge more accessible and public. But counting in itself is not necessarily an unmitigated good, nor is putting it on public display. We have already introduced the idea of the paradox of exposure where people are harmed by being made visible to a system. But because system designers from dominant groups do not experience the harms of being counted or of being made visible without consent—this is the privilege hazard, once again—they rarely anticipate these needs or account for them in the design process. This is the reason that questions about counting must be accompanied by questions about consent, as well as of personal safety, cultural dignity, and historical context.

It's Facebook, once again, that helps to prove this point. Information studies scholars Oliver Haimson and Anna Lauren Hoffman have studied the effects of the company's "real name" policy, under which the platform determines each user's registered name to be either "real" and authentic or simply "fake."[56] (In our teacher voices, we now say: Does anyone note the problem with this binary thinking here?) Haimson and Hoffman point out that trans and queer people may choose to have multiple online identities, which may be fluid and contextual and possibly necessary to protect themselves. As another example, abuse survivors may need to take steps to make themselves unfindable through search, even as they still want to be connected to their loved ones.

**Figure 4.8** (following two pages)
*Beyond XX and XY* (2017) visualizes the known factors that contribute to sexual differentiation at different stages of human life, from conception to birth to puberty and beyond. Contrary to received wisdom, sex is not a binary that is fixed at birth, but rather a layered and time-based process of differentiation, with more than two possible outcomes. Reproduced with permission. Copyright © 2017 *Scientific American*, a division of Nature America, Inc. All rights reserved.

# BEYOND XX AND XY

A host of factors figure into whether someone is female, male or somewhere in between

**Humans are socially conditioned** to view sex and gender as binary attributes. From the moment we are born—or even before—we are definitively labeled "boy" or "girl." Yet science points to a much more ambiguous reality. Determination of biological sex is staggeringly complex, involving not only anatomy but an intricate choreography of genetic and chemical factors that unfolds over time. Intersex individuals—those for whom sexual development follows an atypical trajectory—are characterized by a diverse range of conditions, such as 5-alpha reductase deficiency (*circled*). A small cross section of these conditions and the pathways they follow is shown here. In an additional layer of complexity, the gender with which a person identifies does not always align with the sex they* are assigned at birth, and they may not be wholly male or female. The more we learn about sex and gender, the more these attributes appear to exist on a spectrum.

—*Amanda Montañez*

*\*The English language has long struggled with the lack of a widely recognized nongendered third-person singular pronoun. A singular form of "they" has grown in widespread acceptance, and many people who do not identify with a binary gender use it.*

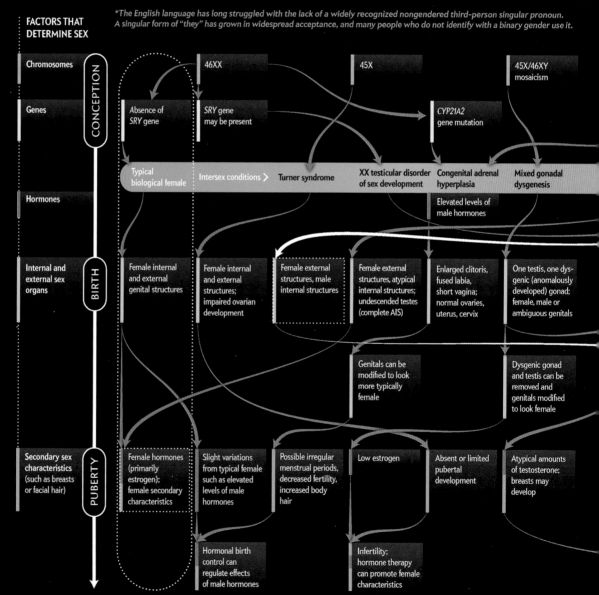

*Graphic by Pitch Interactive and Amanda Montañez*

# The Gender Spectrum

**A transgender woman** is a person who was assigned male at birth based on her anatomy but who identifies as a woman.

**A cisgender woman** is a person who was assigned female at birth based on her anatomy and who also identifies as a woman.

**A nonbinary person** is someone who identifies as neither completely female nor completely male. Such an individual may identify with both genders or neither gender, or they may be gender fluid, meaning their gender fluctuates between female and male.

**A transgender man** is a person who was assigned female at birth based on his anatomy but who identifies as a man.

**A cisgender man** is a person who was assigned male at birth based on his anatomy and who also identifies as a man.

Sexuality refers to an individual's sexual orientation or to the kind of person to whom they are attracted. Sexuality is also a spectrum but is separate from both sex and gender.

**5-alpha reductase deficiency** is an intersex condition that can follow multiple pathways throughout development. Affected individuals have a chromosomal makeup of 46XY, like a typical biological male, but a genetic mutation causes a deficiency of the hormone dihydrotestosterone. Patients' external anatomy can vary, so an individual might be assigned to either sex at birth, but at puberty a surge of testosterone promotes male characteristics. As a result, patients who are raised as girls often end up identifying as male.

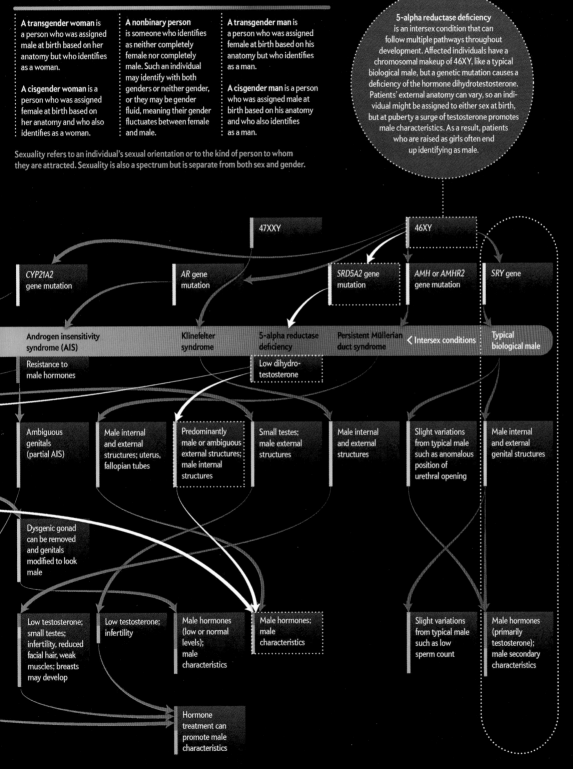

47XXY

46XY

CYP21A2 gene mutation

AR gene mutation

SRD5A2 gene mutation

AMH or AMHR2 gene mutation

SRY gene

Androgen insensitivity syndrome (AIS)
Resistance to male hormones

Klinefelter syndrome

5-alpha reductase deficiency
Low dihydro-testosterone

Persistent Müllerian duct syndrome

< Intersex conditions

Typical biological male

Ambiguous genitals (partial AIS)

Male internal and external structures; uterus, fallopian tubes

Predominantly male or ambiguous external structures; male internal structures

Small testes; male external structures

Male internal and external structures

Slight variations from typical male such as anomalous position of urethral opening

Male internal and external genital structures

Dysgenic gonad can be removed and genitals modified to look male

Low testosterone; small testes; infertility, reduced facial hair, weak muscles; breasts may develop

Low testosterone; infertility

Male hormones (low or normal levels); male characteristics

Male hormones; male characteristics

Slight variations from typical male such as low sperm count

Male hormones (primarily testosterone); male secondary characteristics

Hormone treatment can promote male characteristics

SOURCE: RESEARCH BY AMANDA HOBBS; EXPERT REVIEW BY AMY WISNIEWSKI *University of Oklahoma Health Sciences Center*

Compounding the contextual nature of these factors, Facebook enforces its real name policy algorithmically—flagging names with "too many" words or with unusual capital letters. Haimson and Hoffman note that Facebook's algorithms disproportionately flag Native American names for violation because those names often differ in structure and form from Anglo-Western names (the subject position of the systems' designers, and therefore presumed to be the default; the privilege hazard once again). What's more, users can also report other users for not having real names, resulting in—for example—a single person systematically targeting several hundred drag queens' profiles for removal. Facebook claims that the real name policy exists for safety, but Haimson and Hoffman clearly show that the policy actively imperils the safety of some of the platform's most marginalized users. As we've already begun to suggest, sometimes the most ethical thing to do is to help people be obscure, hidden, and invisible.[57] The example of Facebook demonstrates the fundamental importance of obtaining consent when counting and of enabling individuals to refuse acts of counting and classification in light of potential harms.

Acts of counting and classification, especially as they relate to minoritized groups, must always balance harms and benefits. When data are collected about real people and their lives, risks ranging from exposure to violence are always present. But when deliberately considered, and when consent is obtained, counting can contribute to efforts to increase valuable and desired visibility. The Colored Conventions Project (CCP), led by a team of students and faculty at the University of Delaware, demonstrates how to thoughtfully navigate this balance in the present by looking at the past.[58] Among the goals of the project is to create a machine-readable corpus of meeting minutes from the nineteenth-century Colored Conventions: events in which Black Americans, fugitive and free, gathered to strategize about how to achieve legal, social, economic, and educational justice. These meeting minutes are valuable because they have yet to be counted, so to speak, in the stories commonly told about the movement to end slavery in the nineteenth-century United States. Those stories tend to privilege the actions of white abolitionists because theirs were the stories that were recorded in print. But the Colored Conventions help to document the vital role of the Black activists who were working within their own communities to end slavery and achieve liberation.

The creation of the corpus enables these important activists to be counted, as well as have their words (as recorded in the meeting minutes) analyzed and incorporated into the historical record. But the process of converting the meeting minutes into data strongly recalls the original violence that accompanied the slave trade, when human lives—in fact, the very ancestors of these activists—were reduced to numbers and names. In recognition of this irreconcilable tension, the CCP requires that all those who download the corpus commit to a set of principles, including "a use of data that humanizes and acknowledges the Black people whose collective organizational histories are assembled" in the corpus, and a request to "contextualize and narrate the conditions of the people who appear as 'data' and to name them when possible."[59]

There is a second tension that the CCP navigates in an exemplary fashion, which has to do with the content of the corpus itself. Because it is derived from the conventions' official meeting minutes, it records only the "official" participants in the conventions and the discussions they initiated. These participants were almost exclusively men. To address this disparity, the CCP team asks its teaching partners to sign a Memo of Understanding (MoU) before introducing students to the project. The MoU requests that all instructors introduce a woman involved in the conventions, such as a wife, daughter, sister, or fellow church member, alongside every male delegate who is named (figure 4.9).[60] From this work of recovery, the CCP is creating a second dataset of the women's names—those who would otherwise go uncounted and therefore unrecognized for their work. They are using data collection to make these contributions count.

THE NATIONAL COLORED CONVENTION IN SESSION AT WASHINGTON, D. C.—Sketched by Theo. R. Davis.—[See First Page.]

**Figure 4.9**
An engraving of an 1869 Colored Convention, published in *Harper's Weekly*, showing men at the podium and women seated and standing in the rear. Image courtesy of Jim Casey.

### Counting as Healing, Counting as Accountability

In the nineteenth century, as today, so many of the disparities introduced into datasets had to do with much larger and much more profound asymmetries of power. The asymmetries are often directly reflected in the power dynamics between who is doing the counting and who is being counted. But when a community is counting for itself, about itself, there is the potential that data collection can be not only be empowering but also healing. One example of this that draws from the personal experience of one of the authors of this book. It was 2014, and Catherine was a student and nursing her baby daughter at the time, as well as struggling to pump breastmilk for her in unsavory places like server rooms and bathroom floors. Frustrated, she and six student colleagues came together to publish a call for ideas and stories that could help to improve breast pump technology.[61] These stories led to a research paper about breast pump design, as well as the creation of the Make the Breast Pump Not Suck Hackathon (figure 4.10)—an

**Figure 4.10**
The 2018 Make the Breast Pump Not Suck Hackathon was the second gathering of the community at MIT and focused on racial equity in breastfeeding, as well as shifting paid leave policy in the United States. Photo by Rebecca Rodriguez and Ken Richardson, MIT Media Lab.

ongoing forum for sharing stories, hacking pumps, and reengineering the postpartum ecosystem that surrounds them.[62]

Although innovation spaces had long been holding hackathons for health technology, the 2014 event was one of the first about birth and breastfeeding. As such, it led to participants sharing stories in a space that was (temporarily) free of the stigma surrounding breastfeeding. These stories pointed to common experiences and patterns in the spirit of "the personal is political" consciousness-raising events. Participants recognized these stories as data that could be used—and in fact were used—to demand more from breast pump makers, from workplaces, and from society, to help transform the self-blame that women often experience as a result of difficulties with birth and breastfeeding into collective political action.[63]

But action by whom, and action for whom? Following the 2014 event, we (meaning the organizers) reflected on its successes and its limitations—in particular, its lack of an intersectional approach.[64] In the United States, maternal health carries significant race and class inequities, as discussed in chapter 1. The first hackathon did not consider those inequities; it centered the needs of some the most privileged mothers and produced designs that favored their experiences. We decided to try again. In 2017 and 2018, we multiplied the single event into a participatory research project, a policy summit, and a community innovation program, as well as a hackathon. In all of these, we deliberately centered the needs and the participation of parents of color, low-income parents, and LGBTQ+ parents. When we arrived at the hackathon the second time around, it was the result of over a year of relationship building and identity work on the part of the organizers with our community partners.

Ensuring that the 2018 hackathon would fully welcome the participation of these families required multiple forms of accountability. Guided by Jenn Roberts, our lead organizer for equity and inclusion, we wrote a values statement and convened an advisory board with leaders in breastfeeding, equity, and maternal health. We also developed a set of metrics to shape the demographics of the event.[65] These metrics were designed to prioritize racial diversity, gender diversity, diversity of sexual orientation, geographic diversity, and domain diversity, with additional priority given for young people and newcomers. On the application form, potential participants were encouraged to self-identify their gender and race, specify their location, and choose multiple options from a list of predefined domain expertise categories (like "parent" or "designer/artist"). We also invited them to write about why they wanted to attend, and if they chose to disclose information about their sexual orientation or their financial position, then we considered that information in the process.

Were these categories reductive? Of course they were. No person can fit their whole self into a form, regardless of how many blank text fields are provided. Did the form reflect the true nature of each person's intersecting identities and how those identities impact that person's being in the world? The answer to this question is also unsurprising: of course it did not. But the process of collecting this demographic data—which was, crucially, undertaken voluntarily and from within the community itself—resulted in an event that was indeed guided by the knowledge and experience of the groups that our coalition had hoped to center.[66]

Catherine shared this experience with Lauren as we were beginning to draft this book, and we decided to use a similar process to help hold ourselves accountable to the values that we wanted to inform *Data Feminism* and the criteria by which certain projects and texts would be selected for inclusion. We determined specific numbers and percentages that, in our view, would help keep us accountable to those values, as well as the categories of data collection that would be required to determine whether the metrics had been met. (These are viewable in the appendix, Our Values and Our Metrics for Holding Ourselves Accountable.) At two phases in the process—first when we posted the draft of the manuscript online, and second after we submitted the manuscript for copyediting—one of our research assistants, Isabel Carter, audited the projects and citations of the book. (They describe their research methods in more detail in "Auditing *Data Feminism*," included as another appendix.) As with the hackathon, these metrics were not the only method we employed for holding ourselves accountable. We also interviewed the creators of many of the projects we reference, cleared our quotes and portrayals of their work with them, and published a draft of the book online for open peer review, among other approaches.

Was our method of counting perfect? Of course not. We are certain we have made mistakes. This is among the reasons that we decided to keep our disaggregated data private, even as we published the aggregated results. What about the idea to count people and projects in the first place? Shouldn't that be viewed as contributing to the same reduction in complexity that we have argued against thus far in this book? As this chapter has demonstrated, counting is always complicated. But undertaken deliberately, tailored to specific goals, and with issues of privacy and potential harms always in mind, counting can be used to support accountability—as one method, among many, of working toward a larger goal.

## Rethink Binaries and Hierarchies

Counting and classification can be powerful parts of the process of creating knowledge. But they're also tools of power in themselves. Historically, counting and classification have been used to dominate, discipline, and exclude. This is where the fourth principle of data feminism, *rethink binaries and hierarchies*, enters in. The gender binary offers a key example of how classification systems are constructed by cultures and societies and reflect both their values and their biases. The cases of the TSA airport scanners, Facebook user profiles, and plain old pants show us how gender and sex binaries—along with scientifically incorrect understandings of both gender and sex—get encoded into technical systems (and also jeans)! Those systems, in turn, recirculate erroneous and harmful ideas.

An intersectional feminist approach to counting insists that we examine and, if necessary, rethink the assumptions and beliefs behind our classification infrastructure, as well as consistently probe who is doing the counting and whose interests are served. Counting and measuring do not always have to be tools of oppression. We can also use them to hold power accountable, to reclaim overlooked histories, and to build collectivity and solidarity. When we count within our own communities, with consideration and care, we can work to rebalance unequal distributions of power.

# 5  Unicorns, Janitors, Ninjas, Wizards, and Rock Stars

**Principle: Embrace Pluralism**

*Data feminism insists that the most complete knowledge comes from synthesizing multiple perspectives, with priority given to local, Indigenous, and experiential ways of knowing.*

In Spring 2017, *Bloomberg News* ran an article with the provocative title "America's Rich Get Richer and the Poor Get Replaced by Robots."[1] Using census data, the authors reported that income inequality is widening across the nation. San Francisco is leading the pack, with an income gap of almost half a million dollars between the richest and the poorest twenty percent of residents. As in other places, the wealth gap has race and gender dimensions. In the Bay Area, people of color earn sixty-eight cents for every dollar earned by white people, and 59 percent of single mothers live in poverty.[2] San Francisco also has the lowest proportion of children to adults in any major US city, and—since 2003—an escalating rate of evictions.

Although the San Francisco Rent Board collects data on these evictions, it does not track where people go after they are evicted, how many of those people end up homeless, or which landlords are responsible for systematically evicting major blocks of the city. In 2013, the Anti-Eviction Mapping Project (AEMP) stepped in. The initiative is a self-described collective of "housing justice activists, researchers, data nerds, artists, and oral historians." It is a multiracial group with significant, though not exclusive, project leadership by women. The AEMP is mapping eviction, and doing so through a collaborative, multimodal, and—yes—quite messy process.

If you visit antievictionmap.com, you won't actually find a single map. There are seventy-eight distinct maps linked from the homepage: maps of displaced residents, of evictions, of tech buses, of property owners, of the Filipino diaspora, of the declining numbers of Black residents in the city, and more. The AEMP has a distinct way of working that is grounded in its stated commitment to antiracist, feminist, and decolonial methodologies.[3] Most of the projects happen in collaboration with nonprofits and community-based organizations. Additional projects originate from within the collective. For example, the group is working on producing an atlas of the Bay Area called *Counterpoints: Bay Area Data and Stories for Resisting Displacement*, which will cover topics such as migration and relocation, gentrification and the prison pipeline, Indigenous and colonial histories of the region, and speculation about the future.[4]

One of the AEMP's longest-standing collaborations is with the Eviction Defense Collaborative (EDC), a nonprofit that provides court representation for people who have been evicted. Although the city does not collect data on the race or income of evictees, the EDC does collect those demographics, and it works with 90 percent of the tenants whose eviction cases end up in San Francisco courts.[5] In 2014, the EDC approached the AEMP to help produce its annual report and in return offered to share its demographic data with the organization.[6] Since then, the two groups have continued to work together on EDC annual reports, as well as additional analyses of evictions with a focus on race. The AEMP has also gone on to produce reports with tenants' rights organizations, timelines of gentrification with Indigenous students, oral histories with grants from anthropology departments, and murals with arts organizations, as well as maps and more maps.

Some of the AEMP's maps are designed to leverage the ability of data visualization to make patterns visible at a glance. For example, the *Tech Bus Stop Eviction Map* produced in 2014 plots the locations of three years of Ellis Act evictions.[7] This is a form of "no-fault" eviction in which landlords can claim that they are going out of the rental business. In many cases, this is so that they can convert the building to a condominium and sell the units at significant profit. San Francisco has seen almost five thousand invocations of the Ellis Act since 1994. On the map (figure 5.1), the AEMP plotted Ellis Act evictions in relationship to the location of technology company bus stops. Starting in the 2000s, tech companies with campuses in Silicon Valley began offering private luxury buses as a perk to attract employees who wanted to live in downtown San Francisco but didn't want the hassle of commuting. Colloquially known as "the Google buses" because Google was the most visible company to implement the practice, these vehicles use public bus stops—illegally at first—to shuttle employees to their offices in comfort (and also away from the public transportation system, which would otherwise reap the benefits of their fares).[8] Because a new, wealthy clientele began to seek condos near the bus stops, property values soared—and so did the rate of evictions of long-time neighborhood residents. The AEMP analysis showed that, between 2011 and 2013, 69 percent of no-fault evictions occurred within four blocks of a tech bus stop. The map makes that finding plain.

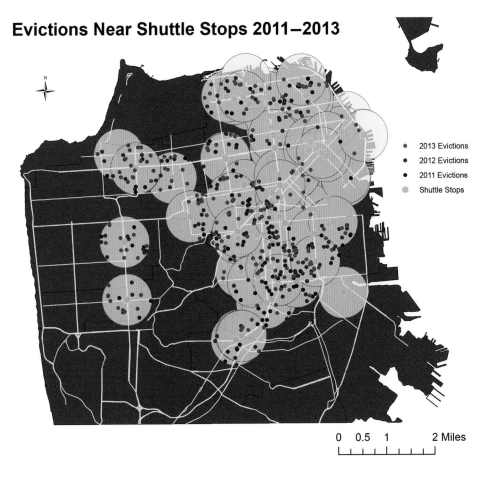

**Figure 5.1**

*Tech Bus Stop Eviction Map* by the Anti-Eviction Mapping Project (2014) plots evictions with "Google bus stops" in San Francisco. The group's analysis showed that 69 percent of no-fault evictions in the city occurred within four blocks of a tech bus stop. Courtesy of the Anti-Eviction Mapping Project.

But other AEMP maps are intentionally designed *not* to depict a clear correlation between evictions and place. In *Narratives of Displacement and Resistance* (figure 5.2a), five thousand evictions are each represented as a differently sized red bubble, so the base map of San Francisco is barely visible underneath.[9] On top of this red sea, sky-blue bubbles dot the locations where the AEMP has conducted interviews with displaced residents, as well as activists, mediamakers, and local historians. Clicking on a blue bubble sets one of the dozens of oral histories in motion—for instance, the story of Phyllis Bowie (figure 5.2b), a resident facing eviction from her one-bedroom apartment. "I was born and raised in San Francisco proudly," she begins. Bowie goes on to recall how she returned from the Air Force and worked like crazy for two years at her own small business, building up an income record that would make her eligible for a lease-to-own apartment in Midtown, the historically Black neighborhood where she had grown up. In 2015, however, the city broke the master lease and took away the rent control on her building. Now tenants like Bowie, who moved there with the promise of homeownership, are facing skyrocketing rents that none of them can afford. Bowie is leading rent strikes and organizing the building's tenants, but their future is uncertain.

This uncertainty is carried over into the design of the map, which uses a phenomenon that is often discouraged in data visualization—*occlusion*—to drive home its point. Occlusion typically refers to the "problem" that occurs when some marks (like the eviction dots) obscure other important features (like the whole geography of the city). But here it underscores the point that there are very few patterns to detect when the entire city is covered in big red eviction bubbles and abundant blue story dots. Put another way, the whole city is a pattern, and that pattern is the problem—much more than a problem of information design.

In this way, the *Narratives* map enacts dissent from San Francisco city policies in the same way that it enacts dissent from the conventions of information design. It refuses both the clarity and cleanliness associated with the best practices of data visualization and the homogenizing and "cleanliness" associated with the forces of gentrification that lead to evictions in the first place.[10] The visual point of the map is simple and exhortative: there are too many evictions. And there are too many eviction stories. The map does not efficiently reveal how evictions data may be correlated with Bay Area Rapid Transit (BART) stops, income, Google bus stops, or any other potential dimensions of the data. Even finding the *Narratives* map is difficult, given the sheer number of maps and visualizations on the AEMP website. There is no "master dashboard" that integrates all the information that the AEMP has collected into a single interface.[11]

**Figure 5.2**

*Narratives of Displacement and Resistance* (2018) by the Anti-Eviction Mapping Project (a), including a detail view from Phyllis Bowie's story (b). Courtesy of the Anti-Eviction Mapping Project. Made in collaboration with the San Francisco Ruth Assawa School of the Arts. The interview was shot by Marianne Maeckelbergh and Brandon Jourdan and edited by students Shilo Arkinson and Avidan Novogrodsky-Godt, and facilitated by Alexandra Lacey and Jin Zhu.

But all these design choices reinforce the main purpose of the AEMP: to document the effects of displacement and to resist it through critical and creative means.

The AEMP thus offers a rich set of examples of feminist counterdata collection and countervisualization strategies. Individually and together, the maps illustrate how Katie Lloyd Thomas, founding member of the feminist art architecture collective *taking place*, envisions "partiality and contestation" taking place in graphical design. "Rather than tell one 'true' story of consensus, [the drawing/graphic] might remember and acknowledge multiple, even contradictory versions of reality," she explains.[12] The *Narratives* map populates its landscape with these contradictory realities. It demonstrates that behind each eviction is a person—a person like Bowie, with a unique voice and a unique story. The voices we hear are diverse, multiple, specific, divergent—and deliberately so.

In so doing, the *Narratives* map, and the AEMP more generally, exemplify the fourth principle of data feminism and the theme of this chapter: *embrace pluralism*. Embracing pluralism in data science means valuing many perspectives and voices and doing so at all stages of the process—from collection to cleaning to analysis to communication. It also means attending to the ways in which data science methods can inadvertently work to suppress those voices in the service of clarity, cleanliness, and control. Many of our received ideas about data science work against pluralistic meaning-making processes, and one goal of data feminism is to change that.

**Strangers in the Dataset**

Data cleaning holds a special kind of mythos in data science. "It is often said that 80% of data analysis is spent on the process of cleaning and preparing the data," writes Hadley Wickham in the first sentence of the abstract of "Tidy Data," his widely cited paper from 2014.[13] Wickham is also the author of the *tidyr* package for R, a popular programming language and statistical computing platform. (*Packages* are prewritten collections of functions, and other forms of code, that can be installed and used in any project.) Wickham's package shares the sentiment expressed in his paper: that data are inherently messy and need to be tidied and tamed.

Wickham is not alone in this belief. Since the release of his tidyr package, Wickham has gone on to create the "tidyverse," an expanded set of packages that formalize a clear workflow for processing data, which have been developed by a team of equally enthusiastic contributors. Articles in the popular and business press corroborate this insistence on tidiness, as well as its pressing need. In *Harvard Business Review*, the work of the data scientist is glorified in terms of this tidying function: "At ease in the digital realm, they are able to bring structure to large quantities of formless data and make analysis possible."[14] Here, the intrepid analyst wrangles orderly tables from unstructured chaos. According to the article, it's "the sexiest job of the 21st century."[15] But for the *New York Times*, the data analyst's work is far less attractive. A 2014 article equated the task of cleaning data to the low-wage maintenance work of "janitors."[16]

Whether or not you think data scientists are sexy (they are), and whether or not you think janitors should be offended by this classist reference (we all should be), certain assumptions and anxieties remain consistent across these different articulations about the need for tidiness, cleanliness, and order, and the qualities of the people who should be doing this work. They must be able to tame the chaos of information overload. They must "scrub" and "cleanse" dirty data. And they must undertake deliberate action—either extraordinary or mundane—to put data back in their proper place.

But what might be lost in the process of dominating and disciplining data? Whose perspectives might be lost in that process? And, conversely, whose perspectives might be additionally imposed? The ideas expressed by Wickham, and by the press, carry the assumption that all data scientists, in all contexts, value cleanliness and control over messiness and complexity. But as the example of the AEMP demonstrates, these are not the requirements, nor the goals, of all data projects.

Before moving forward, we find it important to acknowledge that these ideas about cleanliness and control contain troubling traces of a movement from a prior era: eugenics, the upsetting, nineteenth-century source of much of modern statistics. As we have referenced in previous chapters via the work of Dean Spade and Rori Rohlfs, many of the men often cited as the earliest statistical luminaries, such as Karl Pearson, Adolphe Quetelet, Francis Galton, and Ronald Fisher, were also leaders in the eugenics movement.[17] In her book *Ghost Stories for Darwin: The Science of Variation and the Politics of Diversity*, postcolonial science studies scholar Banu Subramaniam details how over the course of the late nineteenth and early twentieth centuries, as statistics became the preferred language of communication between biologists and social scientists, certain ideas from the eugenics movement also carried over into this broader scientific conversation.[18] While the most odious aspects of these ideas have been largely (and thankfully) stripped away, certain core principles—like a generalized belief in the benefit of control and cleanliness—remain.[19] To be clear: the point here is *not* that anyone who cleans their data is perpetuating eugenics.[20] The point, rather, is that the ideas underlying the belief that data should always be clean and controlled have tainted historical roots. As data scientists, we cannot forget these roots, even as the ideas themselves have been tidied up over time.

This is the long history that library and information studies scholars Katie Rawson and Trevor Muñoz likely allude to in their assertion that "the cleaning paradigm assumes an underlying, 'correct' order." Like Subramaniam, but in the context of cleaning data more specifically, Rawson and Muñoz caution that cleaning can function as a "diversity-hiding trick."[21] In the perceived messiness of data, there is actually rich information about the circumstances under which it was collected. Data studies scholar Yanni Loukissas concurs. Rather than talking about *datasets*, he advocates that we talk about *data settings*—his term to describe both the technical and the human processes that affect what information is captured in the data collection process and how the data are then structured.[22]

As an example of how the data setting matters, Loukissas tells the story of being at a hackathon in Cambridge, Massachusetts, where he began to explore a dataset from the Clemson University Library, located in South Carolina. He stumbled across a puzzling record in which the librarian had noted the location of the item as "upstate." Such a designation is, of course, relational to the place of collection. For South Carolinians, *upstate* is a very legible term that refers to the westernmost region of the state, where Clemson is located. But it does not hold the same meaning for a person from New York, where *upstate* refers to its own northern region, nor does it hold the same meaning for a person sitting at a hackathon in Massachusetts, which does not have an *upstate* part of the state. Had someone at the hackathon written the entry from where they sat, they might have chosen to list the ten or so counties that South Carolinians recognize as *upstate*, so as to be more clearly legible to a wider geographic audience. But there is meaning conveyed by the term that would not be conveyed by other, more general ways of indicating the same region. Only somebody already located in South Carolina would have referred to that region in that way. From that usage of the term, we can reason that the data were originally collected in South Carolina. This information is not included elsewhere in the library record.

It is because of records like this one that the process of cleaning and tidying data can be so complicated and, at times, can be a destructive rather than constructive act. One way to think of it is like chopping off the roots of a tree that connects it to the ground from which it grew. It's an act that irreversibly separates the data from their context.

We might relate the growth of tools like tidyr that help to trim and tidy data to another human intervention into the environment: the proliferation of street names and signs that transformed the landscape of the nineteenth-century United States. Geographer Reuben Rose-Redwood describes how, for example, prior to the Revolutionary War, very few streets in Manhattan had signs posted at intersections.[23] Street names, such as they existed, were vernacular and related to the particularity of a spot—for example, "Take a right at the red house." But with the increased mobility of people and things—think of the postal system, the railroads, or the telegraph—street names needed to become systematized. Rose-Redwood calls this the production of "legible urban spaces." Then, as now, there is high economic value to legible urban spaces, particularly for large corporations (we're looking at you, Amazon) to deliver boxes of anything and everything directly to people's front doors.[24]

The point here is that *one does not need street names for navigation until one has strangers in the landscape.* Likewise, data do not need cleaning until there are *strangers in the dataset.* The Clemson University Library dataset was perfectly clear to Clemson's own librarians. But once hackers in Cambridge get their hands on it, *upstate* started to make a lot less sense, and it was not at all helpful in producing, for instance, a map of all the library records in the United States. Put more generally, once the data scientists involved in a project are not from within the community, once the place of analysis changes, once the scale of the project shifts, or once a single dataset needs to be combined with others—then we have strangers in the dataset.

Who are these strangers? As we've already started to suggest, people who work with data are alternately called *unicorns* (because they are rare and have special skills), *wizards* (because they can do magic), *ninjas* (because they execute complicated, expert moves), *rock stars* (because they outperform others), and *janitors* (because they clean messy data) (figure 5.3). Amazon dropped the "janitor" part in a recent job ad, but it managed to work in a few of these metaphors: "Amazon needs a rockstar engineer ... You are passionate ... You succeed fearlessly ... You are a coding ninja."[25]

These rock stars and ninjas are strangers in the dataset because, like the hackers in Cambridge, they often sit at one, two, or many levels removed from the collection and maintenance process of the data that they work with. This is a *negative externality*—an inadvertent third-party consequence—that arises when working with open data, application programming interfaces (APIs), and the vast stores of training data available online. These data appear available and ready to mobilize, but what they represent is not always well-documented or easily understood by outsiders. Being a stranger in the dataset is not an inherently bad thing, but it carries significant risk of what renowned postcolonial scholar Gayatri Spivak calls *epistemic violence*—the harm that dominant groups like colonial powers wreak by privileging their ways of knowing over local and Indigenous ways.[26]

This problem is compounded by the belief that data work is a solitary undertaking. This is reflected in the fact that unicorns are unique by definition, and wizards, ninjas, and rock stars are also all people who seem to work alone. This is a fallacy, of course; every rock star requires a backing band, and if we've learned anything from *Harry*

## Data Scientists as Unicorns, Wizards, Ninjas, Rock Stars and Janitors
## Mentions in the Media, 2012–2018

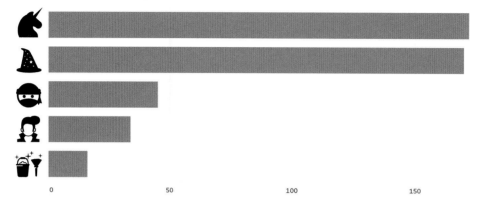

**Figure 5.3**
Searching Media Cloud between 2012 and 2018 shows that *unicorn* is the most commonly referenced metaphor in relation to data scientists, with *wizard* a close second. There are fewer than fifty
articles about data *ninjas*, *rock stars*, and *janitors*, but they appear in high-profile venues like the
*Washington Post* and *Forbes*. The Media Cloud platform at www.mediacloud.org was developed at
the MIT Center for Civic Media and archives just under a million articles and blog posts every day.
Graphic by Catherine D'Ignazio. Data from www.mediacloud.org.

*Potter*, it's that any particular tap of the wand is the culmination of years of education,
training, and support. Wizards, ninjas, rock stars, and janitors each have something
else in common: they are assumed to be men.[27] If you doubt this assertion, try doing a
Google image search and count how many male-presenting wizards and janitors you
see before you get to a single female-presenting one. Or consider why news articles
about "data janitors" don't describe them as doing "cleaning lady" work? Like janito-
rial work, which is disproportionately undertaken by working-class people of color, the
idea of the data ninja also carries racist connotations.[28] And there's even more: shared
among the first four terms—unicorns, wizards, ninjas, rock stars—is a focus on the
individual's extraordinary technical expertise and their ability to prevail when others
cannot. We might have more accurately said "his ability to prevail" because these ideas
about individual mastery and prevailing against steep odds are, of course, also associated with men.

There is a "genius" in the world of eviction data—it is Matthew Desmond, designated as such by the MacArthur Foundation for his work on poverty and eviction

in the United States. He is a professor and director of the Eviction Lab at Princeton University, which has been working for several years to compile a national database of evictions and make it available to the general public. Although the federal government collects national data on foreclosures, there is no national database of evictions, something that is desperately needed for the many communities where housing is in crisis.

Initially, the Eviction Lab had approached community organizations like the AEMP to request their data. The AEMP wanted to know more—about privacy protections and how the Eviction Lab would keep the data from falling into landlord hands. Instead of continuing the conversation, the Eviction Lab turned to a real estate data broker and purchased data of lower quality. But in an article written by people affiliated with the AEMP and other housing justice organizations, the authors state, "AEMP and Tenants Together have found three-times the amount of evictions in California as Desmond's Eviction Lab show."[29]

As Desmond tells it, this decision was due to a change in the lab's data collection strategy. "We're a research lab, so one thing that's important to us is the data cleaning process. If you want to know does Chicago evict more people than Boston, you've got to compare apples to apples."[30] In Desmond's view, it is more methodologically sound to compare datasets that have been already aggregated and standardized. And yet Desmond acknowledges that Eviction Lab's data for the state of California is undercounting evictions; there is even a message on the site that makes that explicit. So here's an ethical quandary: Does one choose cleaner data at a larger scale that is relatively easy and quick to purchase? Or more accurate data at a local scale for which one has to engage and build trust with community groups?

In this case, the priority was placed on speed at the expense of establishing trusted relationships with actors on the ground, and on broad national coverage at the expense of local accuracy. Though the Eviction Lab is doing important work, continued decisions that prioritize speed and comprehensiveness can't help but maintain the cultural status of the solitary "genius," effectively downplaying the work of coalitions, communities, and movements that are—not coincidentally—often led primarily by women and people of color.

What might be gained if we not only recognized but also valued the fact that data work involves multiple voices and multiple types of expertise? What if producing new social relationships—increasing community solidarity and enhancing social cohesion—was valued (and funded) as much as acquiring data? We think this would lead to a multiplication of projects like the AEMP: projects that do demonstrable good with data and do so together with the communities they seek to support.

## On Power, Pluralism, and Process

Although the Anti-Eviction Mapping Project could have handed off its valuable data to a single mapmaking rock star-unicorn-ninja-wizard-janitor, the group made an intentional decision to include many designers in the process, including many nonexperts who experienced the power of making maps for the first time. In addition to the diverse array of data products that resulted, which reflected the diverse voices of the AEMP's various collaborators, this decision also had the (wholly intentional) result of building technical capacity. Slowly and surely, map by map, collaboration by collaboration, local residents strengthened their mapping skills and their relationships with each other. This latter result too was intentional; one of the stated goals of the AEMP is to "build solidarity and collectivity among the project's participants who could help one another in fighting evictions and collectively combat the alienation that eviction produces."[31]

This goal reflects a key tenet of feminist thinking, which is the recognition that a multiplicity of voices, rather than one single loud or technical or magical one, results in a more complete picture of the issue at hand. Feminist philosophers like Donna Haraway, who we introduced in chapter 3, prompted a wave of thinkers who have continued to develop the idea that all knowledge is partial, meaning no single person or group can claim an objective view of the capital-T Truth.[32] But embracing *pluralism*, as this concept is often described today, does not mean that everything is relative, nor does it mean that all truth claims have equal weight. And it most certainly does not mean that feminists do not believe in science. It simply means that when people make knowledge, they do so from a particular standpoint: from a situated, embodied location in the world. More than that, by pooling our standpoints—or positionalities—together, we can arrive at a richer and more robust understanding of the world.[33]

So, how do we begin down the path to this deeper understanding in data science? The first step in activating the value of multiple perspectives is to acknowledge the partiality of your own. This means disclosing your project's methods, your decisions, and—importantly for work that strives to address injustice—your own positionalities. This is called *reflexivity*, and we modeled this in the introduction to this book. You may have heard the phrase coined by David Weinberger, "transparency is the new objectivity."[34] We take this to mean that there is a way to build space for transparency plus reflexivity in data science, rather than undertaking projects that purport to be objective (but, as we've discussed, never really are). Transparency and reflexivity allow the people involved in any particular project to be explicit about the methods behind their project, as well as their own identities.

People in journalism have been doing this for some time—at least as it relates to their data and methods. For example, Bloomberg's interactive visualization "What's Really Warming the World?" (figure 5.4) walks the reader through a range of common arguments that try to explain away global warming with reasons that don't have to do with human industry or behavior.[35] It's a compelling piece in terms of content alone, but another interesting thing about it is that it devotes nearly a third of its screen real estate to describing its data and methods.

Providing access to data and describing the methods employed are becoming conventions in data journalism, just as they are in scientific fields. Although these accounts are presently focused on technical details—where the data were from, what statistical models were developed, and what analysis was performed—there is a seed of possibility for using this same space to reveal additional details: Who was on the team? What were points of tension and disagreement? When did the team need to talk to data stewards or domain experts or local communities? Which hypotheses were pursued but ultimately proved false? There is a story about how every evidence-based argument comes into being, and it is often a story that involves money and institutions, as well as humans and tools. Revealing this story through transparency and reflexivity can be a feminist act.

Reflexivity can sometimes be as simple as being explicit about or even visualizing who is doing the counting and mapping behind the scenes.[36] Take the example of the aerial-mapping image in figure 5.5.[37] The Public Laboratory for Open Technology and Science (Public Lab) is a citizen science group that got its start during the BP oil spill in 2010.[38] It makes high-resolution aerial maps by flying balloons and kites, which dangle cheap digital cameras over the environmental sites they seek to study. The technique is low cost, but the imagery produced is often higher resolution than existing satellite imagery because of the proximity to the ground. As in the image, the mappers themselves are often visible in the final product in the form of little bodies, gathered in boats or standing in clumps on a shoreline, looking up at the camera above them. The balloon string leads the eye back to their forms. Here the creators are not distanced or absent but represented in the final product. Literally.

**Data for "Good" versus Data for Co-liberation**

Embracing the value of multiple perspectives shouldn't stop with transparency and reflexivity. It also means actively and deliberately inviting other perspectives into the data analysis and storytelling process—more specifically, those of the people most marginalized in any given context. Intersectional feminist scholars have long insisted that

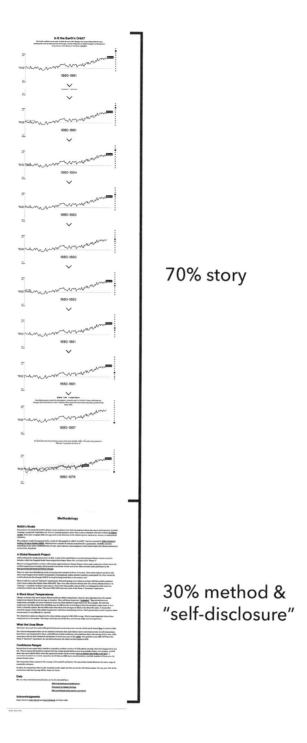

**Figure 5.4**
"What's Really Warming the World?," published in 2015, devotes a third of its real estate to describing the methods for how the authors worked with the data. Graphic by Catherine D'Ignazio, based on reporting by Eric Roston and Blacki Migliozzi for *Bloomberg Businessweek*.

**Figure 5.5**
From a Public Lab research note by Eymund Diegel about mapping sewage flows in the Gowanus Canal in 2012 after Hurricane Sandy. Note the people on boats doing the mapping and the balloon tether that links the camera and image back to their bodies. Courtesy of Eymund Diegel for Public Lab.

we should be creating new knowledge and new designs from the margins. "Marginalized subjects have an epistemic advantage, a particular perspective that scholars should consider, if not adopt, when crafting a normative vision of a just society," as Black feminist scholar Jennifer C. Nash explains.[39]

What does this mean? From a gender perspective, it means beginning with the perspectives of women and nonbinary people. On a project that involves international development data, it means beginning not with institutional goals but with Indigenous standpoints. For the AEMP, it means centering the voices and experiences of those who have been evicted. Follow-up work about designing from the margins argues that designers and engineers shouldn't only be engaging people at the margins but also actively working to dismantle the center/margins distinction in the first place.[40] More recently, the Design Justice Network has transformed this key tenet of intersectional thinking into one of its design principles, stating: "We center the voices of those who are directly impacted by the outcomes of the design process."[41]

How might this work? To begin, it requires a design process in which many actors can participate—people with technical expertise, as well as those with lived expertise, domain expertise, organizing expertise, and community history expertise. It also means shifting the overarching goal of such projects from "doing good with data" to designing for *co-liberation*; remember from chapter 2 that this is an end state in which people from dominant groups and minoritized groups work together to free themselves from oppressive systems. The key differences between data for good and data for co-liberation are highlighted in table 5.1.

Data for good is a frame that is increasingly employed to describe data science projects that are socially engaged and/or undertaken in the public interest. The Bloomberg corporation has been sponsoring Data for Good conferences since 2014. Nonprofit consulting groups like Delta Analytics have sprouted up to match volunteers with technical expertise with mission-driven organizations. In 2019, one such organization, DataKind, received a $20 million gift from a funding collaborative called Data Science for Social Impact, with monies contributed by the Rockefeller Foundation

**Table 5.1**
Features of "data for good" versus data for co-liberation

|  | "Data for good" | Data for co-liberation |
| --- | --- | --- |
| Leadership by members of minoritized groups working in community |  | √ |
| Money and resources managed by members of minoritized groups |  | √ |
| Data owned and governed by the community |  | √ |
| Quantitative data analysis "ground truthed" through a participatory, community-centered data analysis process |  | √ |
| Data scientists are not rock stars and wizards, but rather facilitators and guides |  | √ |
| Data education and knowledge transfer are part of the project design |  | √ |
| Building social infrastructure—community solidarity and shared understanding—is part of the project design |  | √ |

and Mastercard.[42] There are also educational experiments underway, like the Utrecht Data School and the University of Chicago's Data Science for Social Good summer fellowship program. In the latter, aspiring data scientists work with governments and nonprofit organizations to address problems in diverse domains such as education, health, public safety, and economic development.[43] Related efforts in artificial intelligence have sprung up, like the AI4Good Foundation, Project Impact sponsored by Intel Corporation, the women-centered AI Summer Lab at McGill University—the list goes on.

These efforts have had demonstrable social impact. And yet, there remains a nagging fuzziness with respect to what it means to "do good." Whose good are we talking about? What are the terms? Who maintains the databases when the unicorn-wizards leave the community? And who pays for the cloud storage when the development portion of the project is complete?

These issues are beginning to be discussed within the data for good community. Sara Hooker, a deep-learning researcher at Google Brain and the founder of Delta Analytics, has observed that the idea of "data for good" lacks precision.[44] To contribute clarity to the phrase, Hooker proposes a rough taxonomy of this type of work, identifying four distinct flavors: (1) volunteering skilled labor, (2) donating tech, (3) working with nonprofits or governments as partners, and/or (4) running data-education programs in underserved communities.[45] In each of these areas, there remain some thorny issues: the fickleness of volunteer labor, the fact that even committed volunteers often lack local knowledge, the community's capacity (or lack thereof) to perform and/or pay for its own technical maintenance, the fact that work initiated by nonprofits and governments cannot always be assumed to be "good," and so on. Hooker's point is that when the goal is a vague notion of "good," there is no way to address such concerns.

By contrast, a model that positions *co-liberation* as the end goal leads to a very specific set of processes and practices, as well as criteria for success. Co-liberation is grounded in the belief that enduring and asymmetrical power relations among social groups serve as the root cause of many societal problems. Rather than framing acts of technical service as benevolence or charity, the goal of co-liberation requires that those technical workers acknowledge that they are engaged in a struggle for their own liberation as well, even and especially when they are members of dominant groups.

For these reasons, data projects designed with co-liberation in mind must be very specific about the power dynamics involved. They must take preemptive steps to counteract the privilege hazard that comes with work undertaken by members of dominant groups. For example, one such step is to deliberately concentrate strategic leadership, financial resources, and data ownership in the hands of collaborators from minoritized groups.[46] It also recognizes that differential power has a silencing effect and that for a variety of reasons—as discussed in chapter 4—quantitative data can leave people out. To address these almost certain gaps, a model of data for co-liberation would deliberately pair quantitative analyses with inclusive civic processes, resulting in locally informed, ground-truthed insights that derive from many perspectives.

The scope of data for co-liberation is also broader. It explicitly includes two additional outcomes that data for good typically does not: (1) knowledge transfer and (2) building social infrastructure. The first involves a two-way exchange: in one direction, technical capacity building within the community so that any data products can be maintained and/or enhanced without requiring external expertise, and in the other, enhanced understanding of and respect for local knowledge for external collaborators, as well as chipping away at their own individual and institutional privilege hazards. Building social infrastructure involves an explicit focus on cultivating community solidarity through the project. This entails allocating financial and human resources to the community aspects of the project and not only to technical aspects like processing data or building apps. In the co-liberation model, data science projects become community science projects. They take place simultaneously in the database and in public space.

**Data for Co-liberation in Action**

What does data for co-liberation look like in action? Are there examples of feminist data science that value quantitative methods *and* pluralistic processes, data education *and* community solidarity?

Since 2012, Rahul and Emily Bhargava have partnered with community organizations from Belo Horizonte to Boston to create *data murals* in public spaces (figure 5.6). These are large-scale infographics that are both designed by and tell stories about the people who live and work in those spaces. In all cases, the people themselves sought out the collaboration, having recognized a need within their own community space. For example, in 2013, Groundwork Somerville, an urban agriculture nonprofit, approached the Bhargavas because it was in the process of establishing its first urban farm. As Emily recalls, "The site was disorderly—it was behind a used car parts building and hidden between other semi-industrial lots. They had built raised beds and planted for one growing season but passersby were stealing the vegetables."[47] The organization was also running a high school employment program called the Green

**Figure 5.6**
The process of making a data mural involves conversation, building prototypes with craft materials, workshops in data analysis, and actual painting. Courtesy of Data Therapy, Emily and Rahul Bhargava, 2018.

Team but struggling to fully involve the young people in its mission to create healthier communities.

Linking those two challenges together, Groundwork Somerville and the Bhargavas decided to enlist the young people in creating a mural that illustrated the purpose of the farm to the community (figure 5.7). First, they brought together data from several sources: demographic data from the city, geographic information system (GIS) data on unused lots, and internal data from Groundwork Somerville that included growing records, food donations, and attendance logs at community events. Then they worked with the learners over several after-school sessions to review and discuss the data and engage in storyfinding (aka data analysis). By the end of these sessions, the youth had sketched the overall outline and iconography of the resulting mural. Read left to right, the mural frames the problem: A man grasps for a basket of veggies, but it says "healthy food is hard to get." The claim is backed up by data that show the cost of healthy food and the number of people with prediabetes. Additional data document the number of unused lots in the city and the amount of land that has been reclaimed for urban farming, pointing to future opportunity. In the next section, the mural depicts a Groundwork Somerville truck bringing affordable produce to the neighborhood and includes the fact that it employs over four hundred youth residents. The data mural ends with a vision for a unified and healthy community, "Together Livin' Better."

On July 30, 2013, the mayor of Somerville and other community leaders attended the ribbon-cutting to officially launch the renovated garden. Emily describes the visit:

**Figure 5.7**
The Groundwork Somerville data mural, painted by youth, staff and volunteers at Groundwork Somerville, and the Bhargavas in 2013. Courtesy of Data Therapy, Emily and Rahul Bhargava.

"The youth, having just spent weeks looking at the data, painting the mural together, and building relationships with staff and volunteers, were able to talk about the story in great detail to their elected officials."[48]

The partnership represents some of the best aspects of data projects designed for co-liberation. But murals are just one possible output from this type of pluralistic, community-centered process.[49] For example, Digital Democracy works with Indigenous groups around the world to defend their rights through collecting data and making maps.[50] In the process, they have developed SMS services with domestic violence groups in Haiti and supported the Wapichana people in Guyana to make a data-driven case for land rights to the government. In another example, the Westside Atlanta Land Trust (WALT) has worked toward affordable housing in Atlanta, Georgia, through participatory data collection.[51] Disappointed in what it termed "triple-incorrect" county-level data, the organization set out to collect its own spatial dataset about property abandonment and disinvestment and has used it to push municipal policymakers for change.[52]

Each of these projects—the data mural, the land rights claim, and the abandoned property map—could be said to exemplify the data for co-liberation model. Each originates from a need articulated by the community, rather than projected onto it by more powerful institutions. Each relies upon methods of data collection and processing. And each also incorporates an explicit process of knowledge transfer from external collaborators (consultants, academics, nonprofit specialists) to the community itself. More importantly, none of those external collaborators envision themselves as unicorns, janitors, ninjas, wizards, or rock stars. "At Digital Democracy, we try to fight the superhero narrative," says director Emily Jacobi. "We are sidekicks rather than superheroes."[53]

Along these lines, each of these projects is explicitly designed to build community solidarity and shared understanding around a civic issue. As Amanda Meng and Carl DiSalvo explain with respect to the WALT project, "Data collection became a way of generating not simply awareness of the conditions, but a sensitivity to the conditions and to [community members'] own experience of and situation of being blinded to these conditions."[54] Data collection and participatory analysis become a gathering technology—a kind of campfire—where information is exchanged, social cohesion is enhanced, and future actions are co-conspired.[55] The campfire model also has the effect of challenging the deficit narratives that so often surround underserved communities.[56] Instead, as social movement scholar Maribel Casas-Cortés has asserted, cultivating community solidarity contributes to a sense of collective agency and transformative possibility, what she describes as "an innovative sense of political participation and re-invigorating political imaginaries."[57]

## Does the Campfire Scale?

Data for co-liberation is simultaneously more specific than data for good and larger in scope. By advocating for community-based leadership and resources, qualitative investigations to complement quantitative analyses, and intentional project design in areas that exceed the purview of data for good projects, such as knowledge transfer and building community solidarity, data for co-liberation enables the participation of many actors and agents. But can this pluralistic model scale up? What would "big data for co-liberation" look like? Would it be possible at all?

Questions about big data versus little data or quantitative data versus qualitative data are far too often framed as false binaries. And as with all false binaries, there are paths through those distinctions that combine and multiply options, rather than reduce or divide the world into fake choices. The key question to keep in mind is how we can scale up data for co-liberation in ways that remain careful, community-based, and complex.[58]

One real-world example of this intentional practice is the Global Atlas of Environmental Justice (EJ Atlas), a large-scale research project and open archive (figure 5.8) directed by scholars Leah Temper and Joan Martinez-Alier.[59] Initiated in 2012 by a team of researchers at the Universitat Autónoma de Barcelona, in Spain, the EJ Atlas represents a systematic collection of global ecological conflicts. It was created in part as a response to assertions that environmental justice scholars were focusing too much attention on local case studies at the expense of making global connections. For the

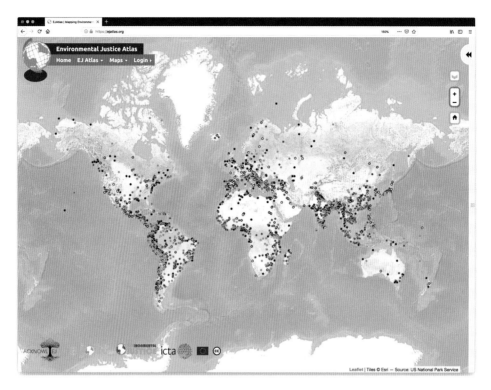

**Figure 5.8**
The Global Atlas of Environmental Justice (EJ Atlas; https://ejatlas.org/) works in partnership with
activists, civil society organizations, and social movements to systematically document ecological
conflicts around the globe. The scope and scale of the EJ Atlas enables activists to connect with
others and facilitates researchers studying conflicts in a quantitative and comparative context,
without sacrificing a commitment to a pluralistic process and the dignity of local and community
knowledges. Courtesy of the Global Atlas of Environmental Justice, 2019.

most part, the global ecological conflicts that the EJ Atlas collects are cases straight out
of the matrix of domination, of wealthy people overutilizing natural resources and
displacing environmental risk and degradation to poorer people, who are often also
minoritized because of their gender, indigeneity, race, and/or geography. For example,
one of the cases in the EJ Atlas discusses the efforts of Berta Cáceres and the Lenca
Indigenous people to oppose the construction of a dam and hydropower plant in Honduras.[60] Cáceres won international awards for her community organizing, only to be
tragically assassinated by employees of the company building the dam. (The assassins
were later convicted in a Honduran court of law.[61]) In the EJ Atlas, the entry on Cáceres's organizing efforts includes copious links, photos, and geographic data about the

dam, as well as information about where the incident falls in the atlas's typology of conflict.

The EJ Atlas demonstrates that scale is not incompatible with valuing local knowledges and relationships with local actors. Temper and her colleagues were able to draw on past relationships with partner organizations to assemble their archive, demonstrating how time invested in building relationships at the outset of a project, or of a career, continues to accrue benefits for all parties involved.[62] At the time of writing, the EJ Atlas includes close to three thousand cases of ecological conflicts from around the world. Collaborators have created submaps from the atlas for countries including Colombia, Italy, and Turkey. Activists have used the atlas to draw media and policymaker attention to overlooked environmental conflicts, such as the construction of a ski resort in the middle of a nature park in Kazakhstan, which has disturbed hydrological systems.[63] The atlas has also enabled comparative empirical research, like that undertaken by economist Begüm Özkaynak and several colleagues.[64] Their work employs social network analysis to study the relationships among mining corporations, their financiers, and environmental groups challenging that practice to understand the geography of the actors and their connections to each other.[65]

Scale is emphatically not incompatible with the feminist imperative to value multiple and local knowledges. Research like Özkaynak's and that of her colleagues would not be possible without a large-scale archive like the EJ Atlas. A pluralistic, participatory, iterative process like the EJ Atlas will take longer to scale than the extractive, quantity-at-all-costs approach of conventional big data. But ultimately the data—and the relationships, and the community capacity—will be of higher quality.

## Embrace Pluralism

The fifth principle of data feminism is to *embrace pluralism* in the whole process of working with data, from collection to analysis to communication to decision-making. As we describe in this chapter, data work carries a high risk of enacting what Gayatri Spivak has termed *epistemic violence*, particularly when the people doing the work are strangers in the dataset, when they are one or more steps removed from the local context of the data, and when they view themselves (or are viewed by society) as unicorns, rock stars, and wizards.

Embracing pluralism is a feminist strategy for mitigating this risk. It allows both time and space for a range of participants to contribute their knowledge to a data project and to do so at all stages of that project. In contrast to an underspecified data for good model, embracing pluralism offers a way to work toward a model of data for co-liberation. This means transferring knowledge from experts to communities and explicitly cultivating community solidarity in data work, as we see in the case of the Anti-Eviction Mapping Project. Moreover, embracing pluralism is not incompatible with "bigness" or scale; the EJ Atlas shows how pluralistic processes can be used to assemble a global archive and support empirical work in the service of justice. A single data scientist wizard will never defeat the matrix of domination alone, no matter how powerful their spells might be. But a well-designed, data-driven, participatory process, one that centers the standpoints of those most marginalized, empowers project participants, and builds new relationships across lines of social difference—well, that might just have a chance.

## 6  The Numbers Don't Speak for Themselves

**Principle: Consider Context**

*Data feminism asserts that data are not neutral or objective. They are the products of unequal social relations, and this context is essential for conducting accurate, ethical analysis.*

In April 2014, 276 young women were kidnapped from their high school in the town of Chibok in northern Nigeria. Boko Haram, a militant terrorist group, claimed responsibility for the attacks. The press coverage, both in Nigeria and around the world, was fast and furious. SaharaReporters.com challenged the government's ability to keep its students safe. CNN covered parents' anguish. The *Japan Times* connected the kidnappings to the increasing unrest in Nigeria's northern states. And the BBC told the story of a girl who had managed to evade the kidnappers. Several weeks after this initial reporting, the popular blog *FiveThirtyEight* published its own data-driven story about the event, titled "Kidnapping of Girls in Nigeria Is Part of a Worsening Problem."[1] The story reported skyrocketing rates of kidnappings. It asserted that in 2013 alone there had been more than 3,608 kidnappings of young women. Charts and maps accompanied the story to visually make the case that abduction was at an all-time high (figure 6.1).

Shortly thereafter, the news website had to issue an apologetic retraction because its numbers were just plain wrong. The outlet had used the Global Database of Events, Language and Tone (GDELT) as its data source. GDELT is a big data project led by computational social scientist Kalev Leetaru. It collects news reports about events around the world and parses the news reports for actors, events, and geography with the aim of providing a comprehensive set of data for researchers, governments, and civil society. GDELT tries to focus on conflict—for example, whether conflict is likely between two countries or whether unrest is sparking a civil war—by analyzing media reports. However, as political scientist Erin Simpson pointed out to *FiveThirtyEight* in a widely cited Twitter thread, GDELT's primary data source is *media reports* (figure 6.2).[2] The project is not at a stage at which its data can be used to make reliable claims about *independent*

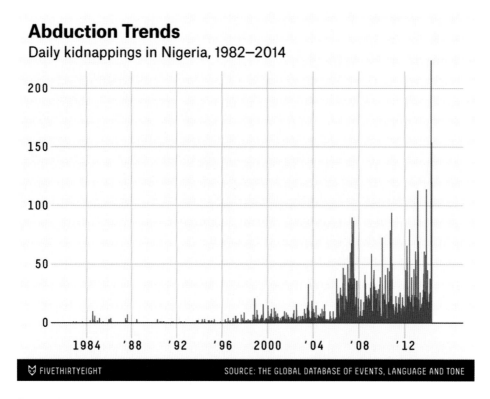

## Abduction Trends
Daily kidnappings in Nigeria, 1982–2014

**Figure 6.1**
In 2014, *FiveThirtyEight* erroneously charted counts of "daily kidnappings" in Nigeria. The news site failed to recognize that the data source it was using was not counting *events*, but rather *media reports about events*. Or some events and some media reports. Or it was counting something, but we are still not sure what. Image by *FiveThirtyEight*.

*cases* of kidnapping. The kidnapping of schoolgirls in Nigeria was a single event. There were thousands of global media stories about it. Although GDELT de-duplicated some of those stories to a single event, it still logged, erroneously, that hundreds of kidnapping events had happened that day. The *FiveThirtyEight* report had counted each of those GDELT pseudoevents as a separate kidnapping incident.

The error was embarrassing for *FiveThirtyEight*, not to mention for the reporter, but it also helps to illustrate some of the larger problems related to data found "in the wild." First, the hype around "big data" leads to projects like GDELT wildly overstating the completeness and accuracy of its data and algorithms. On the website and in publications, the project leads have stated that GDELT is "an initiative to construct a catalog

**Figure 6.2**
Two tweets by Erin Simpson in response to *FiveThirtyEight*'s erroneous interpretation of the GDELT dataset. Tweets by Erin Simpson on May 13, 2014.

of human societal-scale behavior and beliefs across all countries of the world, connecting every person, organization, location, count, theme, news source, and event across the planet into a single massive network that captures what's happening around the world, what its context is and who's involved, and how the world is feeling about it, every single day."[3] That giant mouthful describes no small or impotent big data tool. It is clearly Big Dick Data.

*Big Dick Data* is a formal, academic term that we, the authors, have coined to denote big data projects that are characterized by patriarchal, cis-masculinist, totalizing fantasies of world domination as enacted through data capture and analysis. Big Dick Data projects ignore context, fetishize size, and inflate their technical and scientific capabilities.[4] In GDELT's case, the question is whether we should take its claims of big data at face value or whether the Big Dick Data is trying to trick funding organizations into giving the project massive amounts of research funding. (We have seen this trick work many times before.)

The GDELT technical documentation does not provide any more clarity as to whether it is counting media reports (as Simpson asserts) or single events. The database *FiveThirtyEight* used is called the GDELT Event Database, which certainly makes it sound like it's counting events. The GDELT documentation states that "if an event has been seen before it will not be included again," which also makes it sound like it's counting events. And a 2013 research paper related to the project confirms that GDELT is indeed counting events, but only events that are unique to specific publications. So it's counting events, but with an asterisk. Compounding the matter, the documentation offers no guidance as to what kinds of research questions are appropriate to ask the database or what the limitations might be. People like Simpson who are familiar with the area of research known as *event detection*, or members of the GDELT community, may know to not believe (1) the title of the database, (2) the documentation, and (3) the marketing hype. But how would outsiders, let alone newcomers to the platform, ever know that?

We've singled out GDELT, but the truth is that it's not very different from any number of other data repositories out there on the web. There are a proliferating number of portals, observatories, and websites that make it possible to download all manner of government, corporate, and scientific data. There are APIs that make it possible to write little programs to query massive datasets (like, for instance, all of Twitter) and download them in a structured way.[5] There are test datasets for network analysis, machine learning, social media, and image recognition. There are fun datasets, curious datasets, and newsletters that inform readers of datasets to explore for journalism or analysis.[6] In our current moment, we tend to think of this unfettered access to information as an inherent good. And in many ways, it *is* kind of amazing that one can just google and download data on, for instance, pigeon racing, the length of guinea pig teeth, or every single person accused of witchcraft in Scotland between 1562 and 1736—not to mention truckloads and truckloads of tweets.[7]

And though the schooling on data verification received by *FiveThirtyEight* was rightly deserved, there is a much larger issue that remains unaddressed: the issue of context. As we've discussed throughout this book, one of the central tenets of feminist thinking is that all knowledge is *situated*. A less academic way to put this is that *context matters*. When approaching any new source of knowledge, whether it be a dataset or dinner menu (or a dataset of dinner menus), it's essential to ask questions about the social, cultural, historical, institutional, and material conditions under which that knowledge was produced, as well as about the identities of the people who created it.[8] Rather than seeing knowledge artifacts, like datasets, as raw input that can be simply fed into a statistical analysis or data visualization, a feminist approach insists on connecting data back to the context in which they were produced. This context allows us, as data scientists, to better understand any functional limitations of the data and any associated ethical obligations, as well as how the power and privilege that contributed to their making may be obscuring the truth.

## Situating Data on the Wild Wild Web

The major issue with much of the data that can be downloaded from web portals or through APIs is that they come without context or metadata. If you are lucky you *might* get a paragraph about where the data are from or a data dictionary that describes what each column in a particular spreadsheet means. But more often than not, you get something that looks like figure 6.3.

The data shown in the figure—open budget data about government procurement in São Paulo, Brazil—do not look very technically complicated. The complicated part is figuring out how the business process behind them works. How does the government run the bidding process? How does it decide who gets awarded a contract? Are all the bids published here, or just the ones that were awarded contracts? What do terms like *competition, cooperation agreement*, and *terms of collaboration* mean to the data publisher? Why is there such variation in the publication numbering scheme? These are only a few of the questions one might ask when first encountering this dataset. But without answers to even some of these questions—to say nothing of the local knowledge required to understand how power is operating in this particular ecosystem—it would be difficult to even begin a data exploration or analysis project.

This scenario is not uncommon. Most data arrive on our computational doorstep context-free. And this lack of context becomes even more of a liability when accompanied by the kind of marketing hype we see in GDELT and other Big Dick Data projects. In fact, the 1980s version of these claims is what led Donna Haraway to propose the concept of situated knowledge in the first place.[9] Subsequent feminist work has drawn on the concept of situated knowledge to elaborate ideas about ethics and responsibility in relation to knowledge-making.[10] Along this line of thinking, it becomes the responsibility of the person evaluating that knowledge, or building upon it, to ensure that its "situatedness" is taken into account. For example, information studies scholar Christine Borgman advocates for understanding data in relation to the "knowledge infrastructure" from which they originate. As Borgman defines it, a *knowledge infrastructure* is "an ecology of people, practices, technologies, institutions, material objects, and relationships."[11] In short, it is the context that makes the data possible.

| Nr. Publicação | Licitador | Modalidade | Dt. Abertura | Objeto |
|---|---|---|---|---|
| 01-PREF/SECOM/2018 | Secretaria do Governo Municipal - SGM | CONCORRÊNCIA | 10/06/2019 14:00 | Contratação de empresa para prestação de serviços de assessoria de imprensa e comunicação para a PREF/SECOM |
| 03/SGM-2019 | SGM - Administração de Compras e Contratos | CONCORRÊNCIA | 03/06/2019 10:30 | ALIENAÇÃO DO IMÓVEL MUNICIPAL SITUADO NA AVENIDA PROFESSOR ALCEU MAYNARD ARAÚJO, NO DISTRITO DE SANTO AMARO. |
| 01/SMPED/2019 | Secretaria Municipal da Pessoa com Deficiência - SMPED | TOMADA DE PREÇOS | 31/05/2019 10:30 | Contratação de empresa especializada em produção e atualização de material didático orientador e informativo com produção de conteúdo em versão digital acessível, visando a subsidiar a capacitação do público alvo dos cursos e eventos oferecidos pela Secretaria Municipal da Pessoa com Deficiência - SMPED. |
| 19/SME/2019 | Secretaria Municipal de Educação - SME | PREGÃO ELETRÔNICO | 30/05/2019 10:30 | Registro de preços para aquisição de alimentos não perecíveis açúcar refinado. |
| 007/2019 | São Paulo Transporte S/A | PREGÃO ELETRÔNICO | 27/05/2019 10:00 | OBJETO: AQUISIÇÃO DE 6 (SEIS) EQUIPAMENTOS APPLIANCE DO TIPO UTM?s, COM LICENÇAS DE SEGURANÇA, INSTALAÇÃO E SUPORTE TÉCNICO, PELO PERÍODO DE 24 (VINTE E QUATRO) MESES |
| 109/SMADS/2019 | Secretaria Municipal de Assistência e Desenvolvimento Social - SMADS | TERMO DE COLABORAÇÃO - EDITAL | 24/05/2019 10:00 | C J |
| 108/SMADS/2019 | Secretaria Municipal de Assistência e Desenvolvimento Social - SMADS | TERMO DE COLABORAÇÃO - EDITAL | 24/05/2019 10:00 | Centro de Acolhida com Inserção Produtiva para Adultos em Situação de Rua |
| 001/2018/SEHAB | Secretaria Municipal de Habitação - SEHAB - GABINETE | CONCORRÊNCIA | 24/05/2019 10:00 | EXECUÇÃO DE OBRAS DE CONSTRUÇÃO DE EMPREENDIMENTO HABITACIONAL DE INTERESSE SOCIAL E DE USO MISTO, DENOMINADO COLISEU, NO ÂMBITO DA OPERAÇÃO URBANA CONSORCIADA FARIA LIMA |
| 002/SVMA/2019 | Secretaria Municipal do Verde e Meio Ambiente - SVMA | CONCORRÊNCIA | 23/05/2019 10:30 | CONTRATAÇÃO DE SERVIÇOS TÉCNICOS ESPECIALIZADOS PARA A ELABORAÇÃO DO PLANO DE MANEJO DA ÁREA DE PROTEÇÃO AMBIENTAL (APA) BORORÉ-COLÔNIA |
| 070/18 | São Paulo Turismo - SPTURIS | PREGÃO ELETRÔNICO | 22/05/2019 10:00 | Contratação de empresa, sob o regime de empreitada por preço unitário, para prestação de serviços de BOMBEIRO PROFISSIONAL CIVIL, por um período de 12 (doze) meses, prorrogáveis por iguais ou menores períodos, conforme bases, especificações e condições do Edital e seus Anexos. |
| 093/2019-SMS.G | Secretaria Municipal de Saúde - SMS | PREGÃO ELETRÔNICO | 22/05/2019 09:00 | Registro de preços para o fornecimento de PAPEL CREPADO E SWAB, ALCOOL 70% PARA ANTI-SEPSIA. |
| 121/2019-SMS.G | Secretaria Municipal de Saúde - SMS | PREGÃO ELETRÔNICO | 21/05/2019 10:30 | Registro de preços para o fornecimento de KIT PARA IDENTIFICAÇÃO QUALITATIVA PARA O COMPLEXO M. TUBERCULOSIS. |
| 18/SME/2019 | Secretaria Municipal de Educação - SME | PREGÃO ELETRÔNICO | 21/05/2019 10:30 | Registro de preço para aquisição de Item A: Sardinha em óleo comestível e Item B: Atum em pedaços em conserva. |
| 166/2019 | Autarquia Hospitalar Municipal - AHM | PREGÃO ELETRÔNICO | 21/05/2019 09:30 | AQUISIÇÃO DE SULFAMETOXAZOL 80 MG/ML + TRIMETOPRIMA 16 MG/ML 5 ML, PARA AS UNIDADES DA AUTARQUIA HOSPITALAR MUNICIPAL. |
| 119/2019-SMS.G | Secretaria Municipal de Saúde - SMS | PREGÃO ELETRÔNICO | 20/05/2019 10:30 | Registro de preços para o fornecimento de ETIQUETA TÉRMICA CONTINUA, AUTOADESIVA, PARA IMPRESSAO TÉRMICA ? 62MM X 15M. |
| 117/2019-SMS.G | Secretaria Municipal de Saúde - SMS | PREGÃO ELETRÔNICO | 20/05/2019 09:30 | Aquisição de MATERIAL ODONTOLÓGICO - FÓRCEPS PARA USO ODONTOLÓGICO. |
| 047/2019-HMEC | Hospital Municipal Maternidade-Escola Dr. Mario de Moraes Altenfelder Silva | PREGÃO ELETRÔNICO | 20/05/2019 09:00 | BERACTANTO SUSPENSÃO INTRA-TRAQUEAL 25 MG/ML FAM 8,0 ML ? FAM |
| 103/2019-SMS.G | Secretaria Municipal de Saúde - SMS | PREGÃO ELETRÔNICO | 17/05/2019 10:30 | Registro de preços para o fornecimento de MATERIAL DE LABORATÓRIO - COLETOR UNIVERSAL ESTÉRIL, PIPETA DE TRANSFERÊNCIA E SWAB DE RAYON. |
| 055/2019-HMEC | Hospital Municipal Maternidade-Escola Dr. Mario de Moraes Altenfelder Silva | PREGÃO ELETRÔNICO | 17/05/2019 10:00 | PLACA DESCARTAVEL PARA ELETROCIRURGIA |
| 002/2019 | São Paulo Obras - SP Obras | TOMADA DE PREÇOS | 17/05/2019 09:30 | Contratação de empresa especializada em engenharia e arquitetura para execução das obras de reforma para implantação do DESCOMPLICA SP ? UNIDADE SÃO MATEUS . |

**Figure 6.3**

Open budget data about procurement and expenses from the São Paulo prefecture in Brazil. Although Brazil has some of the most progressive transparency laws on the books, the data that are published aren't necessarily always accessible or usable by citizens and residents. In 2013, researcher Gisele Craveiro worked with civil society organizations to give this open budget data more context. Images from SIGRC for the Prefecture of São Paulo, Brazil.

Ironically, some of the most admirable aims and actions of the open data movement have worked against the ethical urgency of providing context, however inadvertently. *Open data* describes the idea that anyone can freely access, use, modify, and share data for any purpose. The open data movement is a loose network of organizations, governments, and individuals. It has been active in some form since the mid-2000s, when groups like the Open Knowledge Institute were founded and campaigns like Free Our Data from the *Guardian* originated to petition governments for free access to public records.[12] The goals are good ones in theory: economic development by building apps and services on open data; faster scientific progress when researchers share knowledge; and greater transparency for journalists, citizens, and residents to be able to use public information to hold governments accountable. This final goal was a major part of the framing of former US president Obama's well-known memorandum on transparency and open government.[13] On his very first day in office, Obama signed a memorandum that directed government agencies to make all data open by default.[14] Many more countries, states, and cities have followed suit by developing open data portals and writing open data into policy. As of 2019, seventeen countries and over fifty cities and states have adopted the International Open Data Charter, which outlines a set of six principles guiding the publication and accessibility of government data.[15]

In practice, however, limited public funding for technological infrastructure has meant that governments have prioritized the "opening up" part of open data—publishing spreadsheets of things like license applications, arrest records, and flood zones—but lack the capacity to provide any context about the data's provenance, let alone documentation that would allow the data to be made accessible and usable by the general public. As scholar Tim Davies notes, raw data dumps might be good for starting a conversation, but they cannot ensure engagement or accountability.[16] The reality is that many published datasets sit idle on their portals, awaiting users to undertake the intensive work of deciphering the bureaucratic arcana that obscures their significance. This phenomenon has been called *zombie data*: datasets that have been published without any purpose or clear use case in mind.[17]

Zombies might be bad for brains, but is zombie data really a problem? *Wired* magazine editor Chris Anderson would say, emphatically, "No." In a 2008 *Wired* article, "The End of Theory," Anderson made the now-infamous claim that "the numbers speak for themselves."[18] His main assertion was that the advent of big data would soon allow data scientists to conduct analyses at the scale of the entire human population, without needing to restrict their analysis to a smaller sample. To understand his claim, you need to understand one of the basic premises of statistics.

Statistical inference is based on the idea of sampling: that you can infer things about a population (or other large-scale phenomenon) by studying a random and/ or representative sample and then mapping those findings back on the population (or phenomenon) as a whole. Say that you want to know who all of the 323 million people in the US will vote for in the coming presidential election. You couldn't contact all of them, of course, but you could call three thousand of them on the phone and then use those results to predict how the rest of the people would likely vote. There would also need to be some statistical modeling and theory involved, because how do you know that those three thousand people are an accurate representation of the whole population? This is where Anderson made his intervention: at the point at which we have data collected on the entire population, we no longer need modeling, or any other "theory" to first test and then prove. We can look directly at the data themselves.

Now, you can't write an article claiming that the basic structure of scientific inquiry is obsolete and not expect some pushback. Anderson wrote the piece to be provocative, and sure enough, it prompted numerous responses and debates, including those that challenge the idea that this argument is a "new" way of thinking in the first place (e.g., in the early seventeenth century, Francis Bacon argued for a form of inductive reasoning, in which the scientist gathers data, analyzes them, and only thereafter forms a hypothesis).[19] One of Anderson's major examples is Google Search. Google's search algorithms don't need to have a hypothesis about *why* some websites have more incoming links—other pages that link to the site—than others; they just need a way to determine the number of links so they can use that number to determine the popularity and relevance of the site in search results. We no longer need causation, Anderson insists: "Correlation is enough."[20] But what happens when the number of links is also highly correlated with sexist, racist, and pornographic results?

The influence of racism, sexism, and colonialism is precisely what we see described in *Algorithms of Oppression*, information studies scholar Safiya Umoja Noble's study of the harmful stereotypes about Black and Latinx women perpetuated by search algorithms such as Google's. As discussed in chapter 1, Noble demonstrates that Google Search results do not simply correlate with our racist, sexist, and colonialist society; that society *causes* the racist and sexist results. More than that, Google Search reinforces these oppressive views by ranking results according to how many other sites link to them. The rank order, in turn, encourages users to continue to click on those same sites. Here, correlation without context is clearly not enough because it recirculates racism and sexism and perpetuates inequality.[21]

There's another reason that context is necessary for making sense of correlation, and it has to do with how racism, sexism, and other forces of oppression enter into the environments in which data are collected. The next example has to do with sexual assault and violence. If you do not want to read about these topics, you may want to skip ahead to the next section.

In April 1986, Jeanne Clery, a student at Lehigh University, was sexually assaulted and murdered in her dorm room. Her parents later found out that there had been thirty-eight violent crimes at Lehigh in the prior three years, but nobody had viewed that as important data that should be made available to parents or to the public. The Clerys mounted a campaign to improve data collection and communication efforts related to crimes on college campuses, and it was successful: the Jeanne Clery Act was passed in 1990, requiring all US colleges and universities to make on-campus crime statistics available to the public.[22]

So we have an ostensibly comprehensive national dataset about an important public topic. In 2016, three students in Catherine's data journalism class at Emerson College—Patrick Torphy, Michaela Halnon, and Jillian Meehan—downloaded the Clery Act data and began to explore it, hoping to better understand the rape culture that has become pervasive on college campuses across the United States.[23] They soon became puzzled, however. Williams College, a small, wealthy liberal arts college in rural Massachusetts, seemed to have an epidemic of sexual assault, whereas Boston University (BU), a large research institution in the center of the city, seemed to have strikingly few cases relative to its size and population (not to mention that several high-profile sexual assault cases at BU had made the news in recent years).[24] The students were suspicious of these numbers, and investigated further. After comparing the Clery Act data with anonymous campus climate surveys (figure 6.4), consulting with experts, and interviewing survivors, they discovered, paradoxically, that the truth was closer to the *reverse* of the picture that the Clery Act data suggest. Many of the colleges with higher reported rates of sexual assault were actually places where more institutional resources were being devoted to support for survivors.[25]

As for the colleges with lower numbers, this is also explained by context. The Clery Act requires colleges and universities to provide annual reports of sexual assault and other campus crimes, and there are stiff financial penalties for not reporting. But the numbers are self-reported, and there are also strong financial incentives for colleges *not* to report.[26] No college wants to tell the government—let alone parents of prospective students—that it has a high rate of sexual assault on campus. This is compounded by

Clery report data and anonymous survey results leave vastly different impressions of rape culture on college campuses.
Boston University
Climate Survey
18%
Clery Report
0.04%
0
2
4
6
8
10
12
14
16
18
Boston University surveyed its students in 2015, with a response rate of 22 percent. Nearly one in five respondents reported experiencing some type of sexual harassment or assault during their time at Boston University, compared to one in 2500 who reported assault in 2014.
Emerson College
Climate Survey
9%
Clery Report
0.15%
0
2
4
6
8
10
12
14
16
18
Emerson College surveyed its students in 2015, with a 32 percent response rate. About one in 10 respondents said they experienced nonconsensual sexual contact on-campus during their time at Emerson, compared to one in 666 students that reported forcible sex offenses in 2014.

**Figure 6.4**

Data journalism students at Emerson College were skeptical of the self-reported Clery Act data and decided to compare the Clery Act results with anonymous campus climate survey results about nonconsensual sexual contact. Although there are data-quality issues with both datasets, the students assert that if institutions are providing adequate support for survivors, then there will be less of a gap between the Clery-reported data and the proportion of students that report nonconsensual sexual conduct. Courtesy of Patrick Torphy, Michaela Halnon, and Jillian Meehan, 2016.

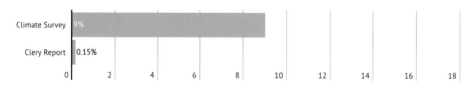

the fact that survivors of sexual assault often do not want to come forward—because of social stigma, the trauma of reliving their experience, or the resulting lack of social and psychological support. Mainstream culture has taught survivors that their experiences will not be treated with care and that they may in fact face more harm, blame, and trauma if they do come forward.[27]

There are further power differentials reflected in the data when race and sexuality are taken into account. For example, in 2014, twenty-three students filed a complaint against Columbia University, alleging that Columbia was systematically mishandling cases of rape and sexual violence reported by LGBTQ students. Zoe Ridolfi-Starr, the lead student named in the complaint, told the *Daily Beast*, "We see complete lack of knowledge about the specific dynamics of sexual violence in the queer community, even from people who really should be trained in those issues."[28]

Simply stated, there are imbalances of power in the *data setting*—to use the phrase coined by Yanni Loukissas that we discussed in chapter 5—so we cannot take the numbers in the dataset at face value. Lacking this understanding of power in the collection environment and letting the numbers "speak for themselves" would tell a story that is not only patently false but could also be used to reward colleges that are systematically underreporting and creating hostile environments for survivors. Deliberately undercounting cases of sexual assault leads to being rewarded for underreporting. And the silence around sexual assault continues: the administration is silent, the campus culture is silent, the dataset is silent.[29]

## Raw Data, Cooked Data, Cooking

As demonstrated by the Emerson College students, one of the key analytical missteps of work that lets "the numbers speak for themselves" is the premise that data are a *raw input*. But as Lisa Gitelman and Virginia Jackson have memorably explained, data enter into research projects already fully cooked—the result of a complex set of social, political, and historical circumstances. "'Raw data' is an oxymoron," they assert, just like "jumbo shrimp."[30] But there is an emerging class of "data creatives" whose very existence is premised on their ability to *context-hop*—that is, their ability to creatively mine and combine data to produce new insights, as well as work across diverse domains. This group includes data scientists, data journalists, data artists and designers, researchers, and entrepreneurs—in short, pretty much everyone who works with data right now. They are the strangers in the dataset that we spoke of in chapter 5.

Data's new creative class is highly rewarded for producing work that creates new value and insight from mining and combining conceptually unrelated datasets.

Examples include Google's now defunct Flu Trends project, which tried to geographi-
cally link people's web searches for flu symptoms to actual incidences of flu.[31] Or a project of the *Sun Sentinel* newspaper, in Fort Lauderdale, Florida, which combined police
license plate data with electronic toll records to prove that cops were systematically and
dangerously speeding on Florida highways.[32] Sometimes these acts of creative synthe-
sis work out well; the *Sun Sentinel* won a Pulitzer for its reporting and a number of the
speeding cops were fired. But sometimes the results are not quite as straightforward.
Google Flu Trends worked well until it didn't, and subsequent research has shown that
Google searches cannot be used as 1:1 signals for actual flu phenomena because they
are susceptible to external factors, such as what the media is reporting about the flu.[33]

Instead of taking data at face value and looking toward future insights, data scientists can first interrogate the context, limitations, and validity of the data under use.
In other words, one feminist strategy for considering context is to consider the cooking process that produces "raw" data. As one example, computational social scientists
Derek Ruths and Jürgen Pfeffer write about the limitations of using social media data
for behavioral insights: Instagram data skews young because Instagram does; Reddit
data contains far more comments by men than by women because Reddit's overall
membership is majority men. They further show how research data acquired from
those sources are shaped by sampling because companies like Reddit and Instagram
employ proprietary methods to deliver their data to researchers, and those methods
are never disclosed.[34] Related research by Devin Gaffney and J. Nathan Matias took on
a popular corpus that claimed to contain "every publicly available Reddit comment."[35]
Their work showed the that the supposedly complete corpus is missing at least thirty-
six million comments and twenty-eight million submissions.

Exploring and analyzing what is missing from a dataset is a powerful way to gain
insight into the cooking process—of both the data and of the phenomenon it purports
to represent. In some of Lauren's historical work, she looks at actual cooks as they are
recorded (or not) in a corpus of thirty thousand letters written by Thomas Jefferson, as
shown in figure 6.5.[36] Some may already know that Jefferson is considered the nation's
"founding foodie."[37] But fewer know that he relied upon an enslaved kitchen staff to
prepare his famous food.[38] In "The Image of Absence," Lauren used named-entity recognition, a natural language processing technique, to identify the places in Jefferson's
personal correspondence where he named these people and then used social network
analysis to approximate the extent of the relationships among them. The result is a
visual representation of all of the work that Jefferson's enslaved staff put into preparing his meals but that he did not acknowledge—at least not directly—in the text of the
letters themselves.

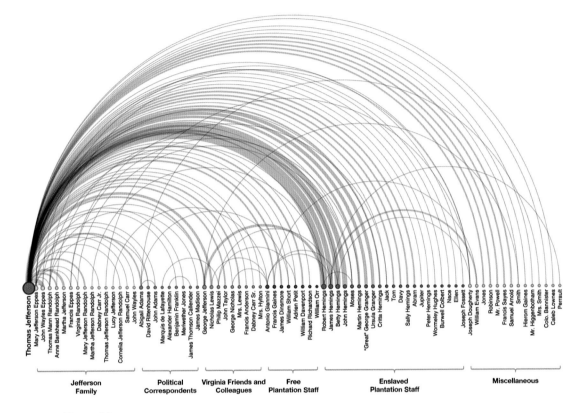

**Figure 6.5**
In "The Image of Absence" (2013), Lauren used machine learning techniques to identify the names of the people whom Thomas Jefferson mentioned in his personal correspondence and then visualized the relationships among them. The result demonstrates all of the work that his enslaved staff put into preparing Jefferson's meals but that was not directly acknowledged by Jefferson himself. Visualization by Lauren F. Klein.

On an even larger scale, computer scientists and historians at Stanford University used word embeddings—another machine learning technique—to explore gender and ethnic stereotypes across the span of the twentieth century.[39] Using several large datasets derived from sources such as the Google Books and the *New York Times*, the team showed how words like *intelligent*, *logical*, and *thoughtful* were strongly associated with men until the 1960s. Since that time, however, those words have steadily increased in association with women. The team attributed this phenomenon to the "women's movement in the 1960s and 1970s," making their work an interesting example of an attempt to quantify the impact of social movements. The paper is also notable for openly acknowledging how their methods, which involved looking at the adjectives surrounding the words *man* and *woman*, limited the scope of their analysis to the gender binary. Furthermore, the researchers did not try to assert that the data represent how women and men "are," nor did they try to "remove the bias" so that they could develop "unbiased" applications in other domains. They saw the data as what they are—cultural indicators of the changing face of patriarchy and racism—and interrogated them as such.

So, how do we produce more work like this—work that understands data as already "cooked" and then uses that data to expose structural bias? Unfortunately for Chris Anderson, the answer is that we need more theory, not less. Without theory, survey designers and data analysts must rely on their intuition, supported by "common sense" ideas about the things they are measuring and modeling. This reliance on "common sense" leads directly down the path to bias. Take the case of GDELT. Decades of research has demonstrated that events covered by the media are selected, framed, and shaped by what are called "news values": values that confirm existing images and ideologies.[40] So what is it really that GDELT is measuring? What events are happening in the world, or what the major international news organizations are focusing their attention on? The latter might be the most powerful story embedded in the GDELT database. But it requires deep context and framing to draw it out.

Refusing to acknowledge context is a power play to avoid power. It's a way to assert authoritativeness and mastery without being required to address the complexity of what the data actually represent: the political economy of the news in the case of GDELT, entrenched gender hierarchies and flawed reporting environments in the case of the Clery data, and so on. But deep context and computation are not incompatible. For example, SAFElab, a research lab at Columbia run by scholar and social worker Desmond Patton, uses artificial intelligence to examine the ways that youth of color navigate violence on and offline. He and a team of social work students use Twitter data to understand and prevent gang violence in Chicago. Their data are big, and they're also complicated in ways that are both technical and social. The team is acutely aware of the history of law enforcement agencies using technology to surveil Black people, for example, and acknowledges that law enforcement continues to do so using Twitter itself. What's more, when Patton started his research, he ran into an even more basic problem: "I didn't know what young people were saying, period."[41] This was true even though Patton himself is Black, grew up in Chicago, and worked for years in many of these same neighborhoods. "It became really clear to me that we needed to take a deeper approach to social media data in particular, so that we could really grasp culture, context and nuance, for the primary reason of not misinterpreting what's being said," he explains.[42]

Patton's approach to incorporating culture, context, and nuance took the form of direct contact with and centering the perspectives of the youth whose behaviors his group sought to study. Patton and doctoral student William Frey hired formerly gang-involved youth to work on the project as domain experts. These experts coded and categorized a subset of the millions of tweets, then trained a team of social work students to take over the coding. The process was long and not without challenges. It required that Patton and Frey create a new "deep listening" method they call the *contextual analysis of social media* to help the student coders mitigate their own bias and get closer to the intended meaning of each tweet.[43] The step after that was to train a machine learning classifier to automatically label the tweets, so that the project could categorize all of the millions of tweets in the dataset. Says Patton, "We trained the algorithm to think like a young African American man on the south side of Chicago."[44]

This approach illustrates how context can be integrated into an artificial intelligence project, and can be done with an attention to *subjugated knowledge*. This term describes the forms of knowledge that have been pushed out of mainstream institutions and the conversations they encourage. To explain this phenomenon, Patricia Hill Collins gives the example of how Black women have historically turned to "music, literature, daily conversations, and everyday behavior" as a result of being excluded from "white male-controlled social institutions."[45] These institutions include academia, or—for a recent example raised by sociologist Tressie McMillan Cottom—the op-ed section of the *New York Times*.[46] And because they circulate their knowledge in places outside of those mainstream institutions, that knowledge is not seen or recognized by those institutions: it becomes *subjugated*.

The idea of subjugated knowledge applies to other minoritized groups as well, including the Black men from Chicago whom Patton sought to understand. An approach that did not attend to this context would have resulted in significant errors. For example, a tweet like "aint kill yo mans & ion kno ya homie" would likely have been classified as aggressive or violent, reflecting its use of the word "kill." But drawing on the knowledge provided by the young Black men they hired for the project, Frey and Patton were able to show that many tweets like this one were references to song lyrics, in this case the Chicago rapper Lil Durk. In other words, these tweets are about sharing culture, not communicating threats.[47]

In the case of SAFElab, as with all research projects that seek to make use of subjugated knowledge, there is also significant human, relational infrastructure required.

Frey and Patton have built long-term relationships with individuals and organizations in the community they study. Indeed, Frey lives and works in the community. In addition, both Frey and Patton are trained as social workers. This is reflected in their computational work, which remains guided by the social worker's code of ethics.[48] They are using AI to broker new forms of human understanding across power differentials, rather than using computation to replace human relationships. This kind of social innovation often goes underappreciated in the unicorn-wizard-genius model of data science. (For more on unicorns, see chapter 5.) As Patton says, "We had a lot of challenges with publishing papers in data science communities about this work, because it is very clear to me that they're slow to care about context. Not that they don't care, but they don't see the innovation or the social justice impact that the work can have."[49] Hopefully that will change in the future, as the work of SAFElab and others demonstrates the tremendous potential of combining social work and data science.

## Communicating Context

It's not just in the stages of data acquisition or data analysis that context matters. Context also comes into play in the framing and communication of results. Let's imagine a scenario. In this case, you are a data journalist, and your editor has assigned you to create a graphic and short story about a recent research study: "Disparities in Mental Health Referral and Diagnosis in the New York City Jail Mental Health Service."[50] This study looks at the medical records of more than forty-five thousand first-time incarcerated people and finds that some groups are more likely to receive treatment, while others are more likely to receive punishment. More specifically, white people are more likely to receive a mental health diagnosis, while Black and Latinx people are more likely to be placed in solitary confinement. The researchers attribute some of this divergence to the differing diagnosis rates experienced by these groups before becoming incarcerated, but they also attribute some of the divergence to discrimination within the jail system. Either way, the racial and ethnic disparities are a product of structural racism.

Consider the difference between the two graphics shown in figure 6.6. The only variation is the title and framing of the chart.

Which one of these graphics would you create? Which one should you create? The first—Mental Health in Jail—represents the typical way that the results of a data analysis are communicated. The title *appears* to be neutral and free of bias. This is a graphic about rates of mental illness diagnosis of incarcerated people broken down by race and ethnicity. The people are referred to as *inmates*, the language that the study used. The

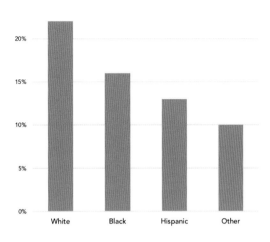

**Mental Health in Jail**
Rate of mental health diagnosis of inmates

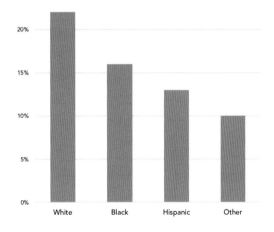

**Racism in Jail**
People of color less likely to get mental health diagnosis

**Figure 6.6**
Two portrayals of the same data analysis. The data are from a study of people incarcerated for the first time in NYC jails between 2011 and 2013. Graphics by Catherine D'Ignazio. Data from Fatos Kaba et al., "Disparities in Mental Health Referral and Diagnosis in the New York City Jail Mental Health Service."

title does not mention race or ethnicity, or racism or health inequities, nor does the title point to what the data mean. But this is where additional questions about context come in. Are you representing only the four numbers that we see in the chart? Or are you representing the context from which they emerged?

The study that produced these numbers contains convincing evidence that we should distrust diagnosis numbers due to racial and ethnic discrimination. The first chart does not simply fail to communicate that but also actively undermines that main finding of the research. Moreover, the language used to refer to people in jail as *inmates* is dehumanizing, particularly in the context of the epidemic of mass incarceration in the United States.[51] So, consider the second chart: Racism in Jail: People of Color Less Likely to Get Mental Health Diagnosis. This title offers a frame for how to interpret the numbers along the lines of the study from which they emerged. The research study was about racial disparities, so the title and content of this chart are about racial disparities. The people behind the numbers are *people*, not *inmates*. In addition, and crucially, the second chart names the forces of oppression that are at work: racism in prison.

Although naming racism may sound easy and obvious to some readers of this book, it is important to acknowledge that fields like journalism still adhere to conventions that resist such naming on the grounds that it is "bias" or "opinion." John Daniszewski, an editor at the Associated Press, epitomizes this view: "In general our policy is to try to be neutral and precise and as accurate as we possibly can be for the given situation. We're very cautious about throwing around accusations of our own that characterize something as being racist. We would try to say what was done, and allow the reader to make their own judgement."[52]

Daniszewski's statement may sound democratic ("power to the reader!"), but it's important to think about whose interests are served by making racism a matter of individual opinion. For many people, racism exists as a matter of fact, as we have discussed throughout this book. Its existence is supported by the overwhelming empirical evidence that documents instances of structural racism, including wealth gaps, wage gaps, and school segregation, as well as health inequities, as we have also discussed. Naming these structural forces may be the most effective way to communicate broad context. Moreover, as the data journalist in this scenario, it is your responsibility to connect the research question to the results and to the audience's interpretation of the results. Letting the numbers speak for themselves is emphatically not more ethical or more democratic because it often leads to those numbers being misinterpreted or the results of the study being lost. Placing numbers in context and naming racism or sexism when it is present in those numbers should be a requirement—not only for feminist data communication, but for data communication full stop.

This counsel—to name racism, sexism, or other forces of oppression when they are clearly present in the numbers—particularly applies to designers and data scientists from the dominant group with respect to the issue at hand. White people, including ourselves, the authors of this book, have a hard time naming and talking about racism. Men have a hard time naming and talking about sexism and patriarchy. Straight people have a hard time seeing and talking about homophobia and heteronormativity. If you are concerned with justice in data communication, or data science more generally, we suggest that you practice recognizing, naming, and talking about these structural forces of oppression.[53]

But our work as hypothetical anti-oppression visualization designers is not over yet. We might have named racism as a structural force in our visualization, but there are still two problems with the "good" visualization, and they hinge on the wording of the subtitle: People of Color Less Likely to Get Mental Health Diagnosis. The first problem is that this is starting to look like a deficit narrative, which we discuss in chapter 2—a narrative that reduces a social group to negative stereotypes and fails to portray them with creativity and agency. The second issue is that by naming racism and then talking about people of color in the title, the graphic reinforces the idea that race is an issue for people of color only. If we care about righting the balance of power, the choice of words matters as much as the data under analysis. In an op-ed about the language used to describe low-income communities, health journalist Kimberly Seals Allers affirms this point: "We almost always use a language of deficiency, calling them disadvantaged, under-resourced and under-everything else. ... It ignores all the richness those communities and their young people possess: the wealth of resiliency, tenacity and grit that can turn into greatness if properly cultivated."[54]

So let's give it a third try, with the image in figure 6.7.

In this third version, we have retained the same title as the previous chart. But instead of focusing the subtitle on what minoritized groups lack, it focuses on the unfair advantages that are given to the dominant group. The subtitle now reads, White People Get More Mental Health Services. This avoids propagating a deficit narrative that reinforces negative associations and clichés. It also asserts that white people have a race, and that they derive an unfair advantage from that race in this case.[55] Finally, the title is proposing an interpretation of the numbers that is grounded in the context of the researchers' conclusions on health disparities.

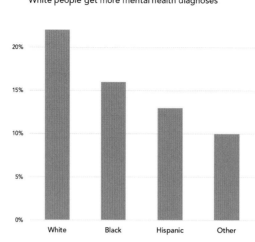

**Figure 6.7**
A third portrayal of the same data, with only the framing title and subtitle changed. *Source:* Data from Kaba et al., "Disparities in Mental Health Referral and Diagnosis in the New York City Jail Mental Health Service." Graphic by Catherine D'Ignazio. Data from Fatos Kaba et al., "Disparities in Mental Health."

## Restoring Context

Three iterations on a single chart title might feel excessive, but it also helps to under-score the larger point that considering context always involves some combination of interest and time. Fortunately, there is a lot of energy around issues of context right now, and educators, journalists, librarians, computer scientists, and civic data publishers are starting to develop more robust tools and methods for keeping context attached to data so that it's easier to include in the end result.

For example, remember figure 6.3, that confusing chart of government procure-ments in São Paulo that we discussed earlier in this chapter? Gisele Craveiro, a professor at the University of São Paulo, has created a tool called Cuidando do Meu Bairro (Caring for My Neighborhood) to make that spending data more accessible to citizens by add-ing additional local context to the presentation of the information.[56] In the classroom, Heather Krause, a data scientist and educator, has developed the concept of the "data biography."[57] Prior to beginning the analysis process, Krause asks people working with data, particularly journalists, to write a short history of a particular dataset and answer five basic questions: Where did it come from? Who collected it? When? How was it collected? Why was it collected? A related but slightly more technical proposal advocated by researchers at Microsoft is being called *datasheets for datasets*.[58] Inspired by the datasheets that accompany hardware components, computer scientist Timnit Gebru and colleagues advocate for data publishers to create short, three- to five-page documents that accompany datasets and outline how they were created and collected, what data might be missing, whether preprocessing was done, and how the dataset will be maintained, as well as a discussion of legal and ethical considerations such as whether the data collection process complies with privacy laws in the European Union.[59]

Another emerging practice that attempts to better situate data in context is the development of *data user guides*.[60] Bob Gradeck, manager of the Western Pennsylvania Regional Data Center, started writing data user guides because he got the same questions over and over again about popular datasets he was managing, like property data and 311 resident reports in Pittsburgh. Reports Gradeck, "It took us some time to learn tips and tricks. ... I wanted to take the stuff that was in my head and put it out there with additional context, so other data users didn't have to do it from scratch."[61] Data user guides are simple, written documents that each contain a narrative portrait of a dataset. They describe, among other things, the purpose and application of the data; the history, format, and standards; the organizational context; other analyses and stories that have used the dataset; and the limitations and ethical implications of the dataset. This is similar to the work that data journalists are doing to compile datasets and then make them available for reuse. For example, the Associated Press makes comprehensive national statistics about school segregation in the United States available for purchase.[62] The spreadsheets are accompanied by a twenty-page narrative explainer about the data that includes limitations and sample story ideas.

These developments are exciting, but there is further to go with respect to issues of power and inequality that affect data collection environments. For example, professor of political science Valerie Hudson has worked for decades to trace the links between state security and the status of women. "I was interested in whether forms of oppression or subordination or violence against women were related to national, and perhaps international, instability and conflict," she explains. She and geographer Chad Emmett started the project WomanStats as a modest Excel spreadsheet in 2001. It has since grown to a large-scale web database with over a quarter of a million data points, including over 350 variables ranging from access to health care to the prevalence of rape to the division of domestic labor.[63]

Notably, their sources are qualitative as well as quantitative. Says Hudson, "If you want to do research on women, you have to embrace qualitative data. There's no two ways about it, because the reality of women's lives is simply not captured in quantitative statistics. Absolutely not."[64] At the present, WomanStats includes two types of qualitative variables: practice variables are composed from women's reports of their lived experiences, and law variables are coded from the legal frameworks in a particular country. Indeed, the WomanStats codebook is a context nerd's dream that outlines measurement issues and warns about the incompleteness of its own data, especially with respect to difficult topics.[65] In regard to the data that records reports of rape, for example—a topic upsetting enough to even consider, let alone contemplate its scale and scope in an entire country—the codebook states: "CAVEAT EMPTOR! Users are warned that this scale only reflects reported rape rates, and for many, if not most, countries, this is a completely unreliable indicator of the actual prevalence of rape within a society!"[66] Instead of focusing on a single variable, users are directed to WomanStats's composite scales, like the Comprehensive Rape Scale, which look at reported prevalence in the context of laws, whether laws are enforced, reports from lived experience, strength of taboos in that environment, and so on.

So tools and methods for providing context are being developed and piloted. And WomanStats models how context can also include an analysis of unequal social power. But if we zoom out of project-level experiments, what remains murky is this: Which actors in the data ecosystem are responsible for providing context?

Is it the end users? In the case of the missing Reddit comments, we see how even the most highly educated among us fail to verify the basic claims of their data source. And datasheets for datasets and data user guides are great, but can we expect individual people and small teams to conduct an in-depth background research project while on a deadline and with a limited budget? This places unreasonable expectations and responsibility on newcomers and is likely to lead to further high-profile cases of errors and ethical breaches.

So is it the data publishers? In the case of GDELT, we saw how data publishers, in their quest for research funding, overstated their capabilities and didn't document the limitations of their data. The Reddit comments were a little different: the dataset was provided by an individual acting in good faith, but he did not verify—and probably did not have the resources to verify—his claim to completeness. In the case of the campus sexual assault data, it's the universities who are responsible for self-reporting, and they are governed by their own bottom line.[67] The government is under-resourced to verify and document all the limitations of the data.

Is it the data intermediaries? Intermediaries, who have also been called *infomediaries*, might include librarians, journalists, nonprofits, educators, and other public information professionals.[68] There are strong traditions of data curation and management in library science, and librarians are often the human face of databases for citizens and residents. But as media scholar Shannon Mattern points out, librarians are often left out of conversations about smart cities and civic technology.[69] Examples of well-curated, verified and contextualized data from journalism, like the Associated Press database on school segregation or other datasets available in ProPublica's data store, are also promising.[70] The nonprofit Measures for Justice provides comprehensive and contextualized data on criminal justice and incarceration rates in the United States.[71] Some data intermediaries, like Civic Switchboard in Pittsburgh, are building their own local data ecosystems as a way of working toward sustainability and resilience.[72] These intermediaries who clean and contextualize the data for public use have potential (and have fewer conflicts of interest), but sustained funding, significant capacity-building, and professional norms-setting would need to take place to do this at scale.

Houston, we have a public information problem. Until we invest as much in providing (and maintaining) context as we do in publishing data, we will end up with public information resources that are subpar at best and dangerous at worst. This ends up getting even more thorny as the sheer quantity of digital data complicates the verification, provenance, and contextualization work that archivists have traditionally undertaken. Context, and the informational infrastructure that it requires, should be a significant focus for open data advocates, philanthropic foundations, librarians, researchers, news organizations, and regulators in the future. Our data-driven lives depend on it.

## Consider Context

The sixth principle of data feminism is to *consider context*. The bottom line for numbers is that they cannot speak for themselves. In fact, those of us who work with data must actively prevent numbers from speaking for themselves because when those numbers derive from a data setting influenced by differentials of power, or by misaligned collection incentives (read: pretty much all data settings), and especially when the numbers have to do with human beings or their behavior, then they run the risk not only of being arrogantly grandiose and empirically wrong, but also of doing real harm in their reinforcement of an unjust status quo.

The way through this predicament is by considering context, a process that includes understanding the provenance and environment from which the data was collected, as well as working hard to frame context in data communication (i.e., the numbers should not speak for themselves in charts any more than they should in spreadsheets). It also includes analyzing social power in relation to the data setting. Which power imbalances have led to silences in the dataset or data that is missing altogether? Who has conflicts of interest that prevent them from being fully transparent about their data? Whose knowledge about an issue has been subjugated, and how might we begin to recuperate it? The energy around context, metadata, and provenance is impressive, but until we fund context, then excellent contextual work will remain the exception rather than the norm.

# 7   Show Your Work

## Principle: Make Labor Visible

*The work of data science, like all work in the world, is the work of many hands. Data feminism makes this labor visible so that it can be recognized and valued.*

If you work in software development, chances are that you have a GitHub account. As of June 2018, the online code-management platform had over twenty-eight million users worldwide. By allowing users to create web-based repositories of source code (among other forms of content) to which project teams of any size can then contribute, GitHub makes collaborating on a single piece of software or a website or even a book much easier than it has ever been before.

Well, easier if you're a man. A 2016 study found that female GitHub users were less likely to have their contributions accepted if they identified themselves in their user profiles as women. (The study did not consider nonbinary genders.)[1] Critics of GitHub's commitment to inclusivity, or the lack thereof, also point to the company's internal politics. In 2014, GitHub's cofounder was forced to resign after allegations of sexual harassment were brought to light.[2] More recently, in 2018, Agnes Pak, a former top attorney at GitHub, sued the company for allegedly altering her performance reviews after she complained about her gender and race contributing to a lower compensation package, giving them the grounds to fire her.[3] Pak's suit came only shortly after transgender software developer Coraline Ada Ehmke, in 2017, declined a significant severance package so that she could talk publicly about her negative experience of working at GitHub.[4] Clearly, GitHub has several major issues of corporate culture that it must address.

But a corporate culture that is hostile to women does not necessarily preclude other feminist interventions. And here GitHub makes one important one: its platform helps *show the work* of writing collaborative code. In addition to basic project management tools, like bug tracking and feature requests, the GitHub platform also generates visualizations of each team member's contributions to a project's codebase. Area charts, arranged in small multiples, allow viewers to compare the quantity, frequency, and duration of any particular member's contributions (figure 7.1a). A line graph reveals patterns in the day of the week when those contributions took place (figure 7.1b). And a flowchart-like diagram of the relationships between various branches of the project's

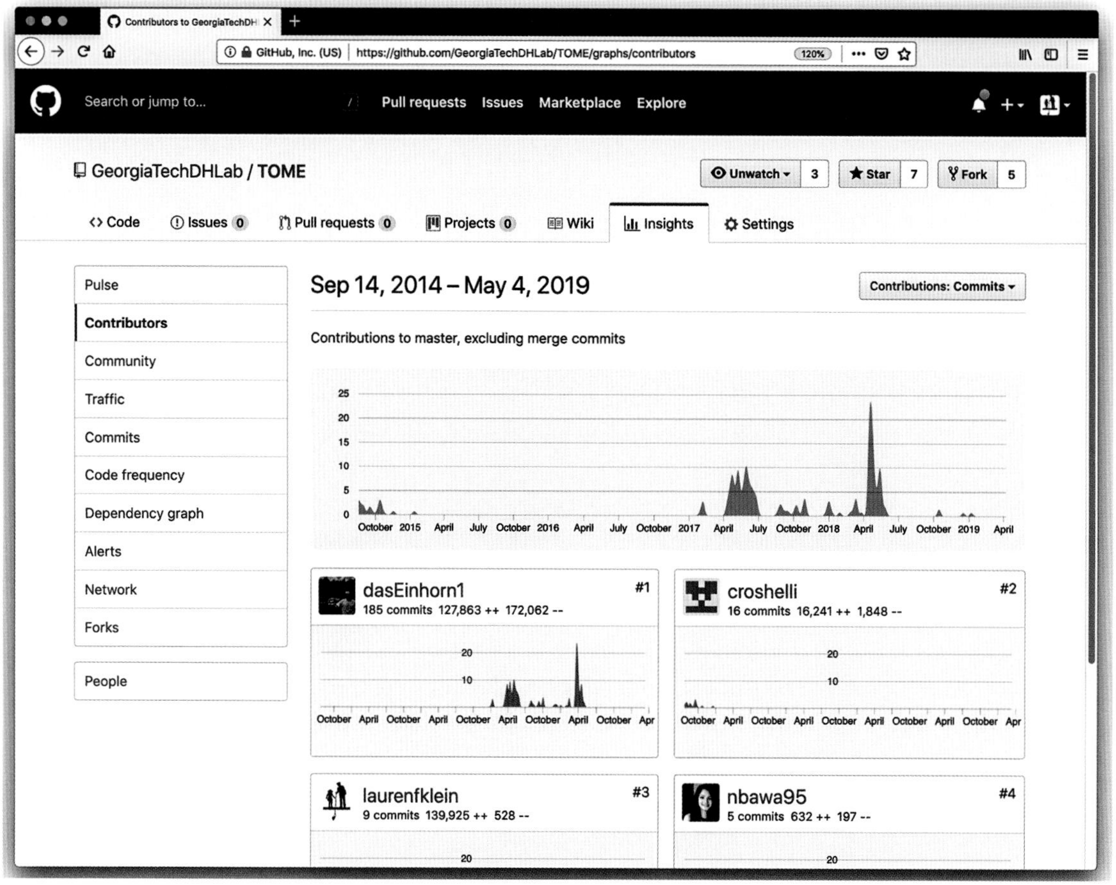

**Figure 7.1**

(a) The first of three visualizations of the code associated with a project from Lauren's research group, showing the significant contributions of student researchers between the years 2014 and 2019. (b) A bar chart shows the frequency of code commits over time, and a line graph shows any patterns in the day of the week when the commits were made. Screenshot by Lauren F. Klein. (c) A flowchart-like diagram documents the relationships between the various branches of the project's codebase. Screenshots by Lauren F. Klein.

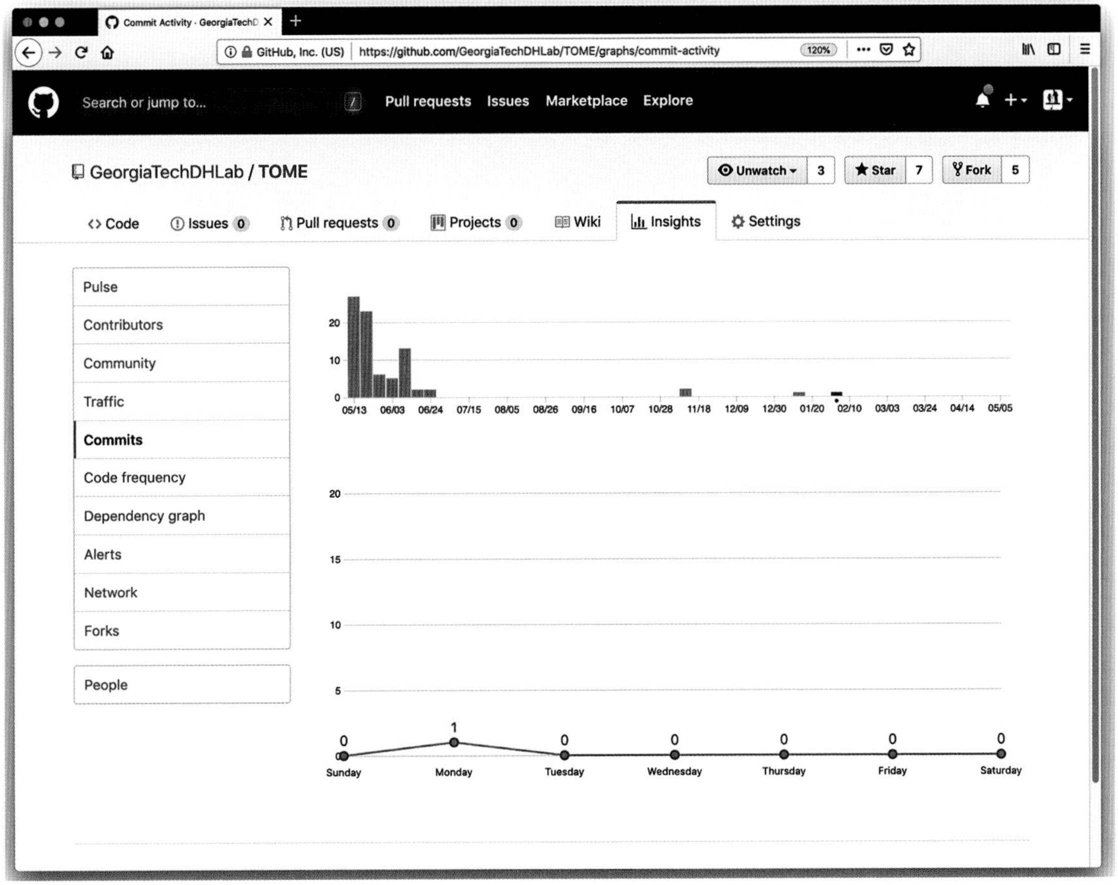

**Figure 7.1** (continued)

code helps to acknowledge any sources for the project that might otherwise go uncredited, as well as any additional projects that might build upon the project's initial work (figure 7.1c).

Coding is work, as anyone who's ever programmed anything knows well. But it's not always work that is easy to see. The same is true for collecting, analyzing, and visualizing data. We tend to marvel at the scale and complexity of an interactive visualization like the *Ship Map*, in figure 7.2, which plots the paths of the global merchant fleet over the course of the 2012 calendar year.[5] By showing every single sea voyage, the *Ship Map* exposes the networks of waterways that constitute our global product supply chain. But we are less often exposed to the networks of processes and people that

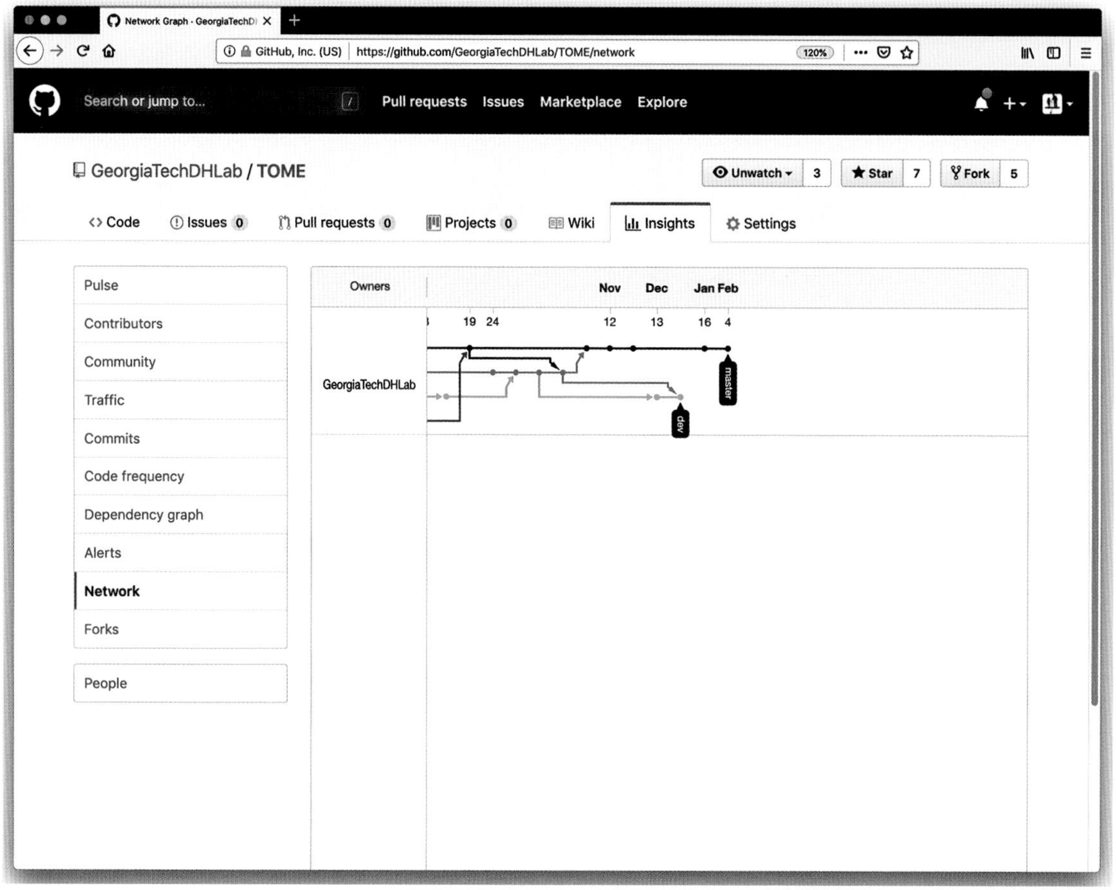

**Figure 7.1** (continued)

help constitute the visualization itself—from the seventy-five corporate researchers at Clarksons Research UK who assembled and validated the underlying dataset, to the academic research team at University College London's Energy Institute that developed the data model, to the design team at Kiln that transformed the data model into the visualization that we see. And that is to say nothing of the tens of thousands of commercial ships that served as the source of data in the first place. Visualizations like the *Ship Map* involve the work of many hands.

Unfortunately, however, when releasing a data product to the public, we tend not to credit the many hands who perform this work. We often cite the source of the dataset, and the names of the people who designed and implemented the code and graphic elements. But we rarely dig deeper to discover who created the data in first

**Figure 7.2**
A time-based visualization of global shipping routes, designed by Kiln in 2016, based on data from
the University College London Energy Institute (UCL EI). The Ship Map website was created by
Duncan Clark and Robin Houston from Kiln, and the dataset was compiled by Julia Schaumeier
and Tristan Smith from the UCL EI. The website also includes a soundtrack: Bach's Goldberg
Variations, played by Kimiko Ishizaka.

place, who collected the data and processed them for use, and who else might have labored to make creations like the *Ship Map* possible. Admittedly, this information is sometimes hard to find. And when project teams (or individuals) are already operating at full capacity, or under budgetary strain, this information can—ironically—simply be too much additional work to pursue.[6] Even in cases in which there are both resources and desire, information about the range of the contributors to any particular project sometimes can't be found at all. But the various difficulties we encounter when trying to acknowledge this work reflects a larger problem in what information studies scholar Miriam Posner calls our *data supply chain*.[7] Like the contents of the ships visualized on the *Ship Map*, about which we only know sparse details—the map can tell us *if* a shipping container was loaded onto the boat, but not *what* the shipping container contains—the invisible labor involved in data work, as Posner argues, is something that corporations have an interest in keeping out of public view.

To put it more simply, it's not a coincidence that much of the work that goes into designing a data product—visualization, algorithm, model, app—remains invisible and uncredited. In our capitalist society, we tend to value work that we can see. This is the result of a system in which the cultural worth of any particular form of work is directly connected to the price we pay for it; because a service costs money, we recognize its larger value. But more often than not, the reverse also holds true: we fail to recognize the larger value of the services we get for free. When, in the early 1970s, the International Feminist Collective launched the Wages for Housework campaign, it was this phenomenon of *invisible labor*—labor that was unpaid and therefore unvalued—that the group was trying to expose (figure 7.3).[8] The precise term they used to describe this work was *reproductive labor*, which comes from the classical economic distinction between the paid and therefore economically *productive* labor of the marketplace, and the unpaid and therefore economically *unproductive* labor of everything else. By reframing this latter category of work as reproductive labor, rather than simply (and inaccurately) unproductive labor, groups like the International Feminist Collective sought to emphasize how the range of tasks that the term encompassed, like cooking and cleaning and child-rearing, were precisely the tasks that enabled those who performed "productive" labor, like office or factory work, to continue to do so.

The Wages for Housework movement began in Italy and migrated to the United States with the help of labor organizer and theorist Silvia Federici. It eventually claimed chapters in several American cities, and did important consciousness-raising work.[9] Still, as prominent feminists like Angela Davis pointed out, while housework might have been unpaid for white women, women of color—especially Black women in the United States—had long been paid, albeit not well, for their housework in other people's homes: "Because of the added intrusion of racism, vast numbers of Black women

**Figure 7.3**
A Wages for Housework march, 1977. Photograph by Bettye Lane. Courtesy of the Schlesinger Library, Radcliffe Institute/Bettye Lane.

have had to do their own housekeeping and other women's home chores as well."[10] Here, Davis is making an important point about *racialized labor*: just as housework is structured along the lines of gender, it is also structured along the lines of race and class. The domestic labor of women of color was and remains *underwaged labor*, as feminist labor theorists would call it, and its low cost was what permitted (and continues to permit) many white middle- and upper-class women to participate in the more lucrative *waged labor* market instead.[11]

Since the 1970s, the term *invisible labor* has come to encompass the various forms of labor, unwaged, underwaged, and even waged, that are rendered invisible because they take place inside of the home, because they take place out of sight, or because they lack physical form altogether.[12] Visit WagesforFacebook.com and you'll find a version of the Wages for Housework argument updated for a new form of invisible work. This invisible labor can be found all over the web, as digital labor theorists such as Tiziana Terranova have helped us to understand.[13] "They call it sharing. We call it stealing," is one of the statements that scrolls down the screen in large black type. The word *it* refers to work that most of us perform every day, in the form of our Facebook likes, Instagram posts, and Twitter tweets. The point made by Laurel Ptak, the artist behind Wages for Facebook—a point also made by Terranova—is that the invisible unpaid labor of our likes and tweets is precisely what enables the Facebooks and Twitters of the world to profit and thrive.

## The Invisible Labor of Data Science

The world of data science is able to profit and thrive because of unpaid invisible labor as well. How did Netflix improve its movie recommendation algorithm? The company crowdsourced it.[14] How did the *Guardian*, the British newspaper, determine which among two million leaked documents might contain incriminating information about government misspending? The paper crowdsourced it.[15] The optical character recognition (OCR) error correction performed on the dataset of early modern books that you downloaded for your text-analysis project? That was crowdsourced, too.[16]

Each of these crowdsourcing projects were framed as acts of benevolence (and, in the case of Netflix, an opportunity to win a million-dollar prize). People should want to contribute to these projects, their proponents claimed, since their labor would further the public good.[17] However, Ashe Dryden, the software developer and diversity consultant, points out that people can only help crowdsource if they have the inclination and the time.[18] Think back to that study of GitHub. If you were a woman and you knew your contributions to a programming project were less likely to be accepted than if you were a man, would that motivate you to contribute the project? Or, for another example, consider Wikipedia. Although the exact gender demographics of Wikipedia contributors are unknown, numerous surveys have indicated that those who contribute content to the crowdsourced encyclopedia are between 84 percent and 91.5 percent men.[19] Why? It could be that there, too, edits are less likely to be accepted if they come from women editors.[20] It could also be attributed to Wikipedia's exclusionary editing culture and technological infrastructure, as science and technology studies (STS) scholars Heather Ford and Judy Wajcman have argued.[21] And there is also reason to go back to the housework argument. Dryden cites a 2011 study showing that women in twenty-nine countries spend more than twice as much time on household tasks than men do, even when controlling for women who hold full-time jobs.[22] The study did not consider nonbinary genders or same-sex (or other non-hetero-typical) households. But even as a rough estimate, it seems that women simply don't have as much time.[23]

In capitalist societies, it's very often the case that time is money. But it's also important to remember to ask whose time is being spent and whose money is being saved. The premise behind Amazon's Mechanical Turk—or MTurk, as the crowdsourcing platform is more commonly known—is that data scientists want to save their own time and their own bottom line.[24] The MTurk website touts its access to a global marketplace of "on-demand Workers," who are advertised as being more "scalable and cost-effective" than the "time consuming [and] expensive" process of hiring actual employees.[25] But the data-entry and data-processing tasks performed by these workers earn them less than minimum wage, even as a recent study by the Pew Research Center showed that 51 percent of US-based Turkers, as they are known, hold college degrees, and 88 percent are below the age of fifty, among other metrics that would otherwise rank them among the most desired demographic for salaried employees.[26] This form of underwaged work is also increasingly outsourced from the United States to countries with fewer (or worse) labor laws and fewer (or worse) opportunities for economic advancement. A 2010 University of California, Irvine study measured a 20 percent drop in the number of US-based Turkers over the eighteen months that it monitored.[27] This trend has continued, the real-time MTurk tracker shows. (The gender split, interestingly, has evened out over time.)

Even at resource-rich companies like Amazon and Google, the work of data entry is profoundly undervalued in proportion to the knowledge it helps to create. Andrew Norman Wilson's 2011 documentary *Workers Leaving the Googleplex* (figure 7.4) exposes how the workers tasked with scanning the books for the Google Books database are hired as a separate but unequal class of employee, with ID cards that restrict their access to most of the Google campus and that prevent them from enjoying the company's famed employee perks.[28] (Evidently, working overtime to preserve the world's cultural heritage still does not entitle you to a free lunch, let alone a free class on how to cook Pad Kee Mao.)[29]

Wilson also observes that Google's book-scanning workers are disproportionately women and people of color—a fact that would not surprise the long line of women of color scholar-activists, including Angela Davis, Patricia Hill Collins, and Evelyn Nakano Glenn, who have insisted that economic oppression be recognized as a vector that cuts across the matrix of domination as a whole. Information studies scholar Lilly Irani confirms that "today's hierarchy of data labor echoes older gendered, classed, and raced technology hierarchies."[30] Here, Irani compares the hierarchy of data labor to the hierarchy encountered by the first generation of female computers, like Christine Darden, whom we discussed in this book's introduction.[31] But Irani's own research also considers contemporary digital labor practices, and in particular, Amazon's Mechanical Turk, the people it employs, and the people it exploits. In 2008, Irani and collaborators built a web tool called the Turkopticon, which enabled Turkers to anonymously report unfair labor conditions, as well as any additional information that might help

**Figure 7.4**
Andrew Norman Wilson's *Workers Leaving the Googleplex* (2011) documents the hidden inequities at Google's Mountain View headquarters. Still courtesy of Andrew Norman Wilson.

them decide whether to accept future tasks.[32] Irani envisioned the Turkopticon as a worker-led project. However, the same unfair labor conditions that necessitated the tool also ultimately limited its reach. In 2018, after ten years of service, its all-volunteer team of moderators called it quits. "We're all burned out," they wrote on Twitter.[33] And amid tagging images and correcting error-laden text, no additional Turkers could find the time.

The people who perform this *cultural data work*, as Irani terms it, are not only found on the MTurk platform, however. They're also increasingly the people on whom the entire information economy depends. Cultural data workers are responsible for the invisible labor involved in moderating the veritable deluge of content produced online every day, ensuring that your Facebook feed is free of, for example, child pornography and violent propaganda videos. When a 2014 exposé in *Wired* magazine documented the emotional costs of this labor, performed by some of the least empowered of these workers—women in the Global South—it was met with an outpouring of shock and outrage.[34] But subsequent studies like *Ghost Work*, by anthropologist Mary Gray and computer scientist Siddharth Suri, have documented the existence of a large "global underclass" performing this work of content moderation, transcription, and captioning.[35] They make the point that the so-called automation of artificial intelligence relies on a vast number of human beings in the loop.[36] Moreover, while the demographics of Silicon Valley tech workers remain steadily young, white, and male, these global "ghost workers" are often older women of color, and always required to accept precarious labor conditions.

Those who study the human costs of global capitalism would be quick to point out that this exploitation of precarious, racialized, colonial labor has a long history, one that has its roots in the original form of human exploitation: slavery. Slavery and capitalism are closely connected, after all, and one infamous story is often told to illustrate this point: in 1781, the British slave ship *Zong* made a series of navigational errors while crossing the Atlantic, resulting in a shortage of drinking water for the seventeen crew members and 133 captives on board.[37] After performing a cost-benefit analysis, the captain decided to throw the crew's enslaved human "cargo" overboard so that the crew members could consume all of the remaining water and rations themselves. The decision was made because the captain calculated that he could collect enough insurance money on his captives' loss of life to come out ahead, even if he couldn't sell them once they landed ashore. He was thinking about human lives solely in terms of their market value—the notion that capitalism holds in highest regard.

The stark inhumanity of this calculation has prompted numerous scholars and artists to return to the *Zong* as they reckon with what Christina Sharpe calls, in the language of the ship, the "wake" of slavery.[38] Poet M. NourbeSe Philip, for example, composed a book-length poem, *Zong!*, using only the words of the legal case that serves as the sole documentation of the original event.[39] Written over the course of many years and published in 2011, Philip's poem plucks words and short phrases out of the language of the court case, arranging them across the printed page. Philip's poem regularly shifts tenses from past to present, and from present to past, lending an additional voice to Sharpe's claim that the effects of that originary crime—the exploitation of Black bodies for white financial gain—are far from resolved.[40]

Our present technological infrastructure follows this same pattern of exploitation. In the United States, the scarcely paid or altogether unpaid labor of those who endure a contemporary form of enslavement—incarceration—has been used for everything from packaging Windows software to cleaning up the 2010 BP oil spill.[41] In a global colonial context, we might consider how the cobalt required to produce the lithium-ion batteries that power our cell phones and laptops is associated with significant human rights violations, including coercing labor from Congolese children as young as seven.[42] The unregulated disposal of electronics has resulted in roadside salvage and repair shops in places like Agbogbloshie, just outside of Accra, Ghana, which have long served as sites of invention and ingenuity, being transformed into toxic "e-waste" sites, with profound consequences for the health of those who live and work there, as well as for the environment.[43] The humanitarian and ecological stakes of our attachments to data and technology cannot be higher, nor can their source be any more clear: the capitalist and colonial forces that encourage the exploitation of Black and brown bodies so that white bodies can thrive.[44]

## Examining Data Production

The forces of global capitalism can feel overwhelming. And as people who use data and technology in our everyday work, we are each complicit to varying degrees. But there are certain small things we can do in our work, and in our work with others, to push back against this weight. In prior chapters, we have described some of these possibilities: incorporating an examination of power into a data analysis project (chapters 1 and 2); pushing back against false binaries and hierarchies (chapter 4); including multiple and marginalized voices in the design process (chapter 5); and contextualizing data so that they are not imagined to "speak for themselves" (chapter 6).

Along with these starting points, we can also begin to carve out additional space for the scholars, journalists, and other researchers who are explicitly studying the labor of data science—those who are examining and challenging power by tracing visualizations and algorithms and bots back to their human and material sources. This growing area of research might be called *data production studies*, borrowing a rubric from the field of production studies that currently sits at the intersection of film and media studies and labor studies. The primary focus of production studies, as it relates to film and media, is how media artifacts are produced. Media studies scholar Miranda Banks has asserted that "production studies is a feminist methodology" because it pays particular attention to the power differentials involved in the media production process, as well as the material conditions of media workers.[45] Work focused on data production is already happening in fields like STS, the digital humanities, library and information science, and archival studies, among others.[46] It looks at the production process of datasets, algorithms, and models, and traces those products back to the people and conditions that enabled their creation.

As an example of work in this emerging area, we might consider "Anatomy of an AI System," a project by technology researcher Kate Crawford and design scholar Vladlan Joler that seeks to describe and diagram the human labor, data dependencies, and material resources that contribute to a single Amazon Echo. The project was published online as a diagram of Borgesian proportions, too big to view in its entirety on a standard laptop screen (figure 7.5a); it was accompanied by a nine-thousand-word essay. Viewers are first introduced to the mineral extraction required to produce the electronics components for the device and made aware of the hard labor (and sometimes child labor) this task requires. The chart (and narrative) proceeds through processes of refining, assembling, and distributing these components, then transporting them physically, then transporting them virtually—through the infrastructure of the internet. Once within the Amazon corporate boundary, the chart depicts the layers of workers who provide everything from network maintenance to training datasets (figure 7.5b). Crawford and Joler also diagram patterns in the organization of Amazon's labor force, which they describe in terms of "fractal chains of production and exploitation." But what is required for this replication is people: "At every level contemporary technology is deeply rooted in and running on the exploitation of human bodies," the essay concludes.[47]

"Anatomy of an AI System" is an investigation and exposé of the invisible labor involved in making a single product on a global scale. In this way, it is an ambitious example of the seventh principle of data feminism: *show the work*. Behind the magic and marketing of data products, there is always hidden labor—often performed by women and people of color, which is both a cause and effect of the fact that this labor is both underwaged and undervalued. Data feminism seeks to make this labor visible so that it can be acknowledged and appropriately valued, and so that its truer cost—for people and for the planet—can be recognized.

## Crediting Data Work

The emphasis on giving formal credit for a broad range of work derives from feminist practices of citation. Feminist theorist Sara Ahmed describes this practice as a way of resisting how certain types of people—usually cis and white and male—"take up spaces by screening out others."[48] When those other people are screened out, they become

**Figure 7.5** (following three pages)
Overview (a) and detail (b) of "Anatomy of an AI System" (2018)—a diagram and essay by Kate Crawford and Vladan Joler that attempts to chart all of the human labor, data, and planetary resources used to create an Amazon Echo device. Courtesy of Kate Crawford and Vladan Joler.

# Anatomy of an AI system

An anatomical case study of the Amazon echo as a artificial intelligence system made of human labor

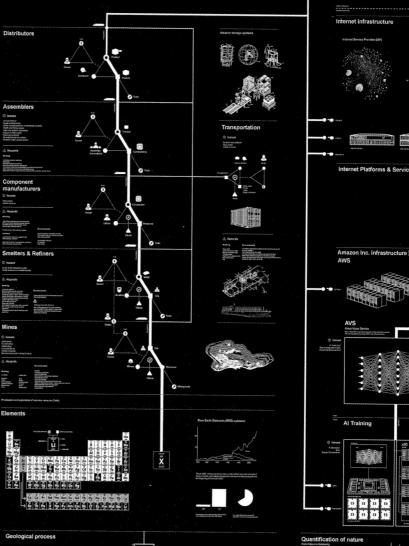

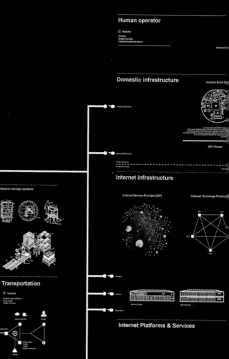

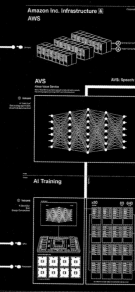

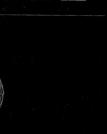

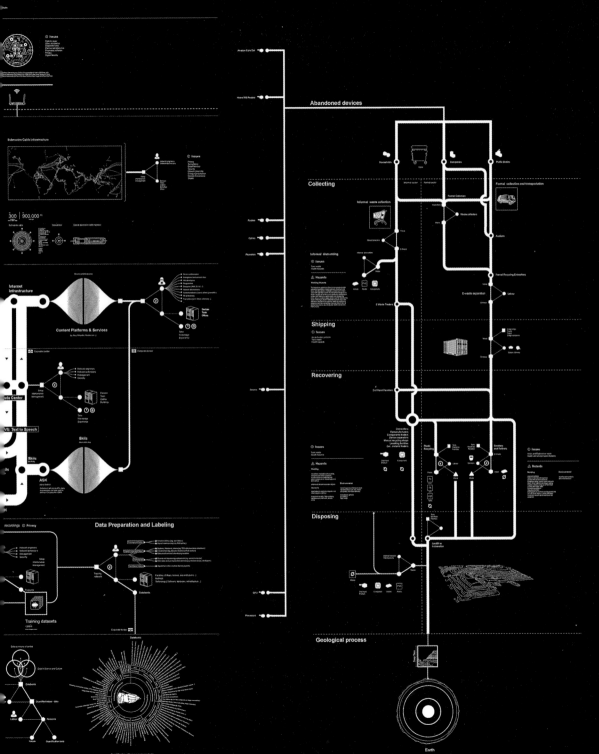

Earth

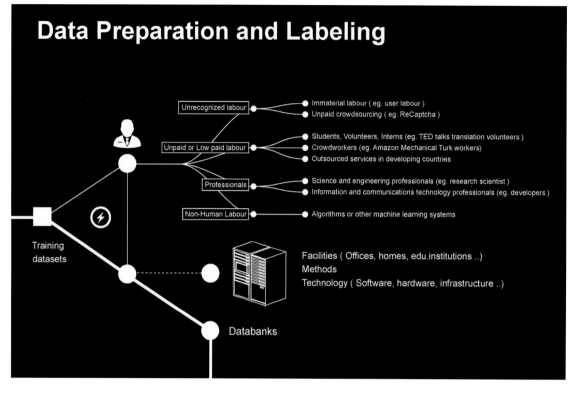

**Figure 7.5** (continued)

invisible, and their contributions go unrecognized. The *screening techniques* that lead to their erasure, as Ahmed terms them, are not always intentional, but they are, unfortunately, self-perpetuating. Ahmed gives the example of sinking into a leather armchair that is comfortable because it's molded to the shape of your body over time. You probably wouldn't notice how the chair would be uncomfortable for those who haven't spent time sitting in it—those with different bodies or with different demands on their time. Which is why those of us who occupy those comfortable leather seats—or, more likely in the design world, molded plastic Eames chairs—must remain vigilant in reminding ourselves of the additional forms of labor, and the additional people, that our own data work rests upon.

This gets complicated quickly even on the scale of a single data science project. The names of all the people and the work they perform are not always easy to locate—if they can be located at all. But taking steps to document all the people who work on a particular project at the time that it is taking place can help to ensure that a record of that work remains after the project has been completed. In fact, this is among the four core principles that comprise the Collaborators' Bill of Rights, a document developed by an interdisciplinary team of librarians, staff technologists, scholars, and postdoctoral fellows in 2011 in response to the proliferation of types of positions, at widely divergent ranks, that were being asked to contribute to data-based (and other digital) projects.[49]

When designing data products from a feminist perspective, we must similarly aspire to show the work involved in the entire lifecycle of the project. This remains true even as it can be difficult to name each individual involved or when the work may be collective in nature and not able to be attributed to a single source. In these cases, we might take inspiration from the Next System Project, a research group aimed at documenting and visualizing alternative economic systems.[50] In one report, the group compiled information on the diversity of community economies operating in locations as far-ranging as Negros Island, in the Philippines; Quebec province, in Canada; and the state of Kerala, in India. The report employs the visual metaphor of an iceberg (figure 7.6), in which wage labor is positioned at the tip of the iceberg, floating above the water, while dozens of other forms of labor—informal lending, consumer cooperatives, and work within families, among others—are positioned below the water, providing essential economic ballast but remaining out of sight.

With the idea of underwater labor in mind, we might return to the example of GitHub, which began this chapter, to ask what additional forms of labor might contribute to the production of code but cannot be represented by the visualization scheme that GitHub currently employs. We might think of the work of the project manager, which is not directly expressed in a particular number or size or frequency of contributions, but nevertheless ensures the quality and consistency of all project code. We might wonder about the work of the designer on a project or of the technical writer—both of whom might have helped to shape the project in its initial phases, but who have likely moved on to other tasks. In the case of a consumer-facing project, we might also consider the contributions of the customer support teams. Or in a community-oriented project, we might include organizers who have spent years developing strong relationships with community members. These forms of labor, both productive and reproductive, are essential to the success of any project but are not currently rendered visible, nor could they ever be easily visualized, by a scheme that considers project contributions to consist of code alone.[51]

But in more instances than you might think, the labor associated with data work can be surfaced through the data themselves. For instance, historian Benjamin Schmidt, whose research centers on the role of government agencies in shaping public knowledge, decided to visualize the metadata associated with the digital catalog of the US Library of Congress, the largest library in the world (figure 7.7).[52] Schmidt's initial goal was to understand the collection and the classification system that structured the

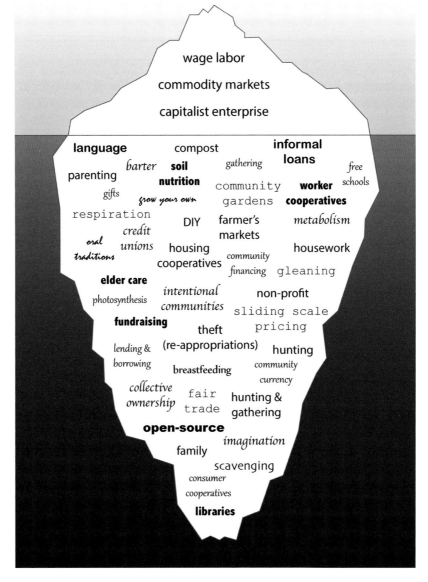

**Figure 7.6**
The "Diverse Economies Iceberg" (2017), a diagram of multiple labor practices created by the Next System Project for a report on cultivating community economies. Image courtesy of J. K. Gibson-Graham, Jenny Cameron, Kelly Dombrowski, Stephen Healy, and Ethan Miller for the Next System Project.

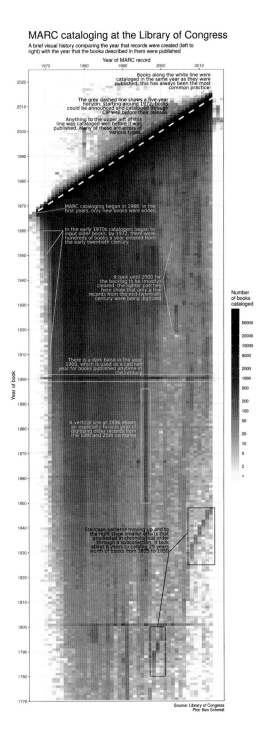

**Figure 7.7**
"A Brief Visual History of MARC Cataloging at the Library of Congress" (2017) visualizes when books at the Library of Congress entered their digital catalog. Image courtesy of Benjamin M. Schmidt.

catalog. But in the process of visualizing the catalog records, he discovered something else: a record of the labor of the cataloguers themselves. When he plotted the year that each book's record was created against the year that the book was published, he saw some unusual patterns in the image: shaded vertical lines, step-like structures, and dark vertical bands that didn't match up with what one might otherwise assume would be a basic two-step process of (1) acquire a book and (2) enter it in.

The shaded vertical lines, Schmidt soon realized, showed the point at which the cataloguers began to turn back to the books that had been published before the library went digital, filling in the online catalogue with older books. The step-like patterns indicated the periods of time, later in the process, when the cataloguers returned to specific subcollections of the library, entering in the data for the entire set of books in a short period of time. And the horizontal lines? Well, given that they appear only in the years 1800 and 1900, Schmidt inferred that they indicated missing publication information, as best practices for library cataloguing dictate that the first year of the century be entered when the exact publication date is unknown.

With an emphasis on showing the work, these visual artifacts should also prompt us to consider just how much physical work was involved in converting the library's paper records to digital form. The darker areas of the chart don't just indicate a larger number of books entered into the catalog, after all. They also indicate the people who typed them all in. (Schmidt estimates the total number of records at ten million and growing.) Similarly, the step-like formations don't just indicate a higher volume of data entry. They indicate strategic decisions made by library staff to return to specific parts of the collection and reflect those staff members' prior knowledge of the gaps that needed to be filled—in other words, their intellectual labor as well. Schmidt's visualization helps to show how the dataset always points back to the *data setting*—to use Yanni Loukissas's helpful phrase—as well as to the people who labored in that setting to produce the data that we see.[53]

### Crediting Emotional Labor and Care Work

In addition to the invisible labor of data work, there is also labor that remains hidden because we are not trained to think of it as labor at all. This is what is known as *emotional labor*, and it's another form of work that feminist theory has helped to bring to light.[54] As described by feminist sociologist Arlie Hochschild, emotional labor describes the work involved in managing one's feelings, or someone else's, in response to the demands of society or a particular job.[55] Hochschild coined the term in the late 1970s to describe the labor required of service industry workers, such as flight attendants, who are required to manage their own fear while also calming passengers during adverse flight conditions, and generally work to ensure that flight passengers feel cared for and content. In the decades that followed, the notion of emotional labor was supplemented by a related concept, *affective labor*, so that the work of projecting a feeling (the definition of emotion) could be distinguished from the work of experiencing the feeling itself (the definition of affect).[56]

We can see both emotional and affective labor at work all across the technology industry today. Consider, for instance, how call center workers and other technical support specialists must exert a combination of affective and emotional labor, as well as technical expertise, to absorb the rage of irate customers (affective labor), reflect back their sympathy (emotional labor), and then help them with—for instance—the configuration of their wireless router (technical expertise).[57] In the workplace, we might also consider the affective labor required by women and minoritized groups, in all situations, who must take steps to disprove (or simply ignore) the sexist, racist, or otherist assumptions they face—about their technical ability or about anything else. And they must do so while also performing the emotional labor that ensures that they do not threaten those who hold those assumptions, who often also hold positions of power over them.[58] Are there ways to visualize these forms of labor, giving visual presence—and therefore acknowledgement and credit—to these outlays of work?

One example that strives to visualize emotional and affective labor is the Atlas of Caregiving (figure 7.8), an ongoing project that aims to document the work involved in caring for a chronically ill family member. The project's name plays on the concept of the anatomy atlas, a compendium of illustrations of the human body that doctors can consult for information and reference. In this case, the goal was to illustrate the sometimes physical and sometimes emotional or affective work of care. The research team outfitted its participants with a variety of biometric sensors, including accelerometers and heart rate monitors, as well as with body cameras programmed to take a picture every fifteen minutes. They then visualized these data alongside excerpts from personal interviews and from the activity logs they asked the caregivers in the study to complete.

The result is a complex picture of caregiving, one that marshals data in the interest of creating a comprehensive view of the range of labor involved in caregiving work.[59] The stress of serving as a caregiver—a form of affective labor—is broken down into six distinct levels, and then visualized as a gradient (figure 7.8a). The work of caregiving itself is divided into seven subtypes of work, including concrete tasks like healthcare management and household chores, and more abstract forms of labor like being available and social support (figure 7.8b). This, too, helps others recognize the wide range

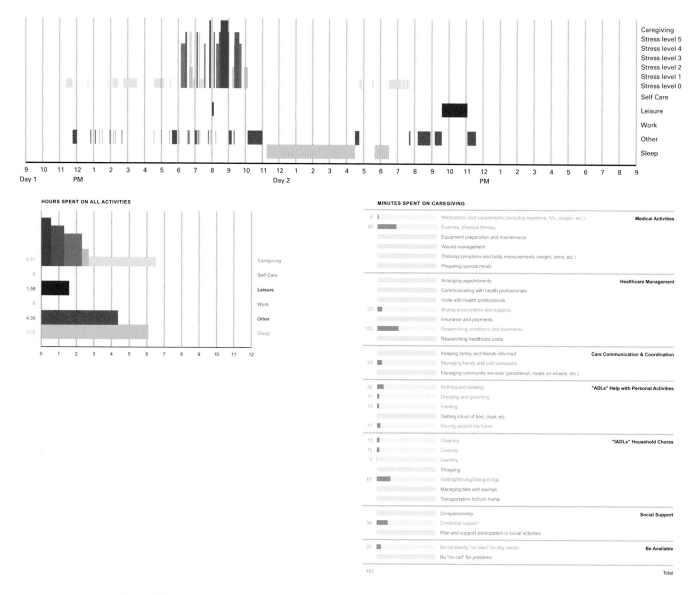

**Figure 7.8**
The Atlas of Caregiving visualizes the labor of caring for chronically ill family members. (a) A thirty-six-hour log of caregiving activities; (b) caregiving activities separated by type; (c) a photo log created during that same time. Image courtesy of the Atlas of Caregiving, 2016.

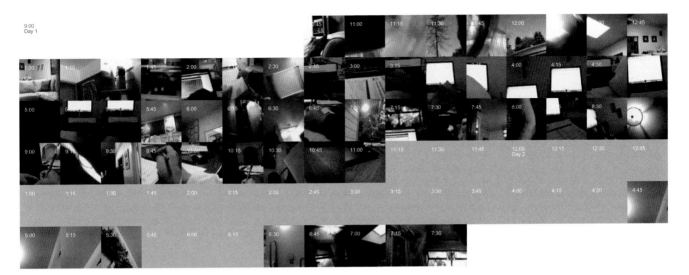

**Figure 7.8** (continued)

of work—indeed, expertise—associated with caregiving. And as some of the study's participants reported, it helped them to recognize that work for themselves.[60]

Of course, a diagram of work is only a proxy for the work itself—and that is to say nothing about the complexity of human feelings. This understanding served as the genesis for "Bruises—the Data we Don't See."[61] This artful visualization, created by designer Giogia Lupi and accompanied by a musical score composed by Kaki King, attempts to get closer to a visual representation of the emotional toll of caregiving (figure 7.9). The project began when King's daughter was diagnosed with a rare autoimmune disease, idiopathic thrombocytopenic purpura (ITP). ITP is described as a "very visual disease," and presents as bruises and burst blood vessels all over the body. For this reason, King was instructed to watch her daughter's skin and record any significant changes. She also recorded her own feelings in terms of hope, stress, and fear, creating subjective data to complement the hard numbers she received from the blood tests her daughter was required to endure.

When Lupi, who knew King from previous collaborations, set out to design her visualization, her goal was to "evoke empathy" and help her audience "feel a part of

**Figure 7.9** (following three pages)
Still from *Bruises—the Data We Don't See* (2018) and the legend that helps decode the data visualization. Image courtesy of Giorgia Lupi and Kaki King.

Giorgia Lupi and Kaki King

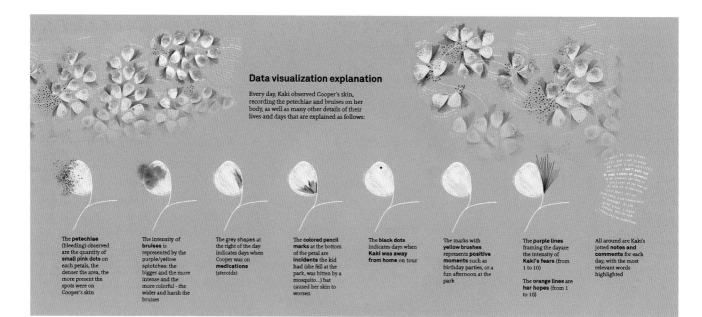

**Data visualization explanation**

Every day, Kaki observed Cooper's skin, recording the petechiae and bruises on her body, as well as many other details of their lives and days that are explained as follows:

The **petechiae** (bleeding) observed are the quantity of **small pink dots** on each petals, the denser the area, the more present the spots were on Cooper's skin

The intensity of **bruises** is represented by the purple/yellow splotches: the bigger and the more intense and the more colorful - the wider and harsh the bruises

The grey shapes at the right of the day indicates days when Cooper was on **medications** (steroids)

The **colored pencil marks** at the bottom of the petal are **incidents** the kid had (she fell at the park, was bitten by a mosquito...) hat caused her skin to worsen

The **black dots** indicates days when **Kaki was away from home** on tour

The marks with **yellow brushes** represents **positive moments** such as birthday parties, or a fun afternoon at the park

The **purple lines** framing the day are the intensity of **Kaki's fears** (from 1 to 10)

The **orange lines** are **her hopes** (from 1 to 10)

All around are Kaki's jotted **notes and comments** for each day, with the most relevant words highlighted

**Figure 7.9** (continued)

a story of a human's life."[62] In contrast to the Atlas of Caregiving, which relies upon standard visualization techniques like radial timelines and Gantt-style charts to legitimate the work of care, Lupi sought alternative visualization strategies to emphasize the particularity and specificity of a single family's situation. She employed a fluid timeline to reflect the subjective nature of what disability studies scholars call *crip time*. With this term, as Ellen Samuels explains it, "Sometimes we just mean that we're late all the time—maybe because we need more sleep than nondisabled people, maybe because the accessible gate in the train station was locked."[63] But it can also mean something more profound, as described by Alison Kafer: "Rather than bend disabled bodies and minds to meet the clock, crip time bends the clock to meet disabled bodies and minds."[64]

In Lupi's depiction of how the clock met King's daughter's body and King's own mind, days became white aspen-shaped leaves, segmented not by weeks or years but by hospital visits. Red dots were employed to indicate platelet counts, with color deployed mimetically to convey the intensity of the bruises, as well as the visuality of the data recorded by King. Lupi also employed color to represent King's record of her feelings, with black corresponding to stress and fear and yellow to signify hope. King's fear and hope were also visualized by hand-drawn lines that reflected each on a scale of one to ten. The result is rendered as an animation that unfolds over time and is set to music, a visually and aurally affecting composition of the affective labor of mothering and care.

Of course, neither Lupi and King nor the Atlas of Caregiving project team are the first to want to identify and make visible the work of care. As early as 1969, shortly after the birth of her own child, artist Mierle Laderman Ukeles penned the Manifesto for Maintenance Art, which called on the art world to elevate the care and maintenance of human life to an art, over and above the solitary creative (male) genius.[65] In the years that followed, *care work* would become a significant topic of interest for feminist scholars—especially after the mid-1990s, when Nancy Folbre formalized the term. Folbre's primary model of care work was the everyday work of caring for a child, although care work, like housework, isn't necessarily performed for free. It can also include the underwaged work performed by daycare workers or home health aids, as well as the waged work of doctors, nurses, physical therapists, mental health professionals, and so on. What binds these forms of work together across economic lines is their motivation. As theorized by Folbre, care work is undertaken out of a sense of compassion or responsibility for others, rather than with a goal of monetary gain. But when it comes to the market, altruism is a double-edged sword. These same professional care workers—who are predominantly women and people of color—are often paid less than they would be in other fields.[66] Why? Because they care.

So how do we "show the work" of care workers? How do we ensure that this work is sufficiently recognized and valued? And can we do anything more to challenge the root cause of this undervalued work? In the academy, groups like the Maintainers have sought to learn from the theories of care developed by feminist labor studies scholars such as Folbre as they attempt to make visible and value the labor of data work.[67] Through workshops, conferences, and publications, the Maintainers seek to counter the current tendency in technology fields to celebrate innovation and discovery alone. The work that maintains and sustains the world we live in today should also be celebrated, they insist. Among their current areas of research are the people they call *Info-Maintainers*: the people who work in libraries and archives and in related preservation fields to ensure that the knowledge of the present remains accessible for generations to come. Because the work of librarians and archivists and curators is focused on facilitating access to future knowledge, the Maintainers argue, it can be viewed as a form of care work too.

Across many technical fields, there is an increasing amount of attention paid to care work, and to other forms of invisible labor, now that so much work is virtual rather than physical; as well as to issues of job insecurity now that white-collar jobs have begun to be outsourced to freelancers as well. In this context, it is important to recall that professional care workers have long dealt with issues of undercompensated and precarious work; and for just as long, they have been involved with efforts to resist and organize against the inequities they have faced. Today, these efforts are being enhanced by data and technology, as unions and other advocacy groups are making use of new platforms and data streams for their work. But they are also being obstructed, as Uber-style apps to connect caregivers and employers increasingly abound. These apps do nothing to solve the systemic problems that caregivers face. A 2016 study of on-demand domestic worker apps by the UK's Overseas Development Institute (ODI) reports that because they displace risk onto workers, these platforms potentially reinforce discrimination and "further entrenchment of unequal power relations within the traditional domestic work sector."[68]

As a corrective, we might look to emerging prototypes that center the needs of workers, those that are developed by and with workers themselves. In the US, for example, the National Domestic Workers Alliance (NDWA) has developed an app, Alia, in order to serve as a portable benefits platform.[69] It allows clients to contribute a small amount into the worker's benefits account each time that worker provides them with a service. Workers can then pool contributions from multiple clients to purchase benefits on-demand, such as paid time off and various forms of insurance. Caveats remain, of course: Shouldn't the government require that all workers receive paid time off as a matter of course? Shouldn't we be advocating for a single-payer healthcare system? Yes and yes. But while the NDWA continues to lobby for systemic change, its app offers one way to provide essential benefits to domestic workers right now. It is a harm reduction strategy, one that can be pursued while simultaneously advocating for more transformative change. Thinking back to Kimberly Seals Allers's app, Irth, discussed in chapter 1, we might also begin to imagine how its successful use would contribute to a dataset that could be used to support future advocacy efforts.

**Show Your Work**

Data work is part of a larger ecology of knowledge, one that must be both sustainable and socially just. Like the ship paths visualized on the *Ship Map* or the source code stored on GitHub or the global assemblage of people and materials that make an Amazon Echo device, the network of people who contribute to data projects is vast and complex. Showing this work is an essential component of data feminism, and it is the reason why "show your work" is the seventh and final principle in this book. An emphasis on labor opens the door to the interdisciplinary area of data production studies: taking a data visualization, model, or product and tracing it back to its material conditions and contexts, as well as to the quality and character of the work and the people required to make it. This kind of careful excavation can be undertaken in academic, journalistic, or general contexts, in all cases helping to make more clearly visible—and therefore to value—the work that data science rests upon.

We can also look to the data themselves in order to honor the range of forms of invisible labor involved in data science. Who is credited on each project? Whose work has been "screened out"? While one strategy is to show the work behind making data products themselves, another strategy for honoring work of all forms is to use data science to show the work of people (mostly women) who labor in other sectors of the economy, those that involve emotional labor, domestic work, and care work. We see this in action in the Atlas of Caregivers, which focuses on legitimizing care work, and the Alia app, which provides more financial security for domestic workers. Designing in solidarity with domestic workers can begin to challenge the structural inequalities that relegate their work to the margins in the first place.

This point brings us back to the ideas about power that began this book. Power imbalances are everywhere in data science: in our datasets, in our data products, and in the environments that enable our data work. Showing the work is crucial to ensure that undervalued and invisible labor receives the credit it deserves, as well as to understand the true cost and planetary consequences of data work.

# Conclusion: Now Let's Multiply

On November 1, 2018, at 11:10 a.m. local time, workers at Google offices in fifty cities around the world closed their browser tabs, shut their laptops, and walked off their jobs.[1] The walkout included both full-time employees and freelancers. It was women-led at a company that, despite years of lip-service to inclusion, only has 31 percent women employees.[2] And it was massive—more than twenty thousand workers participated (figure 8.1). Why did workers at one of the most powerful companies on the planet take to the streets?

One week earlier, the *New York Times* broke a story about the $90 million exit package that Andy Rubin, the creator of Google's Android mobile operating system, had received after he was accused of sexual misconduct (and after an internal investigation had found the claim to be credible).[3] The story mentioned two other executives accused of sexual misconduct whom Google had similarly protected. As journalists Daisuke Wakabayashi and Katie Benner wrote, "In settling on terms favorable to two of the men, Google protected its own interests." Evidently, Rubin's package had been paid out in installments of $2 million per month over the course of four years. His final payment was scheduled for later that month.

As soon as the *New York Times* article was published, additional stories of discrimination faced by women, as well as men and nonbinary people, began pouring out on company email lists and chat channels and in face-to-face forums. The stories pointed to patterns of toxic behavior.[4] Within a week, the massive walkout—initially floated as an idea on a Google moms list—had been planned. "Tech industry business as usual is failing us," said Meredith Whittaker, the founder of Google's Open Research Group. "Google paying $90M to Andy Rubin is one example among thousands, which speak to a company where abuse of power, systemic racism, and unaccountable decision-making are the norm. ... It's clear that we need real structural change, not adjustments to the status quo."[5]

**Figure 8.1**
The Sunnyvale, California, Google campus during the Google Walkout for Real Change on November 1, 2018. Employees turned out en masse to protest the company's handling of sexual misconduct cases. Courtesy of Wikimedia Commons user Grendelkhan.

A group of seven core organizers, including Whittaker, came together to craft five concrete demands, including ending forced arbitration in cases of discrimination and sexual harassment and promoting the chief diversity officer to report directly to the CEO.[6] When November 1 arrived, employees congregated first in indoor atriums, and then in courtyards and on streets. They carried signs that said, "Not OK, Google," "I Reported, He Got Promoted," and "Happy to Quit for $90 Million, No Sexual Harassment Required." Google management started paying attention.

Although the Google Walkout for Real Change, as the protest was formally known, was framed in the media as a milestone for big tech, there are clear precedents for white-collar tech organizing. Historian Mar Hicks has connected the Google walkout to a strike among computer workers—then a workforce that was comprised mainly of women—that took place in the United Kingdom in the 1970s. The strike took down twenty-six government computer centers and disrupted the work of nine others. These were the centers that enabled the government to process its value-added tax (VAT), and without the computers online, the tax couldn't be collected. The government was required to pay attention. Writes Hicks, "Even though many of these workers were women, and limited in their pay, promotion, and work opportunities due to sexism, their proximity to the literal machinery of government gave them a great deal of power."[7]

The organizers of the Google walkout recognized their proximity to another source of power: Google itself. A single worker might have limited power, but their collective organizing drew attention to the proximity of a relatively small number of people—Google employees—to the global digital infrastructure of everyday life. Part of the reason that data and computation have proved to be so lucrative is their ability to scale.

As journalist Moira Weigel points out, "This kind of scale means these companies can make extraordinarily high profits. But it also means the core workers they rely on have an extraordinary amount of bargaining power."[8] They also have messaging power, interruption power, and subversion power.

How might tech workers marshal these strengths to mass-occupy digital infrastructure? To teach algorithms to "work to rule" in the style of assembly-line slow-downs? To slow the flow of everyday capitalism to gather attention? To channel digital solidarities back into physical spaces and human relationships?

There are already many examples that point to how these questions might begin to be answered. In an article about tech organizing in the magazine *n+1*, for instance, an anonymous software developer points out, "If the developers from Slack decided to strike, they could, without too much difficulty, push out a change that made it so that any message that got sent would push a message about the purpose of the strike" to its ten million daily users.[9] And just for a minute, imagine if they did.

Although Slack developers haven't hacked their own platform (yet), collective organizing around data and technology has already taken a range of powerful forms. Groups like the Tech Workers Coalition are building bridges between the programmers who code the search engines and the cafeteria workers who prepare their food. They have also helped popularize the hashtag #TechWontBuildIt to indicate a collective refusal to work on ethically compromised software.[10] Platforms like Coworker.org are helping gig-economy workers, like Uber drivers, get organized. Other organizations, such as Tech Solidarity, are focusing on electoral politics. Some projects are taking explicitly political stands; the Lerna JavaScript library briefly added a clause to its license prohibiting entities that collaborate with US Immigration and Customs Enforcement (ICE) from using it.[11] Individuals are forming worker-owned tech cooperatives in the United States and around the globe and drafting values statements, such as the Design Action Collective's Points of Unity, that guide their work together and help them decide which projects to take on.[12] Other collective organizing efforts are working to draft codes of ethics like the Toronto Declaration[13] and statements of values like those guiding the Canadian government's action plan for open government.[14] Note that these efforts are not limited to white-collar workers, nor to employees of the big five technology companies, nor to large-scale events.

Some groups are using movement-building strategies to effect change across entire industries. For example, Una Lee, Wesley Taylor, Victoria Barnett, Ebony Dumas, Carlos (L05) Garcia, and Sasha Costanza-Chock are coordinating a networked community of practice called *design justice*.[15] The idea for design justice emerged from a workshop at the Allied Media Conference in Detroit in 2015, where thirty people assembled to challenge the idea of "design for good." As co-organizer Una Lee put it, "How could we redesign design so that those who are normally marginalized by it, those who are characterized as passive beneficiaries of design thinking, become co-creators of solutions, of futures?"[16]

Since then, the design justice group has produced dozens of workshops, pop-up educational forums, and scholarly texts. One of its central projects is a set of ten Design Justice Network Principles, which guide designers in navigating inequality and achieving justice through design.[17] Principle 1, for example, reads: "We use design to sustain, heal, and empower our communities, as well as to seek liberation from exploitative and oppressive systems." Principle 5 reads: "We see the role of the designer as a facilitator rather than an expert." The Design Justice Network promotes these principles through its workshops and other events at which designers meet, discuss, and co-conspire. To date, more than 350 additional designers have signed on.

Data for Black Lives (D4BL) is another example of inspired organizing and movement building at a national scale. Founded by veteran organizer Yeshimabeit Milner, who was herself trained by Black Lives Matter organizers, D4BL is "a network [of] over 4,000 scientists and activists working to harness the power of data and technology to make real change in the lives of Black people."[18] D4BL organizes annual conferences, runs online communities, and helps connect people in its network. The group pursues two simultaneous strategies: pushing back against the harmful impacts of data as they are currently deployed, and creating new spaces for organizers, data scientists, and engineers to come together to generate meaningful research questions. The group's emphasis on abolition and liberation, rather than a generic form of social good, leads it to design projects that actively work to overturn the data-driven discrimination experienced in Black communities. Milner's vision is "to make data a tool for profound social change instead of a weapon of oppression."[19]

The vision of D4BL will take time to realize, as is true of all visions that motivate transformative work. The organizers of the Google Walkout for Real Change are discovering this as we write. When they first assembled, they envisioned a world in which executives would listen to the demands of their workers and would undertake immediate measures for change. Although Google publicly expressed support for the workers involved, and the CEO issued a memo that read, "We are taking in all their feedback so we can turn these ideas into action," that action has yet to transpire. Claire Stapleton, the woman who originally floated the idea of taking mass action, stated: "We're almost three months out from the walkout and exactly zero of the five demands have been met." The corporation did end forced arbitration—and it led to other tech companies doing the same—but it was only a partial win because it covered cases of sexual miscon-duct alone, not all discrimination cases. As Amr Gaber, another key organizer, added, "it's also the cheapest thing, the most minor thing they could've done."[20]

These paltry actions, clearly motivated by the bottom line, underscore the unyield-ing influence of profit and power and the need for a feminism that is intersectional as a matter of course. In this book, we have described *intersectional feminism*—a vibrant body of knowledge and action that challenges the unequal distribution of power—and how it can be applied to the field of data science today. In doing this work, we have drawn heavily from the work of Black feminist theorists and activists, to reflect both their central role in defining and elaborating intersectionality and our own position as white scholars and white women in the United States. Here, we want to reiterate our appreciation for this foundational work, as well as to once again acknowledge that we cannot speak directly from the life experiences that motivate it. We hope that you, our readers, will use this work in order to reflect on your own identities, as well as to exam-ine how power and privilege operate in data science and in the world..

As we write this conclusion, in July 2019, issues of power and privilege continue to loom large. The other four and a half demands issued by the organizers of the Google walkout included "a commitment to end pay and opportunity inequity" at all levels of the corporation and the collection of "transparent data on the gender, race and ethnic-ity compensation gap, across both level and years of experience," as well as access to extant sexual misconduct reporting mechanisms by all Google employees, including its contract workers (who make up around half of the company's total employees).[21] Yet the public memo stated blandly that Google would continue to work on "creating a more inclusive culture for everyone."[22] Meanwhile, Google's lawyers have been filing legal documents that urge the US National Labor Relations Board to overturn the 2014 ruling that allows workers to use company email to organize without fear of retalia-tion.[23] If the ruling were overturned, it would seriously impede any efforts to organize future actions at Google, or at any large corporation, because company email lists are the primary way that a distributed workforce can organize across office locations and time zones.

Further complicating future organizing efforts, numerous Google employees, includ-ing lead organizers Whittaker and Stapleton, have faced retaliation and even demotion in the months following the walkout. These internal actions have been documented by *Wired* magazine, *Bloomberg News*, and the tech news site Packt, among other news outlets. For example, Stapleton was told to go on medical leave even though she was not sick, and the decision was only reversed after she hired a lawyer; and Whittaker was told that she would be required to "abandon her work" with the AI Now Institute, an independent research group focused on issues of AI and ethics.[24] Stapleton left Google in June 2019 and Whittaker left in July of that year, two high-profile departures that the *Guardian* surmised would "have a chilling effect" on tech workplace activism.[25]

This is the deployment of the structural and disciplinary domains of the matrix of domination, which we introduced in chapter 1. Google's legal team is well-resourced and has the power to shape both federal laws and company policies. This confirms the need to monitor dominant groups and institutions that wield outsized power in the world (and tend to use it to secure their positions). It also affirms the need to collaborate with the groups most impacted by differentials of power. In chapter 2, and throughout this book, we have attempted to heed our own advice, featuring the voices and ideas of those with direct experience of injustice. In so doing, we have sought to feature the sites of energy that have inspired us in our work—ranging from new activist networks to data journalism startups, from librarians authoring data user guides to engineers interrogating human-reporting bias. We've drawn from the work of sociologists who theorize digital power, artists who challenge technological neutrality, educators who teach statistics in real-world settings, and individuals who are single-handedly compiling spreadsheets of missing data. It is from all these locations, using all these methods, and including all these people—and more—that we can challenge the matrix of domination in data science at its source.

As should now be clear, our definition of data science includes more than quantitative methods, more than "big" data, more than "artificial" intelligence, and more than "neutral" displays. We explored the limitations of such a narrow view of data science and its communication in chapter 3. There and throughout the book, we have argued that an expansive conception of data science is essential if we are to work toward our goal of remaking the world.

Enabling this feminist data science to flourish and thrive will require deliberate interventions in each phase of data work, and in our received ideas about the people and communities who perform it. In chapter 4, we showed how the decisions that are made when first collecting data go on to impact future results. In chapter 5, we debunked the myth that data science is a solo enterprise, undertaken by genius wizards working alone. Data science involves collaboration and community, as well as deep context, as we discussed in chapter 6. Equally important is the acknowledgment, as explored in chapter 7, that data science is the work of many hands.

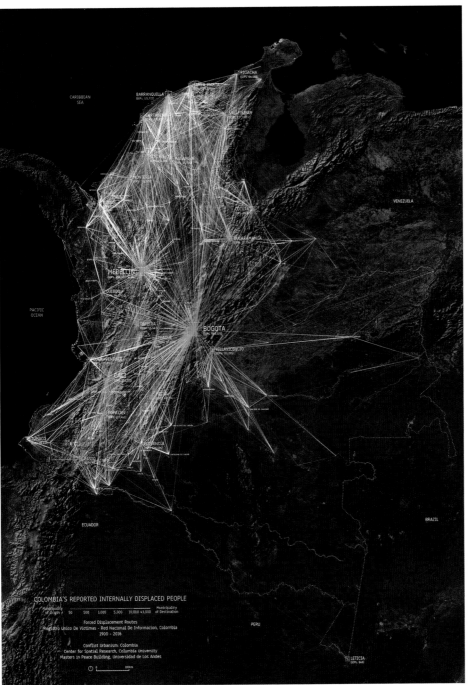

**Figure 8.2**

*Reported Internally Displaced People*, a 2016 map of internally displaced people in Colombia from 1985 to 2015. From the project *Conflict Urbanism: Colombia* by the Center for Spatial Research at Columbia University, which looked at land-use patterns and displacement in Colombia over thirty years of armed internal conflict. The researchers worked with the organization Unidad para la Atención y Reparación Integral a las Víctimas, a massive data collection effort that documented millions of individuals. Courtesy of the Center for Spatial Research, Columbia University.

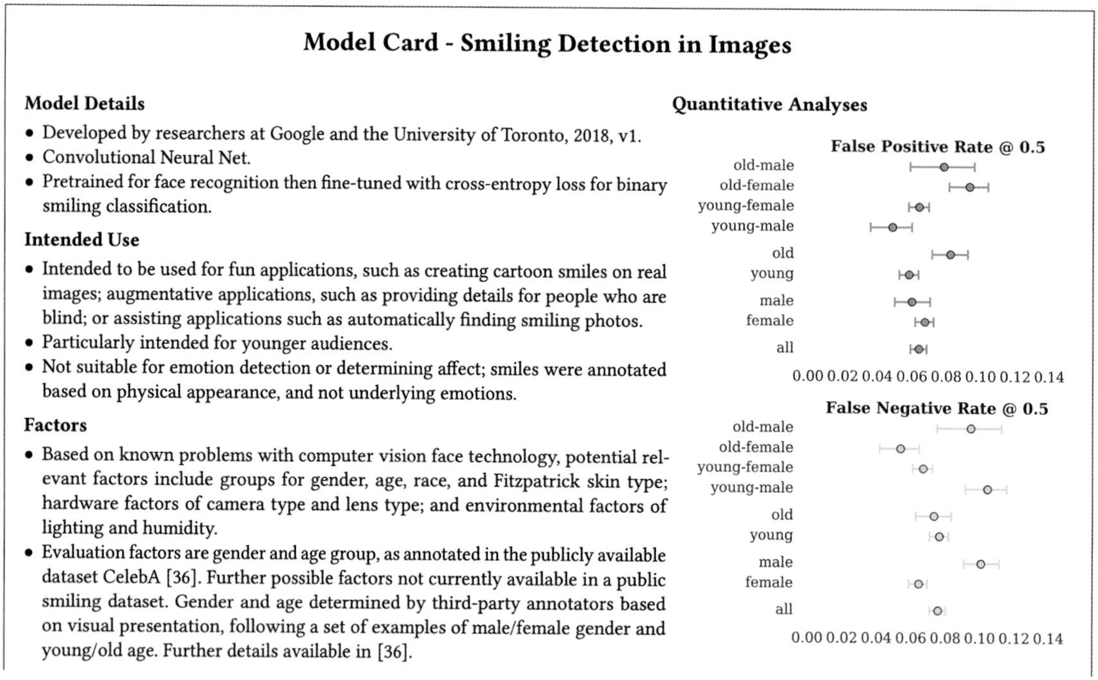

## Model Card - Smiling Detection in Images

**Model Details**

- Developed by researchers at Google and the University of Toronto, 2018, v1.
- Convolutional Neural Net.
- Pretrained for face recognition then fine-tuned with cross-entropy loss for binary smiling classification.

**Intended Use**

- Intended to be used for fun applications, such as creating cartoon smiles on real images; augmentative applications, such as providing details for people who are blind; or assisting applications such as automatically finding smiling photos.
- Particularly intended for younger audiences.
- Not suitable for emotion detection or determining affect; smiles were annotated based on physical appearance, and not underlying emotions.

**Factors**

- Based on known problems with computer vision face technology, potential relevant factors include groups for gender, age, race, and Fitzpatrick skin type; hardware factors of camera type and lens type; and environmental factors of lighting and humidity.
- Evaluation factors are gender and age group, as annotated in the publicly available dataset CelebA [36]. Further possible factors not currently available in a public smiling dataset. Gender and age determined by third-party annotators based on visual presentation, following a set of examples of male/female gender and young/old age. Further details available in [36].

**Quantitative Analyses**

**False Positive Rate @ 0.5**

old-male
old-female
young-female
young-male

old
young

male
female

all

0.00 0.02 0.04 0.06 0.08 0.10 0.12 0.14

**False Negative Rate @ 0.5**

old-male
old-female
young-female
young-male

old
young

male
female

all

0.00 0.02 0.04 0.06 0.08 0.10 0.12 0.14

**Figure 8.3**

Detail of a model card, from a 2019 paper titled "Model Cards for Model Reporting" by AI researcher Margaret Mitchell and coauthors that proposes short documents called *model cards* that would accompany machine learning models as a form of documentation. Model cards detail who developed the model, for what purpose, and how the model performs, including intersectional identity metrics. Model cards would also specify known limitations of a model and use cases for which the model is not suitable. Courtesy of Margaret Mitchell.

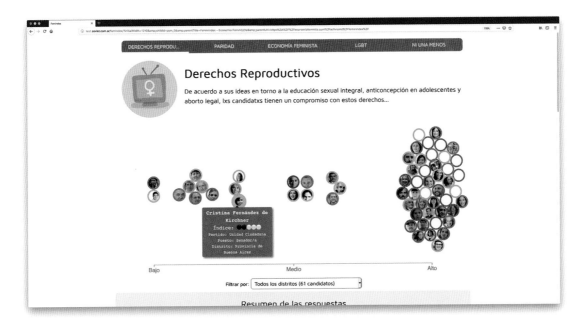

**Figure 8.4**
Feminindex is a civic media project that documents and visualizes where all Argentine political candidates stand on gender and LGBTQ+ issues, including reproductive rights, femicides, and trans rights. The first version was released in 2017 and the second in 2019. Courtesy of Economía Femini(s)ta, including Mercedes D'Alessandro, Andrés Snitcofsky, Lina Castellanos, Aldana Vales, and the Economía Femini(s)ta team. See http://economiafeminita.com/activismo/feminindex/.

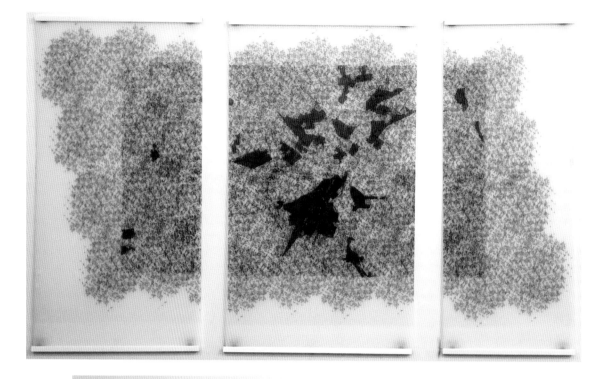

**Figure 8.5**

*Decoding Possibilities* (2017) by Ron Morrison and Treva Ellison, is an artistic examination of redlining's effects in the landscape as well as a celebration of creative resistance to redlining. (a) Contemporary maps of Boston are combined with historic redlining maps, as well as maps created from the Combahee River Collective's writings. (b) The installation includes quotes on the enduring effects of redlining in the landscape. Courtesy of Ron Morrison and Treva Ellison.

Throughout the book, we have described our seven principles of data feminism: *examine power, challenge power, elevate emotion and embodiment, rethink binaries and hierarchies, embrace pluralism, consider context,* and *make labor visible.* We derived these principles from the major ideas that have emerged in the past several decades of intersectional feminist activism and critical thought. At the same time, we welcome the notion that there are many other possible starting points that share the end goal of using data (or refusing data) in order to end oppression.[26]

Those other starting points might come from within the academy. For example, the work of the Center for Spatial Research at Colombia, led by Laura Kurgan, uses a uniquely transdisciplinary approach that includes data science and AI, the humanities, geography, and design to investigate complicated phenomena like urban/rural displacement due to conflict (figure 8.2). Scholars like Dean Spade are using queer theory to challenge the institutions that wield data. And media studies scholars are examining the intersections of race, gender, sexuality, and data, as Shaka McGlotten does through their Black data project.[27] Researchers are writing books about Indigenous statistics and Indigenous data sovereignty,[28] developing decolonial design methods,[29] and leading dynamic conversations about decolonizing data in both the Global North and the Global South.[30] Computer scientists and AI researchers are conducting important studies on bias, as well as developing new ways to promote transparent and responsible use of AI. For example, Margaret Mitchell and her coauthors have recently proposed *model cards* (figure 8.3), a form of documentation that would accompany machine learning models to detail their intended uses and their technical and ethical limitations.[31]

There are many other possible starting points for challenging oppression in data that come out of the arts, activism, community organizing, and consciousness-raising. Cartographer Margaret Pearce's next mapping project indigenizes the Mississippi River map to make new spaces for public dialogue about flood management. Mimi Onuoha and Mother Cyborg's *People's Guide to AI*, designed for newcomers, provides an accessible introduction to the ideas behind artificial intelligence. The activist group Economía Femini(s)ta in Argentina has an ongoing civic accountability project called Feminindex in which the group visualizes where each candidate stands in relation to a range of gender and LGBTQ+ issues (figure 8.4). The group has even produced digital trading cards for politicians, which it circulates on social media. And artist-researchers Ron Morrison and Treva Ellison disrupt Boston redlining maps from 1935 with overlays of "black queer, trans, and feminist geographies" created by the Combahee River Collective (figure 8.5). These require viewers to put on special glasses called Racialized Space Reduction Lenses (RSRL) to see beneath the surface.[32]

These projects are not intended to be exhaustive, and the list could go on. What is most important is not that we all share the same starting point, but rather that we nurture all of these emerging ecosystems and build links between them. We will need all of them for mobilizing resistance to the differentials of power embedded in our current datasets and data systems. And we will also need them for mobilizing courage and creativity—to imagine what data science and artificial intelligence beyond the matrix of domination might look like. The best time for resistance and reimagination is before the norms and structures and regulations of the data economy have been fully determined.

So now let's multiply. Let's multiply now.

## Our Values and Our Metrics for Holding Ourselves Accountable

*A note to readers:* We created this document as we began the writing of this book and included it as part of the manuscript draft that was posted online as part of the open peer review process. We were prompted to write it because of work on a prior project—the Make the Breast Pump Not Suck Hackathon—with equity consultant Jenn Roberts of Versed Education, and because of the values statements (and related statements of principles) published by groups such as the University of Maryland's African American History, Culture, and Digital Humanities Initiative (AADHum) and the University of Delaware's Colored Conventions Project.[1] From these projects, we saw how statements of shared values can become important orientation points, guiding internal decisions at challenging junctures and making ethical commitments public and transparent. The idea to accompany our values with a set of metrics was also proposed by Jenn Roberts for the breast pump hackathon. We discuss that project, and the uses and limits of metrics for accountability, in chapter 4. The metrics below were calculated two times: first on the basis of the draft posted online, and second on the basis of the copyedited book manuscript. Aside from the addition of the second set of metrics, and a short reflection on our successes and failures, the language of the document remains unchanged from the version posted as part of the manuscript draft.

### We Insist on Intersectionality

Feminism has always been multivocal and multiracial, but the movements' diverse voices have not always been valued equally. The women's suffrage movement largely excluded Black women and the abolition of slavery from its agenda. In the 1970s, lesbian feminists were called "the purple menace" by straight feminists. But feminism fails altogether if it is only for elite, white, straight, Christian, Anglo women. The work of activists and scholars, particularly Black feminists, over the past forty years insists on a feminism that is intersectional, meaning it looks at issues of social power related not just to gender, but also to race, class, ability, sexuality, immigrant status, and more. It does so, moreover, by looking to collectives as well as individuals, to structural issues as well as specific instances of injustice.

## We Advocate for Equity

Equity is both an outcome and a process. Future justice must account for an unjust past in which some groups' knowledges have been valued and others have been subjugated, as Patricia Hill Collins teaches us. In the process of achieving equity, those of us in positions of relative power must learn to listen deeper and listen differently—with the ultimate goal of taking action against the status quo that benefits us at the expense of others. For this reason, we listen and give priority in the text to voices who speak from marginalized perspectives, whether because of their gender, ability, race, class, colonial status, or other aspects of their identity.

## We Prioritize Proximity

As Kimberly Seals Allers, women's health advocate, says, "Whatever the question, the answer is in the community." People in a community know its problems intimately, and they know which phenomena go uncounted, underreported, or neglected by institutions in power (or, conversely, who is overly surveilled by institutions in power). They also know what interventions will work to solve those problems. In this book, we try to prioritize voices with closer and more direct experience of issues of injustice over those that study a data injustice from a distance.

## We Acknowledge the Humanity of Data

We recognize that the transformation of human experience into data often entails a reduction in complexity and context. We further acknowledge that there is a long history of data being "all too often wielded as an instrument of oppression, reinforcing inequality and perpetuating injustice," as the group Data for Black Lives explains. We keep these inherent constraints in mind as we write, attempting to introduce context and complexity whenever possible, and acknowledge the limits of the methods we discuss, as well as their strengths.

## We Are Reflexive, Transparent, and Accountable

Acknowledging that our knowledge is shaped by our own perspectives and limitations (see more in the About Us section ahead), we strive to be reflexive, transparent, and accountable for our work. We are on a journey toward justice, and that inevitably involves making mistakes. We are grateful to those who have shown us generosity in letting us learn up to this point. And we respectfully say to our future teachers that you will find in us open listeners: we recognize direct and critical words as a generous offer and a vote of confidence in our ability to hear and be transformed by you.

To that end, we have an evolving table of explicit metrics (table A.1) that will guide us in auditing our citations and the examples that we elevate in the book. We note, here, that our foregrounding of race and racism reflects our location in the United States, where the most entrenched issues of inequality and injustice have racism at their source.

## About Us

Feminist standpoint theory recognizes the value of situated knowledge—acknowledging the perspectives and experiences of the knower and how those have shaped the knowledge they produce. Accordingly, we situate ourselves and the learning contexts in which we work.

Catherine D'Ignazio is an assistant professor at Massachusetts Institute of Technology, a private research university in Cambridge, MA. Before moving to MIT, she worked at Emerson College, a private college in Boston focused on communications and the arts. From a middle-class, Italian-American and Scotch-Irish background, she grew up primarily in the American South, with some formative years spent in Latin America and Europe. She is a mother, an experience that has sharpened her understanding of how women's bodies are stigmatized and underserved by mainstream institutions. Working mainly in urban New England, she experiences significant privilege from her whiteness, ability, institutional affiliation, and education, among other things, and experiences some oppression based on her gender. With decades of professional work in software programming, art/design, and digital media education, she comes to data feminism based on a commitment to democratize information and include more people and professions in contemporary conversations about data and power.

Lauren F. Klein is an associate professor at Emory University, a private university in Atlanta, Georgia, in the Southern United States. Before moving to Emory, she worked at Georgia Tech, a large public research university also in Atlanta. From a middle-class New York Jewish family, she grew

**Table 1**
Aspirational, draft, and final metrics and the structural problems they address

| Structural problem | Aspirational metrics to live our values for this book | Draft metrics (open peer review) | Final metrics (copyedited manuscript) |
|---|---|---|---|
| Racism | • 75 percent of citations of feminist scholarship from people of color<br>• 75 percent of examples of feminist data projects discussed led by people of color | Scholarship: 36 percent from people of color<br>Projects: 49 percent led by people of color | Scholarship: 32 percent from people of color<br>Projects: 42 percent led by people of color |
| Patriarchy | • 75 percent of all citations and examples from women and nonbinary people | 67 percent of citations and examples from women and nonbinary people | 62 percent of citations and examples from women and nonbinary people |
| Cissexism | • Center trans perspectives in discussions of the gender binary<br>• Use transinclusive language throughout the book<br>• Example or theorist in every chapter from a transgender perspective | Three of ten chapters feature transgender example and/or theorist | Nine of nine chapters feature transgender example and/or theorist |
| Heteronormativity | • Resist assumptions about family structure and gender roles<br>• Example or theorist in every chapter that illustrates the power of communal (vs. family) support networks | Ten of ten chapters feature communal example and/or theorist | Nine of nine chapters feature communal example and/or theorist |
| Ableism | • Challenge the dominance of visualization in the presentation of data<br>• Example or theorist in every chapter that employs nonvisual methods of presenting data | Nine of ten chapters feature nonvisual example and/or theorist | Nine of nine chapters feature nonvisual example and/or theorist |

**Table 1** (continued)

| Structural problem | Aspirational metrics to live our values for this book | Draft metrics (open peer review) | Final metrics (copyedited manuscript) |
|---|---|---|---|
| Colonialism | • 30 percent of projects discussed come from the Global South<br>• Example or theorist in every chapter about Indigenous knowledges and/or activism | Projects: 8.5 percent from the Global South<br>Five of ten chapters feature indigenous example and/or theorist | Projects: 7 percent from the Global South<br>Seven of nine chapters feature indigenous example and/or theorist |
| Classism | • Acknowledge that data science, as a field, is premised on economic, educational, and technological privilege<br>• 50 percent of feminist projects discussed come from outside the academy<br>• Example or theorist in every chapter that demonstrates how the ideas can be applied without expensive technology and/or formal training | Projects: 88 percent from outside academy<br>Ten of ten chapters feature nonacademic example and/or theorist | Projects: 78 percent from outside academy<br>Nine of nine chapters feature nonacademic example and/or theorist |
| Proximity | • 50 percent of feminist projects discussed feature and quote people directly impacted by an issue (vs. those who study or report on the phenomena from a distance) | Projects: 49 percent feature people directly impacted | Projects: 34 percent feature people directly impacted |

up in suburban New Jersey and lived in New York for much of her adult life, with some time spent in Boston. Like D'Ignazio, she is also a mother. Working in the US South, she experiences significant privilege from her whiteness, ability, education, and institutional affiliation, among other things, and experiences some oppression based on her gender. She worked in web development before becoming an academic and comes to data feminism through her desire to convert theory into practice and to create more opportunities for humanities research (and researchers) to enter into conversation with communities, activists, organizers, and others working toward justice.

**A Reflection on Our Final Metrics**

There are interesting shifts to note between the metrics of the first draft and those of the final version. Some measures landed closer toward our goals in the final version. For example, we were able to meet our goal of including a trans voice in every chapter, and we included more Indigenous perspectives in the final version than in the first draft. However, we were not able to achieve all of our aspirational metrics in the final draft, and in fact, the proportional representation of women, nonbinary people, and people of color decreased from the first draft to the copyedited manuscript. The final manuscript also had proportionally fewer projects from the Global South, more projects from the academic realm, and less work from people who were directly impacted by each issue.

What explains this outcome that runs counter to our values and stated goals? In our view, there are two explanations. The first has to do with process: we did not keep precise tabs on our citational metrics as we were revising. We hadn't wanted to give the aspirational metrics so much presence in our revision process that we would choose people and projects simply because of their gender, race, or other markers of identity. That seemed tokenizing and like "gaming the system" we had created. We did keep the overall gaps between our draft metrics and our aspirational metrics in mind, and attempted to close those gaps as we introduced new examples and removed old ones And we did in fact introduce more overall references to, for example, scholarship and projects led by people of color (243 vs. 92), and scholarship and projects based in the Global South (41 vs. 20). But we also introduced more references to scholarship and projects led by white people (475 vs. 136), as well as more scholarship and projects based in the Global North (933 vs. 278). In retrospect, since we didn't calculate the second set of metrics until the end of the revision process, we were not aware of the changing proportions of projects and citations, and therefore had no way of rebalancing them after the fact.

An additional pass through the manuscript with this rebalancing in mind would have addressed the issue in the book. Indeed, we wish we had planned for this final pass in our revision process. But the root cause of the issue would have remained unresolved. Put simply: this cause is ourselves and our positions as scholars in the world of higher education, a world dominated by white people and especially white men. To some readers, this answer may seem obvious, but our path to it is worth taking the time to explain.

Many people who participated in our peer review process (both online and anonymously) noted that we should back up our assertions with citations. We agreed with this point, and felt that additional citations would both credit prior work and add legitimacy to our claims. Between the draft and the final version, we added many endnotes and additional references. As a result, the final version of this book is significantly more scholarly than the draft we posted online. But when looking at the history of engagement with a particular idea, or when asking ourselves which notable person in a particular field we should name, we thought less about our values for the book and more about what we already knew about those areas. In so doing, we inadvertently reproduced the biases of academia—ironically, through a mechanism very similar to the privilege hazard we name in the book.

We thought about not including the endnotes in our final accounting; that might have yielded "better" metrics. But that would have misrepresented the contents of the book—and, more importantly, the extent of the work that remains to be done. Instead, our final metrics offer a quantified reminder that "legitimate knowledge" has a race and a gender, as well as a class and a geographic location. These characteristics are inherited from the matrix of domination, and sustained by the matrix of domination as well. Although we challenge that fact in our writing, we readily admit that we fell back on learned habits in our citational practices, particularly when we felt that our scholarly credibility was on the line.

Our takeaway from this process is something we already knew but learned once again: values are not enough. We have to put those values into action and hold ourselves accountable time and time again. This constant emphasis on accountability is not easy, and it is not always successful (case in point). It also takes time. Our final metrics are uncomfortable but in some ways constructive: they serve as evidence of the distance between our ideals and our actions, they help us locate the help we need to bridge those gaps, and they help us persist.

## Auditing *Data Feminism*, by Isabel Carter

In the interest of remaining accountable to the values statement for this book, the authors tasked me with performing an audit of all the individuals, projects, and organizations referenced in *Data Feminism*. Quantifying these references provided important information about which perspectives were being included in the work and to what extent. At the same time, this process presented difficult-to-answer questions about identification and classification. Therefore, this methods statement will serve to explain how we approached those questions and what our answers were.

First, we had to decide what would constitute a "reference" that needed to be recorded in the audit. Every individual mentioned by name was counted as a single reference, as was every project (i.e., Kiln's *Ship Map*). Corporations were mostly excluded from the audit unless they played an active role in an example and were mentioned more than once. For instance, Target is included because its pregnancy-targeted marketing strategy is used as a major example in chapter 1. On the other hand, Instagram is not included despite being mentioned multiple times in the Serena Williams example because in that case the social medium is merely the site where Williams and her fans exchanged stories of their birth experiences.

Each categorization then required its own investigation. For individuals, we attempted to verify their race, gender, country of origin (which fed into our Global South/Global North distinction), and indigeneity. Additional categorizations included whether or not references were from within the academy and if they represented an example of good data practices or "what not to do." References were logged as "communal" if they were community-driven (e.g., Data for Black Lives), and a separate category recorded whether the reference provided a "nonvisual example" of data work or not. Each record was further classified by importance. One reference constituted "passing" importance; two to four references, "more than once;" and beyond that, "central."

Although categories like "importance" and "nonvisual example" were straightforward to assign, others like race and gender were not. We tried to confirm these categorizations through online research, but without the self-identification of the referenced individuals, these categories are obviously subject to further inquiry. It should be noted that only individuals were counted by race or gender, unless an example hinged on a racialized or gendered assertion. For example, General Motors is discussed in relation to its discriminatory treatment of Emma DeGraffenreid, which became the impetus for legal scholar Kimberlé Crenshaw's study of "intersectionality." Discriminatory hiring on the basis of race and gender is the reason for General Motors's inclusion in the book; therefore the company is listed as "white" and "man" in the audit. Also, if upon further research an organization was found to have an all-white board or staff, like Clarksons Research UK, it was categorized as "white."

Future attempts to replicate this audit should take seriously the difficulty of clearly establishing these identity categories without formally consulting with those who are being referenced and therefore classified. Some classifications—such as gender—may not be possible to define, as individuals may choose not to publicly disclose their status as trans or nonbinary to avoid discrimination or simply because they consider it private information. Also, we would like to acknowledge that categorizing references by their social identities and structural affiliations does not inherently guarantee a representative discourse. However, we consider this an important effort to remain accountable to the values that brought forth this entire project, which include intersectional and antihierarchical thought, the honoring of a multiplicity of viewpoints, and a commitment to acknowledging our own positions and limitations.

Isabel Carter is a multimedia reporter who received their master of arts in journalism at Emerson College in 2019.

## Acknowledgment of Community Organizations

Many community organizations are already modeling the principles of data feminism that we have described in this book. As part of our continued efforts to share power, we are redirecting a portion of royalties from this book to two of these organizations that have significance to the authors. We encourage you to seek out these organizations, engage with them on social media and in real life, and consider how you might contribute to their work.

Indigenous Women Rising is committed to honoring Native and Indigenous people's inherent right to equitable and culturally safe health options through accessible health education, resources, and advocacy. Follow Indigenous Women Rising on Twitter at @IWRising, on Instagram at @indigenouswomenrising, and at www.iwrising.org.

Founded in 1996, Charis Circle is the 501c3 nonprofit programming arm of Charis Books, the South's oldest independent feminist bookstore. Charis Circle works with artists, authors, and activists from across the South and around the world to bring innovative, thoughtful, and life-changing programming and events to feminist communities. Charis Circle exists to foster sustainable feminist communities, work for social justice, and encourage the expression of diverse and marginalized voices. Visit Charis Circle at www.chariscircle.org.

# Figure Credits

## Introduction

Figure 0.1: *Source:* Wikipedia: https://en.wikipedia.org/wiki/Christine_Darden#/media/ File:Christine_Darden.jpg. *Credit:* NASA.

Figure 0.2: (a) *Source:* Peter Bright, "Moore's Law Really Is Dead This Time," *Ars Technica*, February 10, 2016, https://arstechnica.com/information-technology/2016/02/moores -law-really-is-dead-this-time/. *Credit:* Gordon Moore. (b) *Source:* https://twitter.com/ TamyEmmaPepin/status/1116014974508371971. Credit: Tamy Emma Pepin / Twitter.

Figure 0.3: (a) *Source:* Stephanie Dinkins, *Not the Only One*, multimedia installation, 2017. (b) *Source:* Ishan Misra, C. Lawrence Zitnick, Margaret Mitchell, and Ross Girshick, "Seeing through the Human Reporting Bias: Visual Classifiers from Noisy Human-centric Labels," in *Proceedings of the IEEE Conference on Computer Vision and Pattern Recognition* (New York: IEEE, 2016), 2930–2939; and Margaret Mitchell, "The Seen and Unseen Factors Influencing Machine Perception of Images and Language," LinkedIn SlideShare, May 19, 2017, https://www.slideshare.net/SessionsEvents/ margaret-mitchell-senior-research-scientist-google-at-mlonf-seattle-2017. (c) *Source:* Hanah Anderson and Matt Daniels, "Film Dialogue," *Pudding*, April 2016, accessed April 3, 2019, https://pudding.cool/2017/03/film-dialogue/.

## Chapter 1

Figure 1.1: *Source:* https://www.facebook.com/SerenaWilliams/videos/1015608613 5726834/. *Credit:* Serena Williams/Facebook.

Figure 1.2: *Source:* Data from Christianne Corbett and Catherine Hill, *Solving the Equation: The Variables for Women's Success in Engineering and Computing* (Washington, DC: American Association of University Women, 2015). *Credit:* Graphic by Catherine D'Ignazio.

Figure 1.3: *Credit:* Courtesy of Joy Buolamwini.

Figure 1.4: *Credit:* Photo by Brandon Schulman.

Figure 1.5: (a) *Source:* https://feminicidiosmx.crowdmap.com/. (b) *Source:* https://www
.google.com/maps/d/u/0/viewer?mid=174IjBzP-fl_6wpRHg5pkGSj2egE&ll=21.3476
09098250942%2C-102.05467709375&z=5. *Credit:* María Salguero.

Figure 1.6: *Source:* Andrew Pole, "How Target Gets the Most out of Its Guest Data to
Improve Marketing ROI," filmed October 2010 at Predictive Analytics World, video,
47:50, https://www.predictiveanalyticsworld.com/patimes/how-target-gets-the-most
-out-of-its-guest-data-to-improve-marketing-roi/6815/.

Figure 1.7: *Source: Networked Nation: The Landscape of the Internet in America*, exhibit,
2013, Center for Land Use Interpretation. *Credit:* Images by the Center for Land Use
Interpretation.

Figure 1.8: *Credit:* Kimberly Seals Allers and the Irth team.

## Chapter 2

Figure 2.1: *Source:* Gwendolyn Warren, "About the Work in Detroit," in *Field Notes No.
3: The Geography of Children, Part II* (East Lansing, MI: Detroit Geographical Expedition and Institute, 1971). *Credit:* Courtesy of Gwendolyn Warren and the Detroit
Geographical Expedition and Institute.

Figure 2.2: *Source:* Robert K. Nelson, LaDale Winling, Richard Marciano, Nathan
Connolly, et al., "Mapping Inequality," in *American Panorama*, ed. Robert K. Nelson
and Edward L. Ayers, accessed May 13, 2019, https://dsl.richmond.edu/panorama/
redlining/#loc=10/42.3475/-83.1365&opacity=0.8&city=detroit-mi.

Figure 2.3: *Source:* Julia Angwin, Jeff Larson, Surya Mattu, and Lauren Kirchner,
"Machine Bias," ProPublica, May 23, 2016, https://www.propublica.org/article/
machine-bias-risk-assessments-in-criminal-sentencing. *Credit:* Courtesy of Julia Angwin, Jeff Larson, Surya Mattu, and Lauren Kirchner for ProPublica, 2016.

Figure 2.4: *Credit:* Courtesy of the City Digits Project Team, including Brooklyn College,
the Civic Data Design Lab at MIT, and the Center for Urban Pedagogy.

Figure 2.5: *Credit:* Courtesy of the City Digits Project Team, including Brooklyn College,
the Civic Data Design Lab at MIT, and the Center for Urban Pedagogy.

Figure 2.6: *Source:* http://104.196.123.131/locallotto#tours-tab. *Credit:* Courtesy of
Emmanuela, Angel, Robert, and Janeva. This work was supported by the National
Science Foundation under Grant No. DRL-1222430.

## Chapter 3

Figure 3.1: *Source:* "United States Gun Death Data Visualization by Periscopic," Periscopic, 2013, accessed March 12, 2019, https://guns.periscopic.com/?year=2013. *Credit:* Images by Periscopic.

Figure 3.2: *Source:* Christopher Ingraham, "FBI: Active Shooter Incidents Have Soared since 2000," *Washington Post*, June 16, 2016, https://www.washingtonpost.com/news/wonk/wp/2016/06/16/fbi-active-shooter-incidents-have-soared-since-2000/?utm_term=.036515c11720. *Credit:* Images by Christopher Ingraham for the *Washington Post*.

Figure 3.3: (a) *Source:* Elizabeth Palmer Peabody, *A Chronological History of the United States* (New York: Sheldon, Blakeman & Company, 1856). (b) *Source:* Lauren Klein, Caroline Foster, Adam Hayward, Erica Pramer, and Shivani Negi, The Shape of History, 2016, http://shape of history.net/#explore. (c) *Credit:* Image by Courney Allen for the Georgia Tech Digital Humanities Lab.

Figure 3.4: *Source:* Mike Bostock, Shan Carter, Amanda Cox, and Kevin Quealy, "One Report, Diverging Perspectives," *New York Times*, October 5, 2012, https://archive.nytimes.com/www.nytimes.com/interactive/2012/10/05/business/economy/one-report-diverging-perspectives.html; as cited in Jonathan Stray, *The Curious Journalist's Guide to Data* (New York: Columbia Journalism School, 2016).

Figure 3.5: *Source: Do Women Have to Be Naked to Get into the Met. Museum?*, 1989, accessed March 13, 2019, https://www.nga.gov/collection/art-object-page.139856.html. *Credit:* Guerrilla Girls.

Figure 3.6: *Source:* "Violin Plot," Data Visualisation Catalogue, accessed March 13, 2019, https://datavizcatalogue.com/methods/violin_plot.html.

Figure 3.7: *Source:* Gregor Aisch, Nate Cohn, Amanda Cox, Josh Katz, Adam Pearce, and Kevin Quealy, "Live Presidential Forecast," *New York Times*, November 9, 2016, https://www.nytimes.com/elections/2016/forecast/president.

Figure 3.8: *Source:* Gregor Aisch, Nate Cohn, Amanda Cox, Josh Katz, Adam Pearce, and Kevin Quealy, "Live Presidential Forecast," *New York Times*, November 9, 2016, https://www.nytimes.com/elections/2016/forecast/president.

Figure 3.9: *Source:* Margaret Wickens Pearce, *Coming Home to Indigenous Place Names in Canada*, Canadian-American Center, 2018, accessed March 13, 2019, https://umaine.edu/canam/publications/coming-home-map/. *Credit:* Map by Margaret W. Pearce; map design copyright 2017 Canadian-American Center, University of

Maine. Place names shared by permission of the following: Alan Corbiere. Hiio Delaronde and Jordan Engel, "Haudenosaunee Country in Mohawk," *The Decolonial Atlas*, decolonialatlas.wordpress.com/2015/02/04/haudenosaunee-country-in -mohawk-2/, by permission of the authors; Charles Lippert and Jordan Engel, "The Great Lakes: An Ojibwe Perspective," *The Decolonial Atlas*, decolonialatlas.word-press.com/2015/04/14/the-great-lakes-in-ojibwe-v2/, by permission of the authors; Kitigan Zibi Anishinabeg; Brian McInnes, *Sounding Thunder: The Stories of Francis Pegahmagabow* (East Lansing: Michigan State University Press, 2016), by permission of Brian McInnes, with gratitude to James Dumont and Wasauksing First Nation; Woodland Cultural Centre, place names from Frances Froman, Alfred Keye, Lottie Keye, and Carrie Dyck, *English-Cayuga/Cayuga-English Dictionary* (Toronto, Ontario: University of Toronto Press); and Marianne Mithun and Reginald Henry, *Wadewayęstanih. A Cayuga Teaching Grammar* (Brantford, Ontario: Woodland Publishing, The Woodland Cultural Centre, 1984), by permission of Amos Key Jr. and Carrie Dyck.

**Chapter 4**

Figure 4.1: *Source:* Joni Seager, *The Women's Atlas*, 5th ed. (Oxford: Penguin Books, 2018).

Figure 4.2: (a) *Source:* Will Oremus, "Here Are All the Different Genders You Can Be on Facebook," *Slate,* February 13, 2014, http://www.slate.com/blogs/future_tense/ 2014/02/13/facebook_custom_gender_options_here_are_all_56_custom_options. html. (b) *Source:* http://www.facebook.com/. *Credit:* Facebook. Screenshot by Lauren F. Klein.

Figure 4.3: *Source:* http://www.facebook.com/. *Credit:* Facebook. Screenshot by Lauren Klein.

Figure 4.4: *Source:* Jan Diehm and Amber Thomas, "Someone Clever Once Said Women Were Not Allowed Pockets," The Pudding, August 2018, https://pudding .cool/2018/08/pockets/.

Figure 4.5: *Source:* Peter Kirwin, "Clinical Outcomes and Experiences of Trans People Accessing HIV Care in England," BHIVA, accessed August 30, 2019, https://www .bhiva.org/file/5ca62e5cf0828/PeterKirwanO22.pdf.

Figure 4.6: *Source:* "Born Equal. Treated Unequally," infographic, *Telegraph*, accessed August 30, 2019, https://www.telegraph.co.uk/women/business/women-mean -business-interactive/. *Credit:* Claire Cohen, Patrick Scott, Ellie Kempster, Richard

Moynihan, Oliver Edgington, Dario Verrengia, Fraser Lyness, George Ioakeimidis, and Jamie Johnson for the *Telegraph*.

Figure 4.7: *Source*: Sam Morris, Juweek Adolphe, and Erum Salam, "Does the New Congress Reflect You?," *Guardian*, June 7, 2019, https://www.theguardian.com/us-news/ng-interactive/2018/nov/15/new-congress-us-house-of-representatives-senate.

Figure 4.8: *Source*: Amanda Montañez, "Visualizing Sex as a Spectrum," *Scientific American*, August 29, 2017, https://blogs.scientificamerican.com/sa-visual/visualizing-sex-as-a-spectrum/. Research by Amanda Hobbs; expert review by Amy Wisniewski University of Oklahoma Health Sciences Center Reproduced with permission. Copyright 2017 *Scientific American*. *Credit:* Pitch Interactive and Amanda Montañez. Reproduced with permission. Copyright © (2017) *Scientific American*, a division of Nature America, Inc. All rights reserved.

Figure 4.9: *Source:* Eve M. Kahn, "Colored Conventions, a Rallying Point for Black Americans before the Civil War," *New York Times*, August 4, 2016, https://www.nytimes.com/2016/08/05/arts/design/colored-conventions-a-rallying-point-for-black-americans-before-the-civil-war.html. *Credit:* Sketched by Theo. R. Davis, published in *Harper's Weekly,* 1869. Image courtesy of Jim Casey.

Figure 4.10: *Credit:* Photo by Rebecca Rodriguez and Ken Richardson, MIT Media Lab.

## Chapter 5

Figure 5.1: *Source:* "Tech Bus Stops and No-Fault Evictions," Anti-Eviction Mapping Project, accessed August 30, 2019, http://www.antievictionmappingproject.net/techbusevictions.html. *Credit:* Anti-Eviction Mapping Project.

Figure 5.2: *Source:* "Narratives of Displacement and Resistance," Anti-Eviction Mapping Project, accessed August 30, 2019, http://www.antievictionmappingproject.net/narratives.html. *Credit:* Anti-Eviction Mapping Project. Made in collaboration with the San Francisco Ruth Assawa School of the Arts. The interview was shot by Marianne Maeckelbergh and Brandon Jourdan and edited by students Shilo Arkinson and Avidan Novogrodsky-Godt, facilitated by Alexandra Lacey and Jin Zhu.

Figure 5.3: *Source:* Data from www.mediacloud.org. *Credit:* Image by Catherine D'Ignazio.

Figure 5.4: *Source:* Catherine D'Ignazio. Original article: Eric Roston and Blacki Miglozzi, "What's Really Warming the World?," *Bloomberg Businessweek*, June 24, 2015, https://www.bloomberg.com/graphics/2015-whats-warming-the-world. *Credit:* Catherine D'Ignazio, based on reporting by Eric Roston and Blacki Migliozzi for *Bloomberg Businessweek*.

Figure 5.5: *Source:* Eymund Diegel, "Mapping Sewage Flows in the Gowanus Canal after Sandy Flood Damages—the Sequel," Public Lab, December 19, 2012, https://publiclab.org/notes/eymund-diegel/12-18-2012/mapping-sewage-flows-gowanus-canal-after-sandy-flood-damages-sequel. *Credit:* Eymund Diegel for Public Lab.

Figure 5.6: *Credit:* Data Therapy, Emily and Rahul Bhargava.

Figure 5.7: *Source:* Rahul Bhargava, "Mural-ing Our Way to Data Literacy," MIT Civic Media, August 6, 2013, https://civic.mit.edu/2013/08/06/mural-ing-our-way-to-data-literacy/. *Credit:* Data Therapy, Emily and Rahul Bhargava.

Figure 5.8: *Source:* Screenshot from https://ejatlas.org/. *Credit:* Global Atlas of Environmental Justice.

## Chapter 6

Figure 6.1: *Source:* Mona Chalabi, "Kidnapping of Girls in Nigeria Is Part of a Worsening Problem," *FiveThirtyEight*, May 6, 2014, https://fivethirtyeight.com/features/nigeria-kidnapping/.

Figure 6.2: *Source:* Erin Simpson (@charlie_simpson), "So if #GDELT says there were 649 kidnappings in Nigeria in 4 months, WHAT IT'S REALLY SAYING is there were 649 news stories abt kidnappings," Twitter, May 13, 2014, 4:04 p.m., https://twitter.com/charlie_simpson/status/466308105416884225; and Erin Simpson (@charlie_simpson), "And never, EVER use #GDELT for reporting of discrete events. That's not what it's for. Not kidnappings, not murders, not suicide bombings," Twitter, May 13, 2014, 1:15 p.m., https://twitter.com/charlie_simpson/status/466310866225217536.

Figure 6.3: *Source:* Prefeitura da Cidade de São Paulo: e-negocios cidadesp, accessed August 30, 2019, http://e-negocioscidadesp.prefeitura.sp.gov.br/. *Credit:* SIGRC for the Prefecture of São Paulo, Brazil.

Figure 6.4: *Source:* Patrick Torphy, Michaela Halnon, and Jillian Meehan, "Reporting Sexual Assault: What the Clery Act Doesn't Tell Us," Atavist, April 26, 2016, https://cleryactfallsshort.atavist.com/reporting-sexual-assault-what-the-clery-act-doesnt-tell-us. *Credit:* Used with permission of Patrick Torphy, Michaela Halnon, and Jillian Meehan.

Figure 6.5: *Source:* Lauren F. Klein, "The Image of Absence: Archival Silence, Digital Humanities, and James Hemings," *American Literature* 85, no. 4 (2013): 661–688. *Credit:* Visualization by Lauren F. Klein.

Figure 6.6: *Source:* Data from Fatos Kaba et al., "Disparities in Mental Health Referral and Diagnosis in the New York City Jail Mental Health Service," *American Journal of Public Health* 105, no. 9 (2015): 1911–1916, https://doi.org/10.2105/AJPH.2015 .302699. *Credit:* Graphics by Catherine D'Ignazio.

Figure 6.7: *Source:* Data from Kaba et al., "Disparities in Mental Health Referral and Diagnosis in the New York City Jail Mental Health Service." *Credit:* Graphic by Catherine D'Ignazio.

## Chapter 7

Figure 7.1: *Source:* https://github.com/GeorgiaTechDHLab/TOME/graphs/contributors; https://github.com/GeorgiaTechDHLab/TOME/graphs/commit-activity; and https:// github.com/GeorgiaTechDHLab/TOME/network. *Credit:* GitHub/Screenshots by Lauren F. Klein.

Figure 7.2: *Source:* https://www.shipmap.org/. *Credit:* Website created by Duncan Clark & Robin Houston from Kiln. Data compiled by Julia Schaumeier & Tristan Smith from the UCL EI. The website also includes a soundtrack: Bach's Goldberg Variations, played by Kimiko Ishizaka.

Figure 7.3: *Source:* "Wages for Housework," Harvard Library archival materials, accessed August 30, 2019, https://hollisarchives.lib.harvard.edu/repositories/8/archival _objects/1438878. *Credit:* Schlesinger Library, Radcliffe Institute/Bettye Lane.

Figure 7.4: *Source:* Still from *Workers Leaving the Googleplex*, dir. Andrew Norman Wilson, video, 12:00. *Credit:* Andrew Norman Wilson.

Figure 7.5: *Source:* Kate Crawford and Vladan Joler, "Anatomy of an AI System: The Amazon Echo as an Anatomical Map of Human Labor, Data and Planetary Resources," AI Now Institute and Share Lab, September 7, 2018, https://anatomyof.ai. *Credit:* Kate Crawford and Vladan Joler.

Figure 7.6: *Source:* J. K. Gibson-Graham and the Community Economies Collective, *Cultivating Community Economies*, Next System Project, February 27, 2017, https:// thenextsystem.org/cultivating-community-economies. *Credit:* J. K. Gibson-Graham, Jenny Cameron, Kelly Dombrowski, Stephen Healy, and Ethan Miller for the Next System Project.

Figure 7.7: *Source:* "A Brief Visual History of MARC Cataloging at the Library of Congress," *Sapping Attention* (blog), May 16, 2017, http://sappingattention.blogspot .com/2017/05/a-brief-visual-history-of-marc.html. *Credit:* Benjamin M. Schmidt.

Figure 7.8: *Source:* "Chantal's Household," Atlas of Caregiving, accessed August 30, 2019, https://atlasofcaregiving.com/studies/chantals-household/chantal/24-hour/. *Credit:* The Atlas of Caregiving.

Figure 7.9: *Source:* Giorgia Lupi, "Bruises—The Data We Don't See," Medium: Neuroscience, January 31, 2018, https://medium.com/@giorgialupi/bruises-the-data-we-dont-see-1fdec00d0036. *Credit:* Giorgia Lupi and Kaki King.

## Conclusion

Figure 8.1: *Source:* https://commons.wikimedia.org/wiki/File:Google_Walkout_For_Real_Change_in_Sunnyvale,_November_1_2018.jpg. License: Creative Commons Attribution—Share Alike 4.0 International. *Credit:* Wikimedia user Grendelkhan.

Figure 8.2: *Credit:* Columbia Center for Spatial Research, 2016.

Figure 8.3: *Source:* Margaret Mitchell, Simone Wu, Andrew Zaldivar, Parker Barnes, Lucy Vasserman, Ben Hutchinson, Elena Spitzer, Inioluwa Deborah Raji, and Timnit Gebru, "Model Cards for Model Reporting," in *Proceedings of the Conference on Fairness, Accountability, and Transparency* (New York: ACM, 2019), 220–229.

Figure 8.4: *Credit:* Economía Femini(s)ta, 2017, http://economiafeminita.com/, including Mercedes D'Alessandro, Andrés Snitcofsky, Lina Castellanos, Aldana Vales, and the Economía Femini(s)ta team.

Figure 8.5: *Source:* Ron Morrison and Treva Ellison, *Decoding Possibilities*, multimedia installation, 2017, https://elegantcollisions.com/decoding-possibilities/. *Credit:* Ron Morrison and Treva Ellison.

## Acknowledgment of Community Organizations

*Credit:* Indigenous Women Rising.

*Credit:* Charis Circle.

# Notes

## Introduction: Why Data Science Needs Feminism

1. The Hampton sit-ins were the first in the state of Virginia and contributed significantly to the dismantling of the Jim Crow era policies of segregation that were still in place at the time. See "Desegregating Hampton: Hampton University Students' Woolworth's Sit In," Humanities for All, National Humanities Alliance, accessed July 23, 2019, https://humanitiesforall.org/projects/oral-history-of-hampton-va-woolworth-s-sit-in.

2. The story about Darden that we tell here derives primarily from Shetterly's account in *Hidden Figures*. Although we have supplemented Shetterly's research with additional sources and reframed the events of Darden's life to emphasize the role that data played in her career, we remain indebted to Shetterly for calling our attention to Darden, as well as her extensive research on Darden's life. For more information on Darden, see the book *Hidden Figures: The Untold True Story of Four African-American Women who Helped Launch Our Nation Into Space* (New York: William Morrow, 2016). Darden does not appear in the film of the same name. For additional scholarship on the history of women in computing, see Jennifer Light, "When Computers Were Women," *Technology and Culture* 40, no. 3 (1999): 455–483; Nathan Ensmenger, *The Computer Boys Take Over: Computers, Programmers, and the Politics of Technical Expertise* (Cambridge, MA: MIT Press, 2010); and Mar Hicks, *Programmed Inequality: How Britain Discarded Women Technologists and Lost Its Edge in Computing* (Cambridge, MA: MIT Press, 2017).

3. *Merriam-Webster Dictionary*, "Feminist," accessed July 22, 2019. In the song "***Flawless" (2014), Beyoncé samples portions of a TED Talk by the Nigerian writer Chimamanda Ngozi Adichie, who cites the Merriam-Webster definition in her remarks. See Chimamanda Ngozi Adichie, "We Should All Be Feminists," filmed December 2012 at TEDxEuston, video, 29:28, https://www.ted.com/talks/chimamanda_ngozi_adichie_we_should_all_be_feminists.

4. Scholars have long described the evolution of feminism in terms of three waves. The first wave is said to have spanned much of the nineteenth and early twentieth centuries, culminating in the United States in 1920 with the passage of the Nineteenth Amendment, which gave women the right to vote. Women's suffrage and related legal issues were the focus of this wave. The second wave, which we reference here, is said to have encompassed the early 1960s to the

early 1980s. This wave was concerned with a wider range of legal and social issues, including the workplace conditions that Friedan describes in her book, as well as reproductive rights, domestic violence, family roles, and issues of sexuality, among others. It is said to have lost cohesion in the 1980s as a result of internal debates within the movement about sexuality and pornography, among others. Feminism's third wave is said to have begun in the 1990s and is characterized by an increased attention to the idea of intersectionality and the emphasis on both individual differences and structural power that the concept entails. Some scholars have proposed that we've entered a fourth wave of feminism, coinciding with the rise of social media in the early 2010s. With all that said, other scholars have rejected the notion of waves altogether for how it elides the longer and more sustained work of organizing and activism that took place before, during, and after these waves—especially by women of color, whose efforts did not often receive as much popular attention as those of their white counterparts. Because we endorse this critique, we attempt to de-emphasize the narrative of waves in this book, employing the terminology of waves only when it helps to establish the context of a particular example, individual, or group.

5. See bell hooks, *Feminist Theory: From Margin to Center* (New York: Routledge, [1984] 2015).

6. In *Black Feminism Reimagined: After Intersectionality* (Durham, NC: Duke University Press, 2019), Jennifer C. Nash references the work of Vivian May, who traces the intellectual origins of intersectionality to nineteenth-century scholar/activist Anna Julia Cooper; see Vivian M. May, "Intellectual Genealogies, Intersectionality, and Anna Julia Cooper," in *Feminist Solidarity at the Crossroads: Intersectional Women's Studies for Transracial Alliance*, edited by Kim Marie Vaz and Gary L. Lemons (New York: Routledge, 2012), 59–71. Brittney Cooper, in *Beyond Respectability: The Intellectual Thought of Race Women* (Urbana: University of Illinois Press, 2017), places Cooper within an intellectual milieu populated by several late nineteenth-century and early twentieth-century figures. P. Gabrielle Foreman, in *Activist Sentiments: Reading Black Women in the Nineteenth Century* (Urbana: University of Illinois Press, 2009), illuminates the intersectional thought and activism of Black women writers (and readers) in the first half of the nineteenth century. These references are not intended to be comprehensive. Rather, they are intended to give a sense of the long history behind the concept of intersectionality.

7. "The Combahee River Collective Statement," 1978, Circuitous.org, accessed April 3, 2019, http://circuitous.org/scraps/combahee.html.

8. For those who want to learn more about the values that have guided our work on this book, as well as the voices and work that we have sought to amplify, please see our values statement included as an appendix.

9. Sandra Johnson, "Interview with Gloria R. Champine," May 1, 2008, NASA Headquarters NACA Oral History Project, https://historycollection.jsc.nasa.gov/JSCHistoryPortal/history/oral_histories/NACA/champinegr.htm.

10. We'll discuss this feeling of "shock" in more detail in later chapters. Here, we will simply make note of it to emphasize how shock is such a common experience for people who occupy dominant group identities when they—and often we, your authors—find out about injustices.

"A sexist promotion structure/dataset/algorithm?!" we say incredulously. "How could that have happened?" Darden's boss, a white man, was surprised because the gender disparity was not something he expected to see, or was even looking for, since it wasn't something that he had personally experienced. Data feminism asks us to deliberately and explicitly apply an intersectional lens to our data science work—even and especially when we represent dominant group positions.

11. "More than 40 Take the Buyout, Retire," *Researcher News*, April 4, 2007, https://www.nasa .gov/centers/langley/news/researchernews/rn_07retirees.html.

12. The scholarship on this subject is vast. Key anthologies include Beverly Guy-Sheftall, *Words of Fire* (New York: New Press, 1995); and Cherríe Moraga and Gloria E. Anzaldúa, *This Bridge Called My Back: Writings by Radical Women of Color* (London: Persephone Press, 1981). For a recent anthology in this tradition, see *The Crunk Feminist Collection*, edited by Brittney C. Cooper, Susana M. Morris, and Robin M. Boylorn (New York: Feminist Press, 2017).

13. One such example is Sojourner Truth (1797–1883), who worked tirelessly over the course of the nineteenth century to advance both women's and civil rights. Truth, a Black woman who escaped slavery as a young mother in 1826, is most famous today for the speech "Ain't I a Woman?" The memorable title line has inspired generations of intersectional activist work. In truth, however, Truth did not state those words in her original speech. She said, "I am a woman's rights." It was a white abolitionist who, a decade later, rewrote Truth's speech in Southern dialect to make the meaning of her statement clearer to other white abolitionists, who would encounter Truth's lines only in print. To scholars today, the example of Sojourner Truth exemplifies both how Black women were on the forefront of intersectional feminist activism and how white women have historically sought to control that narrative, even when it misrepresents ideas and intentions. See the Sojourner Truth Project for the history and context of this example: https:// www.thesojournertruthproject.com/.

14. Scholars often cite two of Crenshaw's early legal essays for the genesis of the term: "Demarginalizing the Intersection of Race and Sex: A Black Feminist Critique of Antidiscrimination Doctrine, Feminist Theory, and Antiracist Politics," *University of Chicago Legal Forum* 8 (1989): 139–167; and "Mapping the Margins: Intersectionality, Identity Politics, and Violence against Women of Color," *Stanford Law Review* 43, no. 6 (1991): 1241–1299.

15. As an accessible entry point into Crenshaw's work, see her TED Talk, "The Urgency of Intersectionality," filmed October 2016 at TEDWomen 2016, video, 18:50. For a compilation of her writings about intersectionality, including her early work referenced in note 8, see *On Intersectionality: Essential Writings* (New York: New Press, 2019). More recently, Crenshaw has been at the forefront of #SayHerName, a campaign to make the gendered dimensions of police violence visible. For more on this effort, see Kimberlé Crenshaw, Andrea J. Ritchie, Rachel Anspach, Rachel Gilmer, and Luke Harris, "Say Her Name: Resisting Police Brutality against Black Women," 2015, African American Policy Forum, Center for Intersectionality and Social Policy Studies, Columbia Law School, http://aapf.org/sayhernamereport.

16. *Positionality* is a term that describes how individuals come to knowledge-making processes from multiple positions, including race, gender, geography, class, ability, and more. Each of these

positions is shaped by culture and context, and they intersect and interact. We, the authors, have a statement about our own positionalities as an appendix in this book.

17. On the popular educational site Everyday Feminism, the comic artist Robot Hugs explains what oppression feels like at the individual level: "[It is] when prejudice and discrimination is supported and encouraged by the world around you. It is when you are harmed or not helped by government, community or society at large because of your identity." "Having Trouble Explaining Oppression? This Comic Can Do It for You," January 30, 2017, https://everydayfeminism .com/2017/01/trouble-explaining-oppression/. Ashley Crossman offers a good explainer of the sociological understanding of oppression in "What Is Social Oppression," ThoughtCo, January 28, 2019, https://www.thoughtco.com/social-oppression-3026593.

18. *Sexism* is discrimination based on a person's sex or gender. *Cissexism* applies to discrimination against transgender people. *Patriarchy* is a term that describes the combination of legal frameworks, social structures, and cultural values that contribute to the continued male domination of society.

19. For more on the racism encoded in demands for "proof," see chapter 2, as well as Candice Lanius, "Fact Check: Your Demand for Statistical Proof Is Racist," Cyborgology, January 12, 2015, https://thesocietypages.org/cyborgology/2015/01/12/fact-check-your-demand-for-statistical -proof-is-racist/. Maya Randolph has connected the points made by Lanius to a famous quote by Toni Morrison, from 1975: "The function of racism is distraction. It keeps you from doing your work. It keeps you explaining, over and over again, your reason for being. Somebody says you have no language so you spend twenty years proving that you do. Somebody says your head isn't shaped properly so you have scientists working on the fact that it is. Somebody says that you have no art so you dredge that up. Somebody says that you have no kingdoms, again you dredge that up. None of that is necessary. There will always be one more thing." "A Humanist View," talk delivered at Portland State University, May 30, 1975, transcribed by Keisha E. McKenzie, accessed July 23, 2019, https://www.mackenzian.com/wp-content/uploads/2014/07/Transcript _PortlandState_TMorrison.pdf. We thank Momin Malik for bringing this quotation to our attention.

20. We discuss these demographics, and the matrix of domination that they create and sustain, in detail in chapter 1.

21. Note that the latter is quite different from *data for good*. We explore these differences in depth in chapter 5.

22. The projects referenced are Terra Incognita (http://civicmediaproject.org/works/civic-media -project/terra-incognita-serendipity-and-discovery-in-the-age-of-personalization), the Border Crossed Us (http://www.kanarinka.com/project/the-border-crossed-us/), Databasic.io (https:// databasic.io/en), and the Make the Breast Pump Not Suck Hackathon (https://makethebreast pumpnotsuck2018.com). To learn more about these and other projects, visit www.kanarinka .com.

23. The projects referenced are Data by Design (https://dhlab.lmc.gatech.edu/data-by-design/), Vectors of Freedom (https://dhlab.lmc.gatech.edu/tome/vectors-of-freedom/), and the Floor

Chart Project (https://dhlab.lmc.gatech.edu/category/floorchart/). To learn more about these and other projects, visit www.lklein.com or read the following publications related to the projects mentioned: Lauren F. Klein, Caroline Foster, Adam Hayward, Erica Pramer, and Shivani Negi, "The Shape of History: Reimagining Elizabeth Palmer Peabody's Feminist Visualization Work," *Feminist Media Histories* 3, no. 3 (Summer 2017): 149–153; Lauren F. Klein, Jacob Eisenstein, and Iris Sun, "Exploratory Thematic Analysis for Historical Newspaper Archives," *Digital Scholarship in the Humanities* 30, no. 1 (December 2015): 130–141; and A. Beall, C. Allen, A. Vujic, and L. Klein, "Reimagining Elizabeth Palmer Peabody's Lost 'Mural Charts,'" in *Digital Humanities 2018 Book of Abstracts* (Mexico City: Association of Digital Humanities Organizations, 2018), 607–609.

24. Mary Poovey, *A History of the Modern Fact: Problems of Knowledge in the Sciences of Wealth and Society* (Chicago: University of Chicago Press, 1998); Miriam Posner and Lauren F. Klein, "Editor's Introduction: Data as Media," *Feminist Media Histories* 3, no. 3 (Summer 2017): 1–8; and Daniel Rosenberg, "Data before the Fact," in *"Raw Data" Is an Oxymoron*, ed. Lisa Gitelman (Cambridge, MA: MIT Press, 2013), 15–40.

25. Lanius, "Fact Check"; and Theodore M. Porter, *Trust in Numbers: The Pursuit of Objectivity in Science and Public Life* (Princeton, NJ: Princeton University Press, 1996).

26. According to the press release for Apple's iPhone XR, its two-core neural engine can perform up to five trillion operations per second. Performance statistics for the IBM System/360 Model 75 were not available, so to calculate this comparison, we employed the performance statistics for the standard IBM System/360 Model 30, which could perform up to 34,500 instructions per second, with memory up to 64 KB. See "System/360 Model 30," IBM Archives, April 17, 1964, accessed April 3, 2019, https://www.ibm.com/ibm/history/exhibits/mainframe/mainframe_PP2030.html.

27. On death tables, see Jacqueline Wernimont, *Numbered Lives: Life and Death in Quantum Media* (Cambridge, MA: MIT Press, 2018); on colonial counting, see Molly Farrell, *Counting Bodies: Population in Colonial American Writing* (Oxford: Oxford University Press, 2016); for a survey of how European nations have collected statistics on minoritized ethnicities, see Patrick Simon, "Collecting Ethnic Statistics in Europe: A Review," *Ethnic and Racial Studies* 35, no. 9 (2012): 1366–1391; and for an in-depth analysis of the politics of the US Census, see Margo J. Anderson, *The American Census: A Social History* (New Haven, CT: Yale University Press, 1988).

28. For a powerful reckoning with this history, see Jessica Marie Johnson, "Markup Bodies: Black [Life] Studies and Slavery [Death] Studies at the Digital Crossroads," *Social Text* 34, no. 4 (2018): 57–79. For a study that focuses on the legacy of eugenics, see Dean Spade and Rori Rohlfs, "Legal Equality, Gay Numbers and the (After?)Math of Eugenics," *Scholar & Feminist Online* 13, no. 2 (Spring 2016). For a transhistorical study of the surveillance of Black people, see Simone Browne, *Dark Matters: On the Surveillance of Blackness* (Durham, NC: Duke University Press, 2015).

29. This was something that community organizations (especially those led by members of targeted groups) and scholars had been saying for years. And there is a whole interdisciplinary field called surveillance studies that theorizes and studies practices of surveillance. Sociologist Simone Browne describes the field in her 2015 study, *Dark Matters*: "Since its emergence, surveillance

studies has been primarily concerned with how and why populations are tracked, profiled, policed, and governed at state borders, in cities, at airports, in public and private spaces, through biometrics, telecommunications technology, CCTV, identification documents, and more recently by way of Internet-based social network sites such as Twitter and Facebook" (13). Rita Raley and other scholars characterize contemporary digital monitoring practices as "dataveillance"; see Rita Raley, "Dataveillance and Countervailance," in *"Raw Data" Is an Oxymoron,* ed. Lisa Gitelman (Cambridge, MA: MIT Press, 2013), 121–145.

30. For example, Logipix claims to sell cameras and software that detect traffic violations, faces, and "suspicious activity"; see "Safe and Smart Cities," Logipix, accessed April 3, 2019, http://www .logipix.com/index.php/safe-and-smart-cities.

31. On segregation, see Ruha Benjamin, *Race after Technology: Abolitionist Tools for the New Jim Code* (New York: Wiley, 2019). On overpolicing, see Cathy O'Neil, *Weapons of Math Destruction* (New York: Broadway Books, 2016). On social services, see Virginia Eubanks, *Automating Inequality: How High-tech Tools Profile, Police, and Punish the Poor* (New York: St. Martin's Press, 2018). For an exploration of how civic "data dashboards" package these data up for the public, see Shannon Mattern, "Mission Control: A History of the Urban Dashboard," *Places Journal*, March 2015, https://placesjournal.org/article/mission-control-a-history-of-the-urban-dashboard/.

32. For evidence of Catherine's exceptionally interesting Maine Coon cat named n00b, see https://www.instagram.com/p/BgxGicVhhTW/.

33. Yoree Koh, "Forget Fingerprints: Car Seat IDs Driver's Rear End," *Wall Street Journal*, January 18, 2012, https://blogs.wsj.com/drivers-seat/2012/01/18/forget-fingerprints-car-seat-ids -drivers-rear-end/.

34. For more information on PredPol, see https://www.predpol.com/technology/. On the racist history of policing, see Browne, *Dark Matters*; Benjamin, *Race after Technology*; and O'Neil, *Weapons of Math Destruction*.

35. "About Us," PredPol, accessed July 23, 2019, https://www.predpol.com/about/.

36. Even after the legal strictures of slavery were lifted with the ratification of the Thirteenth Amendment, in 1865, the proliferation of so-called Black codes abounded. These were laws passed primarily in the South that restricted Black citizens' freedom of movement, access to opportunities, and protection under the law. Although they were challenged by the public and in the in courts, they were difficult to fully dismantle. Indeed, in most Southern states, they were simply replaced with regulations that used vague language to justify the same anti-Black policing and violence. Those regulations were reinforced by the introduction of Jim Crow laws in the late nineteenth and early twentieth centuries, which legalized racial segregation across the South. Scholars such as Michelle Alexander, in *The New Jim Crow: Mass Incarceration in the Age of Colorblindness* (New York: New Press, 2012), and filmmakers such as Ava DuVernay, in *The 13th* (Kandoo Films, 2016), have demonstrated incontrovertible evidence that this history is not over and persists today in mass incarceration and biased police practices.

37. There is now a whole podcast devoted to the question of whether a particular technology should exist or not: see Catarina Fake, "Should This Exist?," accessed April 3, 2019, https://shouldthisexist.com/.

38. To learn more about Data for Black Lives, visit http://d4bl.org/.

39. See the Stop LAPD Spying Coalition report, *"To Observe and to Suspect": A Peoples Audit of the Los Angeles Police Department's Special Order 1*, Stop LAPD Spying Coalition, April 2, 2013, https://stoplapdspying.org/wp-content/uploads/2013/04/PEOPLES-AUDIT-UPDATED-APRIL-2-2013-A.pdf.

40. See Julia Angwin, Jeff Larson, Surya Mattu, and Lauren Kirchner, "Machine Bias," *ProPublica*, May 23, 2016, https://www.propublica.org/article/machine-bias-risk-assessments-in-criminal-sentencing; and Adriana Gallardo, "Lost Mothers: How We Collected Nearly 5,000 Stories of Maternal Harm," *ProPublica*, March 20, 2018, https://www.propublica.org/article/how-we-collected-nearly-5-000-stories-of-maternal-harm. We discuss risk assessment algorithms in detail in chapter 2, and maternal mortality in chapter 1.

41. See, for example, Lize Mogel, *Walking the Watershed* (2016–present), accessed July 23, 2019, http://www.publicgreen.com/projects/watershed.html; and Stephanie Dinkins, *Not the Only One* (2017), accessed July 23, 2019, https://www.stephaniedinkins.com/ntoo.html.

42. People often say that there are two broad kinds of data: quantitative data, consisting of numbers (e.g., how many siblings you have), and qualitative data, consisting of words and categories (e.g., what color is your shirt?). As we will show in chapter 4, any time there is a binary, there is usually also a hierarchy, and in this case it is that quantitative data can be incorrectly perceived as "better" than qualitative data for being more objective, true, generalizable, larger scale, and so on. Feminist researchers have consistently demonstrated the need to collect qualitative data as well, as they can often (but, of course, not always) capture more nuance and detail than numbers.

43. This phenomenon, though new to data science, is unfortunately very, very old. In their now-classic book, *Witches, Midwives, and Nurses: A History of Women Healers*, Barbara Ehrenreich and Deirdre English detail the history of obstetrics in the United States, in which evidence-driven female midwives were replaced by ridiculous-theory-having male obstetricians after the advent of formal medical schools. The same phenomenon can be found in the kitchen, with women performing most home cooking, unpaid altogether, while men attend culinary school to become celebrity chefs. See Barbara Ehrenreich and Deirdre English, *Witches, Midwives, and Nurses: A History of Women Healers* (New York: The Feminist Press, [1973] 2010).

44. See bell hooks, *Feminism Is for Everybody: Passionate Politics* (New York: Pluto Press, 2000).

## 1   The Power Chapter

1. Serena Williams, "Meet Alexis Olympia Ohanian Jr. You have to check out link in bio for her amazing journey. Also check out my IG stories 😍😍♡♡," September 13, 2017, https://www.instagram.com/p/BY-7H9zhQD7/.

2. See Serena Williams, Facebook, January 15, 2018, https://www.facebook.com/SerenaWilliams/videos/10156086135726834/.

3. Nina Martin and Renee Montagne, "Nothing Protects Black Women from Dying in Pregnancy and Childbirth," ProPublica, December 7, 2017, https://www.propublica.org/article/nothing-protects-black-women-from-dying-in-pregnancy-and-childbirth.

4. See New York City Department of Health and Mental Hygiene, *Severe Maternal Morbidity in New York City, 2008–2012* (New York, 2016), https://www1.nyc.gov/assets/doh/downloads/pdf/data/maternal-morbidity-report-08-12.pdf.

5. SisterSong, National Latina Institute for Reproductive Health, and Center for Reproductive Rights, *Reproductive Injustice: Racial and Gender Discrimination in U.S. Health Care* (New York: Center for Reproductive Rights, 2014), https://tbinternet.ohchr.org/Treaties/CERD/Shared%20Documents/USA/INT_CERD_NGO_USA_17560_E.pdf.

6. *USA Today*'s ongoing reporting on maternal mortality can be found at https://www.usatoday.com/series/deadlydeliveries/.

7. Robin Fields and Joe Sexton, "How Many American Women Die from Causes Related to Pregnancy or Childbirth? No One Knows," ProPublica, October 23, 2017, https://www.propublica.org/article/how-many-american-women-die-from-causes-related-to-pregnancy-or-childbirth.

8. In studies that try to infer maternal mortality through other indicators, like hospital records, what has been consistently shown is that the United States is one of the only countries in the world where maternal morbidity is increasing for all races, and increasing even more steeply for Black and brown women. For example, between 2000 and 2014, the CDC reported a 26.6 percent increase in the maternal mortality ratio in the United States. A 2018 report, *Trends and Disparities in Delivery Hospitalizations Involving Severe Maternal Morbidity, 2006–2015*, showed that life-threatening complications increased for all races and ethnicities during that time. In 2015, dying in the hospital was three times more likely for Black mothers than for white mothers. There was no change in the disparity between white mothers and Black mothers during the period that the data covered.

9. According to the 2018 Newspaper Diversity Survey led by researcher Meredith Clarke for the American Society of News Editors, ProPublica's leadership is 89 percent white, with no Black people in leadership positions, and *USA Today*'s leadership is 85 percent white. See https://www.asne.org/diversity-survey-2018.

10. As we wrote this chapter, people were tweeting #believeblackwomen (see https://twitter.com/search?q=believeblackwomen&src=typd) to grieve the death of a young Black woman named Lashonda Hazard who was pregnant and experiencing severe pain. She died at Women and Infants Hospital in Rhode Island after posting on Facebook that medical staff weren't listening to her. In response, a community organization named Sista Fire RI wrote an open letter to the hospital calling for an end to what they characterized as a pattern of racialized gender violence: "In a state that does not put Black women or women of color first, we believe and trust Black women."

See https://docs.google.com/forms/d/e/1FAIpQLSd-B1sBFiip8tB41L-q3j5vu75qwLxVA9a3h5toX-53lMEifFA/viewform, accessed May 11, 2019.

11. Lindsay Schallon, "Serena Williams on the Pressure of Motherhood: 'I'm Not Always Going to Win,'" *Glamour*, April 27, 2018, https://www.glamour.com/story/serena-williams-motherhood -activism-me-too.

12. Schallon, "Serena Williams on the Pressure of Motherhood."

13. Patricia Hill Collins, *Black Feminist Thought: Knowledge, Consciousness, and the Politics of Empowerment* (New York: Routledge, 2008), 21.

14. Even then, Native Americans of all genders were still legally excluded from voting—at least for another few years—since they had yet to be granted US citizenship. The Fourteenth Amendment explicitly excluded Native Americans from US citizenship—another instance of oppression being codified in the structural domain of the matrix of domination. In 1924, the passage of the Indian Citizenship Act granted joint US citizenship to all Native Americans, clearing the path for enfranchisement. But it would take until 1962 for the last US state (New Mexico) to change its laws so that all Native Americans could vote. Even then, obstacles abounded; the 1965 Voting Rights Act offered additional legal language to contest disenfranchisement, but that act is in the process of being dismantled by the Supreme Court (as of 2013, with *Shelby County v. Holder*), which threatens many of its protections. On the subject of voting rights in the United States, it's also worth pointing out that Puerto Rico did not have universal suffrage until 1935, and like other US territories, still does not have voting power in the US Congress or representation in the electoral college.

15. Other disenfranchisement methods devised over the years have included undue wait times for registering to vote, having to pay a tax to vote, or having to take a test about the Constitution. Well through the passage of the Voting Rights Act of 1965, Black and brown people seeking to vote faced threats of bodily harm. Note that the history of voter suppression perpetrated by white people on people of color is not over. One need only consider the 2018 gubernatorial election in Georgia, in which Brian Kemp, secretary of state and a white man, presided over his own gubernatorial race against Stacey Abrams, a Black woman. In his capacity as secretary of state, his actions included purging voter rolls and putting fifty-three thousand voter registrations on hold, 70 percent of which were for voters of color. Long lines and technical problems plagued the election day efforts, and the NAACP and ACLU sued the state of Georgia for voting irregularities. In short, voter suppression—enacted in the disciplinary domain of the matrix of domination—is alive and well. See German Lopez, "Voter Suppression Really May Have Made the Difference for Republicans in Georgia," Vox, November 7, 2018, https://www.vox.com/ policy-and-politics/2018/11/7/18071438/midterm-election-results-voting-rights-georgia-florida.

16. Note that the disciplinary domain does not just have to do with government power and policy, but also with corporate, private, and institutional policies. A particular company prohibiting its workers from leaving early to vote or penalizing those who distribute information about voting on the factory floor is an example of the disciplinary domain.

17. Eleanor Barkhorn, "'Vote No on Women's Suffrage': Bizarre Reasons for Not Letting Women Vote," *Atlantic*, November 6, 2012, https://www.theatlantic.com/sexes/archive/2012/11/vote-no-on-womens-suffrage-bizarre-reasons-for-not-letting-women-vote/264639/.

18. This is a point that Collins underscores: "Oppression is not simply understood in the mind—it is felt in the body in myriad ways," she writes (*Black Feminist Thought*, 293).

19. For further explanation of why *minoritized* makes more sense to use than *minority*, see I. E. Smith, "Minority vs. Minoritized: Why the Noun Just Doesn't Cut It," *Odyssey*, September 2, 2016, https://www.theodysseyonline.com/minority-vs-minoritize; and Yasmin Gunaratnam, *Researching Race and Ethnicity: Methods, Knowledge and Power* (London: Sage, 2003).

20. This role often entails what Sara Ahmed has described as being a "feminist killjoy." As she writes in the first post on her blog, you might be a feminist killjoy if you "have ruined the atmosphere by turning up or speaking up" or "have a body that reminds people of histories they find disturbing" or "are angry because that's a sensible response to what is wrong." The feminist killjoy exposes racism and sexism, but "for those who do not have a sense of the racism or sexism you are talking about, to bring them up is to bring them into existence." In the process of exposing the problem, the feminist killjoy herself becomes a problem. She is "causing trouble" or getting in the way of the happiness of others by bringing up the issue. For example, a personal killjoy moment from the book-writing process happened when Catherine shared the topic of the book with a former professor, who responded that she should stay focused on data literacy and not become one of those "grumpy feminists" who were uncomfortable with their sexuality and sought to make problems for people. For the record, Catherine is not grumpy, feels confident in her sexuality, and is working on the killjoy skills of making more feminist problems for people. Read more about how to navigate being or becoming a feminist killjoy at feministkilljoys.com, or see Sara Ahmed, *Living a Feminist Life* (Durham, NC: Duke University Press, 2017).

21. Feminist methods involve continually asking *who questions*, as AI researcher Michael Muller has observed: By whom, for whom, who benefits, who is harmed, who speaks, who is silenced. Muller articulated what some of the *who questions* are for human-computer interaction in his essay "Feminism Asks the 'Who' Questions in HCI," *Interacting with Computers* 23, no. 5 (2011): 447–449, and in this book we articulate what some of the *who questions* are for data science.

22. "Bureau of Labor Statistics Data Viewer," US Bureau of Labor Statistics, 2019, accessed April 10, 2019, https://beta.bls.gov/dataViewer/view/timeseries/LNU02070002Q.

23. "Data Brief: Women and Girls of Color in Computing," Women of Color in Computing Collaborative, 2018, accessed April 10, 2019, https://www.wocincomputing.org/wp-content/uploads/2018/08/WOCinComputingDataBrief.pdf.

24. Sarah West Myers, Meredith Whittaker, and Kate Crawford. "Discriminating Systems: Gender, Race and Power in AI," AI Now Institute, 2019, https://ainowinstitute.org/discriminatingsystems.pdf.

25. Christianne Corbett and Catherine Hill, *Solving the Equation: The Variables for Women's Success in Engineering and Computing*, American Association of University Women (Washington, DC:

2015). For comparison, 26% women graduates today is the same percentage of women computer science graduates in 1974, and in subfields like machine learning, the proportion of women is far less. As per the points made in this chapter, even knowing the exact extent of the disparity is challenging. According to a 2014 *Mother Jones* report about diversity in Silicon Valley, tech firms convinced the US Labor Department to treat their demographics as a trade secret and didn't divulge any data until after they were sued by Mike Swift of the *San Jose Mercury News*. See Josh Harkinson, "Silicon Valley Firms Are Even Whiter and More Male Than You Thought" *Mother Jones*, May 29, 2014. There are analyses that have obtained the data in other ways. For example, a gender analysis by data scientists at LinkedIn has shown that tech teams at tech companies have far *less* gender parity than tech teams in other industries, including healthcare, education, and government. See Sohan Murthy, "Measuring Gender Diversity with Data from LinkedIn," LinkedIn (blog), June 17, 2015.

26. See Nadya A. Fouad, "Leaning in, but Getting Pushed Back (and Out)," presentation at the American Psychological Association, August 2014, https://www.apa.org/news/press/releases/2014/08/pushed-back.pdf.

27. In the case of a different resume screening tool (not the one developed by Amazon), it was found that the most predictive factors of job performance success were whether someone was named "Jared" and if they had played lacrosse. We might laugh at the absurdity of such random and specific details, but note how they tell us a lot about the group characteristics of who is getting hired: Jared is a mostly men's name, a mostly white name, and lacrosse—in spite of its Native American origins—is an expensive and predominantly elite, white sport. On biased job algorithms, see Dave Gershgorn, "Companies Are on the Hook if Their Hiring Algorithms Are Biased," *Quartz*, October 22, 2018, https://qz.com/1427621/companies-are-on-the-hook-if-their-hiring-algorithms-are-biased/; and Rachel Kraus, "Amazon's Sexist AI Has a Deeper Problem than Code," Mashable, October 10, 2018, https://mashable.com/article/amazon-sexist-recruiting-algorithm-gender-bias-ai/#VSsbMcGmvqqa. On the origins of lacrosse, see Anthony Aveni, "The Indian Origins of Lacrosse," *Colonial Williamsburg Journal* (Winter 2010).

28. Safiya Umoja Noble, *Algorithms of Oppression: How Search Engines Reinforce Racism* (New York: NYU Press, 2018), 80–81.

29. Feminist legal scholar Martha R. Mahoney summarizes the effect of this privilege hazard with respect to race: "A crucial part of the privilege of a dominant group is the ability to see itself as normal and neutral. This quality of being 'normal' makes whiteness and the racial specificity of our own lives invisible as air to white people, while it is visible or offensively obvious to people defined outside the circle of whiteness." The passage appears in "Whiteness and Women, In Practice and Theory: A Response to Catherine McKinnon," *Yale Journal of Law & Feminism* 5, no. 2 (1993): 217–251.

30. From Anita Gurumurthy's keynote address at Data Justice 2018, Cardiff University.

31. Kate Crawford, "Artificial Intelligence's White Guy Problem," *New York Times*, June 26, 2016, https://www.nytimes.com/2016/06/26/opinion/sunday/artificial-intelligences-white-guy-problem.html.

32. *Cis het* is shorthand for *cisgender heterosexual*. These are two dominant group identities: a person is cisgender when their gender identity matches the sex that they were assigned at birth, and a person is heterosexual when they are sexually attracted to people of the opposite sex.

33. Facial analysis software is used to detect faces in a larger image, such as when some digital cameras create outlines around any faces that they detect in a frame. Facial recognition is when the software both detects a face and then matches that face against a database to cross-reference the face with personal information, such as name, demographics, criminal history, and so on.

34. *Blackface* refers to the racist practice of predominantly non-Black performers painting their faces to signal their caricatured representation of Black people. The tradition has a long history, and has directly contributed to the spread of racist stereotypes about Black people. On its history, see Eric Lott, *Love and Theft: Blackface Minstrelsy and the American Working Class* (New York: Oxford University Press, 1993). On some contemporary manifestations, see Lauren Michele Jackson, "We Need to Talk about Digital Blackface in Reaction GIFs," *Teen Vogue*, August 2, 2017, https://www.teenvogue.com/story/digital-blackface-reaction-gifs.

35. See Joy Buolamwini and Timnit Gebru, "Gender Shades: Intersectional Accuracy Disparities in Commercial Gender Classification," *Proceedings of Machine Learning Research* 81 (2018): 1–15, http://proceedings.mlr.press/v81/buolamwini18a/buolamwini18a.pdf.

36. Training automated systems involves using *training data* to teach the model how to classify things. For example, in the case of Buolamwini's work, the training data would consist of images with and without faces, and the model would be trained to detect whether or not there is a face in each image and, if so, to identify the specific location of the face. Once the model is trained, it is evaluated using another dataset—called a *test dataset*—to determine whether the model works only on the training data or whether it is likely to perform well with new data. Finally, once the model has been tested, it is evaluated again with what's called a *benchmarking dataset*. Benchmarking data consists of an agreed upon standard dataset that makes possible to compare different models—so a researcher could say something like, "The facial detection model from X university performed at 90 percent accuracy, whereas the model from Y corporation performed at 87 percent accuracy."

37. For a full list of media outlets that have written about Buolamwini's work, see https://www.poetofcode.com/press.

38. In *Artificial Unintelligence* (Cambridge, MA: MIT Press, 2018), data journalist and professor Meredith Broussard outlines the concept of technochauvinism: the belief that the technological solution to a problem is the right one. She argues that artificial intelligence is often *not* the most efficient, nor most effective, nor even a remotely adequate solution to a given problem at hand.

39. In her book *White Fragility*, DiAngelo goes further, demonstrating how racial innocence can be viewed as a deliberate social strategy for maintaining power and dominance in society. In *Racial Innocence*, literary scholar Robin Bernstein explores its historical roots. See *White Fragility: Why It's So Hard for White People to Talk about Racism* (London: Penguin Books, 2019); and *Racial Innocence: Performing American Childhood from Slavery to Civil Rights* (New York: NYU Press, 2011).

40. Danielle Brown, "Google Diversity Annual Report 2018," Google, 2018, accessed April 10, 2019, https://static.googleusercontent.com/media/diversity.google/en//static/pdf/Google_Diversity_annual_report_2018.pdf; and Catherine D'Ignazio, "How Might Ethical Data Principles Borrow from Social Work?," Medium, September 2, 2018, https://medium.com/@kanarinka/how-might-ethical-data-principles-borrow-from-social-work-3162f08f0353.

41. The paper about DiF states, "For face recognition to perform as desired—to be both accurate and fair—training data must provide sufficient balance and coverage." See Michele Merler, Nalini Ratha, Rogerio Feris, and John R. Smith, "Diversity in Faces," IBM Research, 2019, https://arxiv.org/abs/1901.10436.

42. Amy Hawkins, "Beijing's Big Brother Tech Needs African Faces," *Foreign Policy*, July 24, 2018, https://foreignpolicy.com/2018/07/24/beijings-big-brother-tech-needs-african-faces/.

43. Hawkins, "Beijing's Big Brother Tech."

44. See Os Keyes, Nikki Stevens, and Jacqueline Wernimont, "The Government Is Using the Most Vulnerable People to Test Facial Recognition Software," *Slate*, March 17, 2019, https://slate.com/technology/2019/03/facial-recognition-nist-verification-testing-data-sets-children-immigrants-consent.html.

45. @ShovelRemi, "I hope facial recognition software has a problem identifying my face too. That'd come in handy when the police come rolling around with their facial recognition truck at peaceful demonstrations of dissent, cataloging all dissenter for 'safety and security,'" Twitter, February 12, 2018, 7:58 p.m., https://twitter.com/ShovelRemi/status/963215680559489024.

46. "Research shows facial analysis technology is susceptible to bias and even if accurate can be used in ways that breach civil liberties. Without bans on harmful use cases, regulation, and public oversight, this technology can be readily weaponized, employed in secret government surveillance, and abused in law enforcement," Buolamwini warns. In early 2019, the AJL collaborated with the Center on Technology & Privacy at Georgetown Law to launch the Safe Face Pledge, a set of four ethical commitments that businesses and governments make when using facial analysis technology. Many AI companies and prominent researchers have signed the Safe Face Pledge at the time of this writing. Notably, Amazon, which sells its Rekognition technology to police departments around the country, has not signed and has actively attacked Buolamwini's research. In response, top AI researchers have come to her defense and have called on Amazon to stop selling Rekognition to police departments. See Matt O'Brien, "Face Recognition Researcher Fights Amazon over Biased AI," Associated Press, April 3, 2019, https://apnews.com/24fd8e9bc6bf485c8aff1e46ebde9ec1. For other references, see Joy Buolamwini, "AI Ain't I a Woman?," YouTube video, 3:32, June 2018, https://www.youtube.com/watch?v=QxuyfWoVV98; Joy Buolamwini, "How I'm Fighting Bias in Algorithms," filmed November 2016 in Boston, TED video, 8:34, https://www.ted.com/talks/joy_buolamwini_how_i_m_fighting_bias_in_algorithms?language=en; Federal Trade Commission, "Hearings on Competition and Consumer Protection in the 21st Century," event agenda, November 13–14, 2018, https://www.ftc.gov/news-events/press-releases/2018/10/ftc-announces-agenda-seventh-session-its-hearings-competition; and Soledad O'Brien and Joy

Buolamwini, "Artificial Intelligence Is Biased: She's Working to Fix It," *Matter of Fact*, September 8, 2018, https://matteroffact.tv/artificial-intelligence-is-biased-shes-working-to-fix-it/.

47. Arrianna Planey, "Devalued Lives, & Premature Death: Intervening at the Axes of Social 'Difference,'" *Arrianna Planey's Blog*, March 29, 2019, https://arriannaplaney.wordpress.com/2019/03/29/intervening-at-the-axes-of-social-difference-devalued-lives-premature-death/.

48. Mimi Onuoha, "On Missing Data Sets," GitHub, January 25, 2018, https://github.com/MimiOnuoha/missing-datasets.

49. Mayra Buvinic, Rebecca Furst-Nichols and Gayatri Koolwal, *Mapping Gender Data Gaps* (New York: Data2X, 2014), https://data2x.org/wp-content/uploads/2019/05/Data2X_Mapping GenderDataGaps_FullReport.pdf; and Caroline Criado Perez, *Invisible Women: Exposing Data Bias in a World Designed for Men* (New York: Random House, 2019).

50. See Adriana Gallardo, "How We Collected Nearly 5,000 Stories of Maternal Harm," ProPublica, March 20, 2018, https://www.propublica.org/article/how-we-collected-nearly-5-000 -stories-of-maternal-harm.

51. See https://www.boston.gov/neighborhood/roxbury.

52. Penn Loh, Jodi Sugerman-Brozan, Standrick Wiggins, David Noiles, and Cecelia Archibald, "From Asthma to AirBeat: Community-Driven Monitoring of Fine Particles and Black Carbon in Roxbury, Massachusetts," *Environmental Health Perspectives* 110 (April 2002): 297–301.

53. On counter-data, see Morgan Currie, Britt S. Paris, Irene Pasquetto, and Jennifer Pierre, "The Conundrum of Police Officer-Involved Homicides: Counter-Data in Los Angeles County," *Big Data & Society* 3, no. 2 (2016): 1–14. On data activism, see Stefania Milan and Lonneke Van Der Velden, "The Alternative Epistemologies of Data Activism," *Digital Culture & Society* 2, no. 2 (2016): 57–74. On statactivism, see the introduction to the special issue of *Partecipazione e conflitto. The Open Journal of Sociopolitical Studies* on the topic, edited by Isabelle Bruno, Emmanuel Didier, and Tommaso Vitale: "Statactivism: Forms of Action between Disclosure and Affirmation," *Partecipazione e conflitto: The Open Journal of Sociopolitical Studies* 7, no. 2 (2014): 198–220. There is a large body of literature on citizen science; a good starting point is Sara Ann Wylie, Kirk Jalbert, Shannon Dosemagen, and Matt Ratto, "Institutions for Civic Technoscience: How Critical Making Is Transforming Environmental Research," *Information Society* 30, no. 2 (2014): 116–126.

54. See Ida B. Wells, "A Red Record: Tabulated Statistics and Alleged Causes of Lynchings in the United States, 1892-1893-1894: Respectfully Submitted to the Nineteenth Century Civilization in 'the Land of the Free and the Home of the Brave,'" New York Public Library Digital Collections, accessed July 24, 2019, http://digitalcollections.nypl.org/items/510d47df-8dbd-a3d9 -e040-e00a18064a99.

55. See the About and Data Institute pages of the Ida B. Wells Society website. Since 2016, the Ida B. Wells Society has partnered with ProPublica to offer a two-week data science institute for both journalism students and working reporters. See http://idabwellssociety.org/data-institute/, accessed August 8, 2019.

56. *Femicide* is a term first used publicly by feminist writer and activist Diana Russell in 1976 while testifying before the first International Tribunal on Crimes Against Women. Her goal was to situate the murders of women in a context of unequal gender relations. In this context, men use violence to systematically dominate and exert power over women. And the research bears this out. While male victims of homicide are more likely to have been killed by strangers, a 2009 report published by the World Health Organization and partners notes a "universal finding in all regions" that women are far more likely to have been murdered by someone they know. Femicide includes a range of gender-related crimes, including intimate and interpersonal violence, political violence, gang activity, and female infanticide. Such deaths are often depicted as isolated incidents and treated as such by authorities, but those who study femicides characterize them as a pattern of underrecognized and underaddressed systemic violence. See World Health Organization, *Strengthening Understanding of Femicide: Using Research to Galvanize Action and Accountability* (Washington, DC: Program for Appropriate Technology in Health [PATH], InterCambios, Medical Research Council of South Africa [MRC], and World Health Organization [WHO], 2009), 110.

57. See Maria Salguero's map at https://feminicidiosmx.crowdmap.com/ and https://www .google.com/maps/d/u/0/viewer?mid=174IjBzP-fl_6wpRHg5pkGSj2egE&ll=23.942983359872816 %2C-101.9008685&z=5.

58. Indeed, Marisela Escobedo Ortiz, the mother of one such victim, was herself shot at point-blank range and killed while demonstrating in front of the Governor's Palace in Chihuahua in 2010.

59. The toll now stands at more than 1,500. Three hundred women were killed in Juárez in 2011 alone, and only a tiny fraction of those cases have been investigated. The problem extends beyond Ciudad Juárez and the state of Chihuahua to other states, including Chiapas and Veracruz.

60. *Strengthening Understanding of Femicide* states that "instances of missing, incorrect, or incomplete data mean that femicide is significantly underreported in every region." See World Health Organization, *Strengthening Understanding of Femicide*, 4.

61. After three years of investigating, the commission, chaired by politician Marcela Lagarde, found that femicide was indeed occurring and that the Mexican government was systematically failing to protect women and girls from being killed. Lagarde suggested that femicide be considered, "a crime of the state which tolerates the murders of women and neither vigorously investigates the crimes nor holds the killers accountable." See World Health Organization, *Strengthening Understanding of Femicide*, 11.

62. See Maria Rodriguez-Dominguez, "Femicide and Victim Blaming in Mexico, Council on Hemispheric Affairs, October 2, 2017, http://www.coha.org/wp-content/uploads/2017/10/ Maria-Rodriguez-Femicidio-Mexico-.pdf.

63. Mara Miranda (@MaraMiranda25), "#SiMeMatan es porque me gustaba salir de noche y tomar mucha cerveza ... ," Twitter, May 5, 2017, 11:17 a.m., https://twitter.com/MaraMiranda25/ status/860559096285720581. For an in-depth study of the hashtag and its use in social and political organizing, see Elizabeth Losh, *Hashtag* (New York: Bloomsbury, 2019).

64. Missing data is not a new problem; the fields of critical cartography and critical GIS have long considered the phenomenon of missing data. Contemporary examples of missing data and counterdata collection include "The Missing and Murdered Indigenous Women Database," created by doctoral student Annita Lucchesi, which tracks Indigenous women who are killed or disappear under suspicious circumstances in the United States and Canada (https://www.sovereign-bodies.org/mmiw-database). Jonathan Gray, Danny Lämmerhirt, and Liliana Bounegru also wrote a report which includes case studies of citizen involvement in collecting data on drones, police killings, water supplies, and pollution. See "Changing What Counts: How Can Citizen-Generated and Civil Society Data Be Used as an Advocacy Tool to Change Official Data Collection?," 2016, https://dx.doi.org/10.2139/ssrn.2742871. Environmental health and justice is an area in which communities are out front collecting data when agencies refuse or neglect to do so. The MappingBack Network (http://mappingback.org/home_en/aboutus/) provides mapping capacity and support to Indigenous communities fighting extractive industries, and Sara Wylie, cofounder of Public Lab, works with communities impacted by fracking to measure hydrogen sulfide using low-cost DIY sensors. See Sara Wylie, Elisabeth Wilder, Lourdes Vera, Deborah Thomas, and Megan McLaughlin, "Materializing Exposure: Developing an Indexical Method to Visualize Health Hazards Related to Fossil Fuel Extraction," *Engaging Science, Technology, and Society* 3 (2017): 426–463. Indigenous cartographers Margaret Wickens Pearce and Renee Pualani Louis describe cartographic techniques for recuperating Indigenous perspectives and epistemologies (often absent or misrepresented) into GIS maps. See Margaret Pearce and Renee Louis, "Mapping Indigenous Depth of Place," *American Indian Culture and Research Journal* 32, no. 3 (2008): 107–126. All that said, participatory data collection efforts have their own silences, as Heather Ford and Judy Wajcman show in their study of the "missing women" of Wikipedia: "'Anyone Can Edit,' Not Everyone Does: Wikipedia's Infrastructure and the Gender Gap," *Social Studies of Science* 47, no. 4 (2017): 511–527.

65. Jonathan Stray, *The Curious Journalist's Guide to Data* (New York: Columbia Journalism School, 2016).

66. Virginia Eubanks, *Automating Inequality: How High-Tech Tools Profile, Police, and Punish the Poor* (New York: St. Martin's Press, 2018).

67. "The data are not neutral" is a recurring theme of data feminism. This doesn't mean that data are never useful, just that they are never neutral representations of some sort of essential truth. Examining and understanding the asymmetries of power in the data collection environment (that lead to inequities in the dataset itself) is one of the key responsibilities of the feminist data scientist.

68. Charles Duhigg, "How Companies Learn Your Secrets," *New York Times*, February 19, 2012.

69. Duhigg, "How Companies Learn Your Secrets."

70. The Target "pregnancy prediction score" was more detection than actual prediction because by the time the products were purchased, the customer was likely already pregnant.

71. "Clicking Clean," Greenpeace, May 2015, accessed April 10, 2019, http://www.greenpeace.org/usa/global-warming/click-clean/#top.

72. Joshua S. Hill, "Facebook Los Lunas Data Center Boosted by 100 Megawatts of Solar," Clean-Technica, October 23, 2018, https://cleantechnica.com/2018/10/23/facebook-los-lunas-data -center-boosted-by-100-megawatts-of-solar/.

73. Marie C. Baca, "It's Official: Facebook Breaks Ground in New Mexico Next Month," *Albuquerque Journal*, September 15, 2016, https://www.abqjournal.com/844876/facebook-picks-los-lunas -for-its-data-center.html.

74. On the percentage, see "Women in U.S. Congress 2018," Rutgers Eagleton Institute of Politics, December 13, 2018, https://cawp.rutgers.edu/women-us-congress-2018. On the wealth, see David Hawkings, "Wealth of Congress: Richer than Ever, but Mostly at the Very Top," Roll Call, February 27, 2018, https://www.rollcall.com/news/hawkings/congress-richer-ever-mostly-top.

75. A good visual exploration of the whiteness and the maleness of power across domains can be seen in a photographic data visualization: Haeyoun Park, Josh Keller, and Josh Williams, "The Faces of American Power, Nearly as White as the Oscar Nominees," *New York Times*, February 26, 2016, https://www.nytimes.com/interactive/2016/02/26/us/race-of-american-power.html.

76. See "The World's Most Valuable Resource Is No Longer Oil, but Data," *Economist*, May 6, 2017. For a list of these CEOs, see Michael Haupt, "'Data Is the New Oil'—A Ludicrous Proposition," Medium, May 2, 2016. If you want to hear many people in a row say the phrase, check out the supercut by Neil Perry, "MyDataMyDollars_2018," YouTube, January 28, 2019, https://youtu .be/kKwr1Tp0TBA?t=587.

77. For example, once advertising giants like Facebook and Google have your gender, they can turn around and use it against you. In 2018, Facebook was accused of gender discrimination because it permitted employers to show job ads only to men. Part of this hinges on corporations' reluctance to take responsibility for any of the content that passes through their platforms: Is Facebook discriminating against women? Or merely letting its customers use Facebook data to discriminate? The news article about the suit says that "Facebook said that it was still reviewing the ads but that it generally did not take down job ads that exclude a gender." See Noam Scheiber, "Facebook Accused of Allowing Bias Against Women in Job Ads," *New York Times*, September 18, 2018, https://www.nytimes.com/2018/09/18/business/economy/facebook-job-ads .html. In another example, computer scientists scraped YouTube videos by transgender users and used their images (without consent) to try to train an algorithm to recognize transgender faces. People found their images included in scientific research papers about the technology when they had never granted permission. Because of cissexism, this kind of unethical practice poses severe risk of harm to transgender users in the form of discrimination and violence. See James Vincent, "Transgender YouTubers Had Their Videos Grabbed to Train Facial Recognition Software," *Verge*, August 22, 2017, https://www.theverge.com/2017/8/22/16180080/transgender-youtubers-ai -facial-recognition-dataset. For an example from the government sector with even more severe ethical implications, see Keyes, Stevens, and Wernimont, "The Government Is Using the Most Vulnerable People."

78. In their widely cited paper "Critical Questions for Big Data," danah boyd and Kate Crawford outlined the challenges of unequal access to big data, noting that the current configuration (in

which corporations own and control massive stores of data about people) creates an imbalance of power in which there are "Big Data rich" and "Big Data poor." Boyd and Crawford, "Critical Questions for Big Data: Provocations for a Cultural, Technological, and Scholarly Phenomenon," *Information, Communication & Society* 15, no. 5 (2012): 662–679. Media scholar Seeta Peña Gangadharan has detailed how contemporary data profiling disproportionately impacts the poor, communities of color, migrants, and Indigenous groups. See Seeta Gangadharan, "Digital Inclusion and Data Profiling," *First Monday* 17, no. 5 (April 13, 2012). Social scientist Zeynep Tufekci warns that corporations have emerged as "power brokers" with outsized potential to influence politics and publics precisely because of their exclusive data ownership. See Zeynep Tufekci, "Engineering the Public: Big Data, Surveillance and Computational Politics," *First Monday*, 19, no. 7 (July 2, 2014). And in advancing the idea of *Black data* to refer to the intersection of informatics and Black queer life, Shaka McGlotten states, "How can citizens challenge state and corporate power when those powers demand we accede to total surveillance, while also criminalizing dissent?" See Shaka McGlotten, "Black Data," in *No Tea, No Shade: New Writings in Black Queer Studies* (Durham, NC: Duke University Press, 2016), 262–286.

79. Indeed, four prominent Black maternal health scholars and leaders wrote an essay titled "An Inconvenient Truth: You Have No Answer That Black Women Don't Already Possess." They assert that we should use this moment of increased attention to uplift the work that Black women are already doing, including the "support of Black women in paid, leadership and research roles." Karen A. Scott, Stephanie R. M. Bray, Ifeyinwa Asiodu, and Monica R. McLemore, "An Inconvenient Truth: You Have No Answer That Black Women Don't Already Possess," Black Women Birthing Justice, October 31, 2018, https://www.blackwomenbirthingjustice.org/single-post/2018/10/31/An-inconvenient-truth-You-have-no-answer-that-Black-women-don't-already-possess.

80. See Jochen Profit, Jeffrey B. Gould, Mihoko Bennett, Benjamin A. Goldstein, David Draper, Ciaran S. Phibbs, and Henry C. Lee, "Racial/Ethnic Disparity in NICU Quality of Care Delivery," *Pediatrics* 140, no. 3 (2017) : e20170918; as well as Kelly M. Hoffman, Sophie Trawalter, Jordan R. Axt, and M. Norman Oliver, "Racial Bias in Pain Assessment and Treatment Recommendations, and False Beliefs about Biological Differences between Blacks and Whites," *Proceedings of the National Academy of Sciences* 113, no. 16 (April 4, 2016): 4296–4301, https://doi.org/10.1073/pnas.1516047113.

81. Kimberly Seals Allers, interview by Catherine D'Ignazio, February 26, 2019.

## 2   Collect, Analyze, Imagine, Teach

1. Gwendolyn Warren, "About the Work in Detroit," in *Field Notes No. 3: The Geography of Children, Part II* (East Lansing, MI: Detroit Geographical Expedition and Institute, 1971), 12. The report also included data that Warren and her team collected—and quantified—on factors as specific as amount of broken glass found on playgrounds in white versus Black neighborhoods, as well as essays from other members of the DGEI.

2. Paul Szewczyk, a historian of Detroit, has created a map that overlays demographic information on top of the redlining map to show how all of Detroit's majority Black neighborhoods were colored red. See the blog post authored by Alex B. Hill, "Detroit Redlining Map 1939," *Detroitography*, December 10, 2014, https://detroitography.com/2014/12/10/detroit-redlining-map-1939/.

3. And that's not the end of the cycle: those who could not buy in those neighborhoods but still wanted to own their homes were required to look elsewhere, depriving those neighborhoods of higher-income individuals, as well as those committed to the neighborhood's long-term growth. Fewer higher-income individuals and prospects for long-term growth made those neighborhoods less desirable as locations for business or other developments, and so the cycle continued on, as it does into the present. For a summary of these and other pernicious effects, including their impact into the present on homeownership rates, home values, and credit scores, see Daniel Aaronson, Daniel Hartley, and Bhash Mazumder, "The Effects of the 1930s HOLC 'Redlining' Maps," Working Paper No. 2017-12 (2017), Federal Reserve Bank of Chicago; and Emily Badger, "How Redlining's Racist Effects Lasted for Decades," *New York Times*, August 24, 2017.

4. We use the term *whiteness* here and throughout the book to refer to the social category of whiteness and to distinguish it from any biological or otherwise essentialist conception of race. The concept of whiteness has a long history, just as the concept of Blackness does. Indeed, many have argued that the two are co-constructed. In the early twentieth century, James Weldon Johnson, James Baldwin, and W. E. B. Du Bois devoted significant attention to the relationship between Blackness and whiteness, emphasizing how Black people needed to understand whiteness for their very survival. In more recent years, scholars from across the humanities have taken up this category, offering additional historical context and theoretical importance. See, for example, David Roediger, *The Wages of Whiteness: Race and the Making of the American Working Class* (New York: Verso, 1991) and Toni Morrison, *Playing in the Dark: Whiteness and the Literary Imagination* (New York: Vintage, 1993) for two early works in this tradition; or, more recently, Nell Irvin Painter, *The History of White People* (New York: Norton, 2010).

5. *Buchanan v. Warley*, a 1917 legal case heard before the US Supreme Court, declared that a race-based zoning ordinance in Kentucky was unconstitutional. But many states and cities, as well as private communities, continued to implement other laws and covenants that would effectively exclude certain inhabitants on the basis of race. See Christopher Silver, "The Racial Origins of Zoning in American Cities," in *Urban Planning and the African American Community: In the Shadows*, ed. June Manning Thomas and Marsha Ritzdorf (Thousand Oaks, CA: Sage, 1993), 23–42.

6. On racial capitalism, see Cedric Robinson, *Black Marxism: The Making of the Black Radical Tradition* (1983; Chapel Hill: University of North Carolina Press, 2000); and Jodi Melamed, "Racial Capitalism," *Critical Ethnic Studies* 1, no. 1 (2015): 76–85. On credit scores, see Mikella Hurley and Julius Adebayo, "Credit Scoring in the Era of Big Data," *Yale Journal of Law and Technology* 18, no. 1 (2017), 148-216. On the tax code, see Michael Leachman, Michael Mitchell, Nicholas Johnson, and Erica Williams, "Advancing Racial Equity with State Tax Policy," Center on Budget and Policy Priorities, November 15, 2018.

7. Cheryl I. Harris, "Whiteness as Property," *Harvard Law Review* 106, no. 8 (1993): 1758. Along similar lines, transgender activist and writer Dean Spade and computational biologist Rori Rohlfs, following Michel Foucault, theorize these effects in terms of "life chances." Sorting techniques like redlining distribute life changes differently for different populations, they explain. This "distribution of life chances" is key. Under the matrix of domination, life chances for majoritized bodies are enhanced, multiplied, and secured by new technologies, whereas life chances for minoritized bodies are diminished, divided, and imperiled by new technologies. Spade and Rohlfs, "Legal Equality, Gay Numbers and the (After?)Math of Eugenics."

8. Redlining is still present with us in numerous ways. In late 2018, the US Department of Housing and Urban Development charged Facebook with discrimination for, among other things, enabling housing advertisers to draw a red line around geographic areas where they did not want their housing ads to appear. See Russell Brandom, "Facebook Has Been Charged with Housing Discrimination by the US Government," *Verge*, March 28, 2019, https://www.theverge.com/2019/3/28/18285178/facebook-hud-lawsuit-fair-housing-discrimination. Scholars have also proposed the concepts of *technological redlining*, to describe the ways that technology reinforces oppression and engages in racial profiling (see Safiya Umoja Noble in *Algorithms of Oppression: How Search Engines Reinforce Racism* [New York: New York University Press, 2018]); *digital redlining*, to refer to the unequal distribution of digital services across different geographies, such as the lack of *Pokémon Go* stops in neighborhoods of color (see Allana Akhtar, "Is Pokémon Go Racist? How the App May Be Redlining Communities of Color," *USA Today*, August 9, 2016); and *discursive redlining*, in which online characterizations of physical places, such as Yelp reviews, directly contribute to gentrification processes (see S. Zukin, S. Lindeman, and L. Hurson, "The Omnivore's Neighborhood? Online Restaurant Reviews, Race, and Gentrification," *Journal of Consumer Culture* 17, no. 3 [2017]: 459–479).

9. Warren, "About the Work in Detroit," 12.

10. Warren, "About the Work in Detroit," 10.

11. In fact, Warren had previously led numerous community actions, including school walkouts and protests, before beginning her collaboration with the DGEI.

12. The work of the Detroit Geographic Expedition and Institute inspired a generation of critical cartographers—geographers who would go on to interrogate the role of power in maps and the potential of counterdata and countermapping to challenge that power. But progressive, "critical" people have their own sexism and racism to negotiate. In academic geography, the DGEI's work is almost exclusively portrayed as the work of the progressive academics who worked on the project. Gwendolyn Warren is rarely credited with leading the work; or, if credited, she is used as an example of how the elite academics were successfully able to collaborate with and transfer knowledge to "the disadvantaged Blacks," to quote geographer Ronald Horvath's account of DGEI ("The 'Detroit Geographic Expedition and Institute' Experience," *Antipode* 3, no. 1 [November 1971]: 74). Warren herself challenged this misattribution of the project and mischaracterization of the community in *Field Notes No. 3*. But the white male savior narrative persists to this day: the DGEI map included in this chapter was referred to as "Bill Bunge's map" (which it definitively is

not!) as recently as 2018 in a scholarly paper. Feminist geographers like Cindi Katz have worked to restore credit to Gwendolyn Warren. See, for example, Gwendolyn Warren, Cindi Katz, and Nik Heynen, "Myths, Cults, Memories, and Revisions in Radical Geographic History: Revisiting the Detroit Geographic Expedition and Institute," in *Spatial Histories of Radical Geography: North America and Beyond*, ed. Trevor Barnes and Eric Sheppard (New York: Wiley, 2019), 59–85. In addition, a video of Katz and Warren in conversation at the City University of New York is available here: https://vimeo.com/111159306.

13. Julia Angwin, Jeff Larson, Surya Mattu, and Lauren Kirchner, "Machine Bias," ProPublica, May 23, 2016, https://www.propublica.org/article/machine-bias-risk-assessments-in-criminal-sentencing.

14. Angwin et al., "Machine Bias."

15. "Children in Single-Parent Families by Race in the United States," Annie E. Casey Kids Count Data Center, accessed July 29, 2019, https://datacenter.kidscount.org/data/tables/107-children-in-single-parent-families-by#detailed/1/any/false/867,133,38,35,18/10,9,12,1,185,13/432,431.

16. Walter S. Gilliam, "Implicit Bias in Preschool: A Research Study Brief," Edward Zigler Center in Child Development & Social Policy, September 28, 2016, https://medicine.yale.edu/childstudy/zigler/publications/briefs.aspx.

17. Ruha Benjamin, *Race after Technology: Abolitionist Tools for the New Jim Code* (Medford, MA: Polity, 2019).

18. Surya Mattu offered this quotation in a comment posted on the first draft of this manuscript, viewable at https://bookbook.pubpub.org/pub/7ruegkt6.

19. Ben Green, *The Smart Enough City: Putting Technology in Its Place to Reclaim Our Urban Future* (Cambridge, MA: MIT Press, 2019).

20. Julia Angwin, interview by Catherine D'Ignazio, March 1, 2018.

21. Kristy Holtfreter and Rhonda Cupp, "Gender and Risk Assessment: The Empirical Status of the LSI-R for Women," *Journal of Contemporary Criminal Justice* 23, no. 4 (November 1, 2007): 363–382, https://doi.org/10.1177/1043986207309436.

22. See, for example, Frank Pasquale, *The Black Box Society* (Cambridge, MA: Harvard University Press, 2015).

23. ProPublica was able to obtain risk scores, criminal records, and incarceration history because of Florida's strong open records laws. A complete accounting of ProPublica's methods can be found here: Jeff Larson, Julia Angwin, Lauren Kirchner, and Surya Mattu, "How We Analyzed the COMPAS Recidivism Algorithm," ProPublica, May 23, 2016, https://www.propublica.org/article/how-we-analyzed-the-compas-recidivism-algorithm.

24. Nicholas Diakopoulos, *Algorithmic Accountability Reporting: On the Investigation of Black Boxes* (New York: Tow Center for Digital Journalism, 2013).

25. Nicholas Diakopoulos and Sorelle Friedler, "How to Hold Algorithms Accountable," *MIT Technology Review*, November 17, 2016, https://www.technologyreview.com/s/602933/how-to-hold -algorithms-accountable/.

26. See https://legistar.council.nyc.gov/LegislationDetail.aspx?ID=3137815&GUID=437A6A6D -62E1-47E2-9C42-461253F9C6D0; see also https://www1.nyc.gov/office-of-the-mayor/news/ 251-18/mayor-de-blasio-first-in-nation-task-force-examine-automated-decision-systems-used-by.

27. Another powerful outcome of the ProPublica story is that more than one hundred civil rights groups came together to write a statement against the use of pretrial risk assessment algorithms and released a signed statement: https://civilrights.org/edfund/pretrial-risk-assessments/. Other good work in the vein of algorithm accountability is emerging in activist and scholarly as well as journalistic spaces. For instance, the lack of data on women impacted by police violence in the United States led Kimberlé Crenshaw and the African American Policy Forum to develop the Black Women Police Violence database (http://www.aapf.org/sayhernamewebinar/), designed to challenge the narrative that police violence only affects males of color. The Fairness, Accountability, and Transparency in Machine Learning (FAT/ML) organization and conference examines fairness, accountability, and transparency for machine-learning systems. This growing community of technical researchers looks at how to measure bias in datasets, how to make visible the workings of machine-learning algorithms, and how to align system recommendations with equity and policy goals, among other things.

28. We'll discuss issues of "bigness" in more detail in chapter 6.

29. Candice Lanius, "Fact Check: Your Demand for Statistical Proof is Racist," *Society Pages*, January 12, 2015, https://thesocietypages.org/cyborgology/2015/01/12/fact-check-your-demand -for-statistical-proof-is-racist/.

30. Maggie Walter and Chris Andersen, *Indigenous Statistics: A Quantitative Research Methodology* (London: Routledge/Taylor & Francis Group, 2016).

31. Nina Rabinovitch Blecker, interview by Catherine D'Ignazio, November 29, 2018.

32. Data2X, "Invisible No More? A Methodology and Policy Review of How Time Use Surveys Measure Unpaid Work," March 2018, https://data2x.org/wp-content/uploads/2019/05/Data2X -Invisible-No-More-Volume-1.pdf. For more on invisible labor, see chapter 7.

33. Maggie Walter, "Indigenous Statistics: Doing Numbers Our Way," keynote presentation at the American Indigenous Research Association 2016 Meeting, https://www.american Indigenousresearchassociation.org/wp-content/uploads/2016/11/Walters-AIRA-2016.pdf.

34. Ruha Benjamin, Twitter post, March 19, 2018, https://twitter.com/ruha9/status/97572251877 3403648.

35. There is a well-documented phenomenon called *automation bias* that is described by many researchers, including M. L. Cummings, director of Duke's Humans and Autonomy Laboratory, in which humans are prone to trusting automated systems more than they should. M. L.

Cummings, "Automation Bias in Intelligent Time Critical Decision Support Systems," AAIA First Intelligent Systems Technical Conference, September 2004.

36. See, for instance, the organization (and related conference) for Fairness, Accountability, and Transparency in Machine Learning: http://www.fatml.org/.

37. Funders have been getting behind ethics in data and artificial intelligence in a big way. The Stanford Institute for Human-Centered Artificial Intelligence aims to raise $1 billion for its university-based research center. Reid Hoffman, Pierre Omidyar, and the Knight Foundation created the $27 million Ethics and Governance in AI Fund. The Rockefeller Foundation and the Mastercard Impact Fund recently gifted nonprofit DataKind $20 million to catalyze a "data for social good" ecosystem. But is "ethics" the right framing for this work?

38. Julia Powles, "The Seductive Diversion of 'Solving' Bias in Artificial Intelligence," Medium, December 7, 2018, https://medium.com/s/story/the-seductive-diversion-of-solving-bias-in-artificial-intelligence-890df5e5ef53.

39. We are not suggesting that ethics have no place in data science. There are very valuable contributions to this discussion that come from the emerging field of data ethics—for instance, those of media studies scholar Aristea Fotopoulou, who is working to theorize a feminist data ethics of care, or of information studies scholar Anna Lauren Hoffman, who explores both the uses and limits of data ethics in her research. We should also make clear that there have been many valuable correctives to traditional ethical frameworks that seek to challenge and redefine the notion of ethics through a feminist lens. For example, the idea of an ethics of care that arose from theorizing work in the home emphasizes shared responsibilities rather than individual obligations, the importance of relationships among individuals and groups, the fundamental role of invisible labor, and the importance of placing issues in context. Virginia Held, in *The Ethics of Care: Personal, Political, and Global* (New York: Oxford University Press, 2006), offers one of the most comprehensive accounts of this feminist framework. More recently, Maria Puig de la Bellacasa, in *Matters of Care: Speculative Ethics in Nonhuman Worlds* (Minneapolis: University of Minnesota Press, 2017), extends this ethical conversation to issues of technology and other nonhuman actors. Aristea Fotopoulou extends this framework to data in particular. See "Understanding Citizen Data Practices from a Feminist Perspective: Embodiment and the Ethics of Care" in *Citizen Media and Practice*, ed. H. Stephansen and E. Trere (Oxford: Taylor & Francis/Routledge, forthcoming); and *Feminist Data Studies: Big Data, Critique and Social Justice* (Thousand Oaks, CA: SAGE Publications, forthcoming). For more of Hoffman's work, see D. Greene, A. L. Hoffmann, and L. Stark, "Better, Nicer, Clearer, Fairer: A Critical Assessment of the Movement for Ethical Artificial Intelligence and Machine Learning," presented at the Hawaii International Conference on System Sciences (HICSS), Maui, Hawaii, 2019; and A. L. Hoffmann, "Beyond Distributions and Primary Goods: Assessing Applications of Rawls in Information Science and Technology Literature since 1990," *Journal of the Association for Information Science and Technology* 68, no. 7 (2017): 1601–1618.

40. Sasha Costanza-Chock, "In Defense of Data Discrimination," keynote at Data Justice 2018, Cardiff University, May 2018. A video of the talk can be found here: https://cardiff.cloud

.panopto.eu/Panopto/Pages/Viewer.aspx?id=d132281d-8bbc-4980-8013-a8e8007c788d. For more perspectives on data justice, see special issue, *Information, Communication & Society* 22, no. 7 (2019), ed. Lina Dencik, Arne Hinta, Joanna Redden, and Emiliano Treré.

41. See, for example, the *New York Times*'s extensive coverage of the 2019 college admissions cheating scandal, at https://www.nytimes.com/news-event/college-admissions-scandal.

42. From Sasha Costanza-Chock's keynote at the Data Justice 2018 conference at Cardiff University. We do not delve into restorative justice in depth in this chapter, but interested readers may wish to consult Margaret Urban Walker, "Restorative Justice and Reparations," *Journal of Social Philosophy* 37, no. 3 (Fall 2006): 377–395.

43. For example, the Computer People for Peace (CPP), a late 1960s activist group, strongly criticized the Association for Computing Machinery (ACM), the largest group of computing professionals at the time. The ACM wanted to remain "neutral" on questions of building technology for the Vietnam war. "ACM's neutral position is in fact support for the status quo," wrote the CPP in its May 1969 newsletter. See *Interrupt: The Newsletter of Computer Professionals for Peace*, May 1969, p. 1, https://eli.naeher.name/pdfs/interrupt-7.pdf.

44. If you are a Black woman in the United States, you are intimately familiar with how your simple participation in everyday encounters becomes political. Writes Patricia Hill Collins, "Oppression is not simply understood in the mind—it is felt in the body in myriad ways. Moreover, because oppression is constantly changing, different aspects of an individual U.S. Black woman's self-definitions intermingle and become more salient: Her gender may be more prominent when she becomes a mother, her race when she searches for housing, her social class when she applies for credit, her sexual orientation when she is walking with her lover, and her citizenship status when she applies for a job. In all of these contexts, her position in relation to and within intersecting oppressions shifts." *Black Feminist Thought*, 274–275.

45. See Michael K. Brown, Martin Carnoy, Elliott Currie, David B. Oppenheimer, David Wellman, and Marjorie M. Shultz, *Whitewashing Race: The Myth of a Color-Blind Society* (Berkeley: University of California Press, 2003).

46. Kiddada Green, talk delivered at the 2014 First Food Forum, Kellogg Foundation, March 24, 2014. Viewable online at https://www.youtube.com/watch?v=MXyTrFRGRt4.

47. Scholars like Costanza-Chock are thinking about what it might take to create models and metrics for equity. Their Intersectional Media Equity Index (https://cmsw.mit.edu/media-communication-intersectional-analysis/), for example, is a speculative metric that would quantify "media ownership (who owns the media), employment in media firms (who works in the media), content production (who makes the media), standing (who gets to speak in the media), and attention (who gets listened to)." Although they admit that all sorts of questions remained unanswered—for instance, how the community or identity categories of the index would be determined and what "communication reparations" would look like for groups who have historically been targets of the media—they speculate that "a project to gather and make legible various indicators of equity in the media and communications system would be potentially very

powerful" in light of the limited studies of representation in the media that presently exist. See "Media, Communication, and Intersectional Analysis: Ten Comments for the International Panel on Social Progress," *Global Media and Communication* 14, no. 2 (2018): 201–209.

48. Seeta Pena Gangadharan, Virginia Eubanks, and Solon Barocas, *Data and Discrimination: Collected Essays* (Washington, DC: Open Technology, 2014).

49. Robin DiAngelo, *White Fragility: Why It's So Hard for White People to Talk about Racism* (New York: Beacon, 2018).

50. See the hashtag #TechWontBuildIt: https://twitter.com/hashtag/techwontbuildit?lang=en.

51. For an in-depth discussion of co-liberation, see Ana María León, "Spaces of Co-liberation," in *Dimensions of Citizenship*, ed. Nick Axel, Nikolaus Hirsch, Ann Lui, and Mimi Zeiger (Los Angeles: Inventory Press, 2018), http://dimensionsofcitizenship.org/essays/spaces-of-co-liberation/.

52. Although this quote ended up circulating on the internet as the work of one person—Lilla Watson—Watson herself describes it as the outcome of a collective process, and she desired that it be credited as "Aboriginal activists group, Queensland, 1970s." See Watson, "Attributing Words," *Unnecessary Evils*, November 3, 2008, http://unnecessaryevils.blogspot.com/2008/11/attributing -words.html.

53. See Tawana Petty, "Anti-racism Organizing Has Stalled," *EclectaBlog*, December 2, 2017, https://www.eclectablog.com/2017/12/anti-racism-organizing-has-stalled.html.

54. Seeta Peña Gangadharan, Tawana Petty, Tamika Lewis, and Mariella Saba, *Digital Defense Playbook: Community Power Tools for Reclaiming Data* (Detroit: Our Data Bodies, 2018), https://www.odbproject.org/wp-content/uploads/2019/03/ODB_DDP_HighRes_Single.pdf; see also https://store.alliedmedia.org/products/our-data-bodies-digital-defense-playbook.

55. Gangadharan et al., *Digital Defense Playbook*.

56. Although the ODB project is admirable for how it has forged connections within and across communities, it is not impossible to design for the general public with a goal of co-liberation in mind. For instance, the Appolition app (https://appolition.us/), based on an idea by scholar and filmmaker Kortney Ryan Ziegler and implemented by tech entrepreneur Tiffany Mikell, converts users' spare change into bail money to subvert what Michelle Alexander, author of *The New Jim Crow*, has called "the unconscionable practice of cash bail." Those who can afford to pay bail can go home to await their trial, whereas those who can't remain incarcerated, resulting in the loss of income, job, and other forms of security. The bail system produces a two-tiered system that, once again, privileges the rich.

57. For coverage of the #TechWontBuiltIt hashtag and its results, see "Solidarity Letter: Tech Won't Build it," *Science for the People* (blog), September 25, 2018, https://scienceforthepeople .org/2018/09/25/solidarity-letter-tech-wont-build-it/; Shirin Ghaffary, "Microsoft Workers Are Demanding the Company Cancel Its $480 Million Contract with the US Military," *Vox*, February 22, 2019, https://www.recode.net/2019/2/22/18236290/microsoft-military-contract-augmented -reality-ar-vr; Drew Harwell, "Amazon Met with ICE Officials over Facial-Recognition Systems

that Could Identify Immigrants," *Washington Post*, October 23, 2018, https://www.washington post.com/technology/2018/10/23/amazon-met-with-ice-officials-over-facial-recognition-system -that-could-identify-immigrants/; and Kate Conger, "Google Plans Not to Renew Its Contract for Project Maven, a Controversial Pentagon Drone AI Imaging Program," *Gizmodo*, June 1, 2018, https://gizmodo.com/google-plans-not-to-renew-its-contract-for-project-mave-1826488620.

58. This is not to say that these problematic alliances have disappeared. Google, for instance, continues to work on several other DoD projects. See Jill Aitoro, "Forget Project Maven: Here Are a Couple Other DoD Projects Google Is Working On," *C4ISRNET*, March 13, 2019, https://www.c4isrnet.com/it-networks/2019/03/13/forget-project-maven-here-are-a-couple -other-dod-projects-google-is-working-on/.

59. See Corinne Iozzio, "The *Playboy* Centerfold that Helped Create the JPEG," *Atlantic*, February 9, 2016, https://www.theatlantic.com/technology/archive/2016/02/lena-image-processing -playboy/461970/.

60. Individual software engineers have begun rejecting recruiter efforts from large companies whose values are not aligned with theirs and then publishing their responses to Twitter with the hashtag #TechWontBuildIt (see note 58). For example, engineer Anna Geiduschek posted her response to an Amazon recruiter: "Thanks for reaching out. I'm sure you're working on some interesting problems over there at AWS, however, I would never consider working for Amazon until you drop your AWS contract with Palantir." (Palantir provides software to US Immigration Customs and Enforcement, the agency responsible for separating thousands of young children from their parents at the US-Mexico border in 2018.). At an industry-wide level, the #MoreThanCode research report recently published an inventory of organizations that mobilize technology in combination with social justice values: https://morethancode.cc/orglist/.

61. Horace Mann is credited as one of the first advocates of universal public education; his ideals have been utilized again and again, as recently as by the Obama administration to illustrate its commitment to poverty reduction through public education. But as was true of so many nineteenth-century figures we will meet throughout this book, Mann was severely constrained by the ideas of his time. In public speeches, Mann argued that men and women could not be treated as equals in the education system because their anatomy is different: "There is not one single organ in structure, position and function alike in man and woman, and therefore there can be no equality between the sexes." So, the radical part of Mann's social imaginary for the time was that he imagined an education system that treated all white, Anglo-Saxon, Christian men, regardless of class background, as worthy of education. But women, nonbinary people, people of color, immigrants, disabled people, non-Christians, and others remained excluded from the equalizing. For a recent assessment of Mann's legacy, including a discussion of the lines cited above, see David Rhode, Kristina Cooke, and Himanshu Ojha, "The Decline of the 'Great Equalizer,'" *Atlantic*, December 19, 2012, https://www.theatlantic.com/business/archive/2012/12/ the-decline-of-the-great-equalizer/266455/.

62. From S. M. Wes, M. Whittaker, and K. Crawford, *Discriminating Systems: Gender, Race and Power in AI* (April 2019, AI Now Institute), https://ainowinstitute.org/discriminatingsystems.html.

63. This does not mean there are *no* data ethics courses, only that it is not the norm to address these concerns in introductory coursework. There is a list compiled by social computing researcher Casey Fiesler of hundreds of courses that specifically address ethics in technical fields at http://bit.ly/tech-ethics-syllabi.

64. An April 2019 report from the AI Now Institute, has an excellent characterization of pipeline research and its shortcomings. See Sarah Myers West, Meredith Whittaker, and Kate Crawford, *Discriminating Systems: Gender, Race, and Power in AI*, https://ainowinstitute.org/discriminating systems.pdf.

65. In their paper "'Anyone Can Edit,' Not Everyone Does," Heather Ford and Judy Wajcman offer an excellent summary of the feminist research into women in STEM. See Ford and Wajc-man, "'Anyone Can Edit,' Not Everyone Does: Wikipedia's Infrastructure and the Gender Gap," *Social Studies of Science* 47, no. 4 (March 2017): 511–527. Although this chapter focuses on the transformations that need to happen in individual classrooms and workshops, we should also think about feminist interventions at the scale of programs and institutions. The single most important action an educational institution could take to interrupt the Man Factory model is to *not* house its data science programs in exclusively technical disciplines. Computer science, statistics, and engineering are extremely important fields for data science, but the humanities and social sciences have far more sophisticated and current models for dealing with the social, political, legal, and ethical concerns that arise in all the human and environmental application areas of data science. The solution to the "ethics issue" in the data science curriculum will not be found by adding ethics courses or by funding big AI ethics initiatives that sit in computer science departments. Rather, it will be found when foundations and institutions begin to value and integrate knowledge in a transdisciplinary model. In practice, this might mean that data science is housed outside of the academic department system and placed within its own institute or transdisciplinary center. In this way, the institution could legitimize historical, local, and domain-specific knowledge, even as some of the disciplinary assets of computation are (and should remain) its methods for abstraction and scale.

66. These concepts ranged from basic topics like ratios and probability to more advanced ideas about combinatorics and modeling For more on the City Digits curriculum, see "City Digits: Local Lotto," Center for Urban Pedagogy, accessed July 30, 2019, http://welcometocup.org/Projects/CityStudies/CityDigits. Their curriculum was aligned with the Common Core standards for the state of New York.

67. For recent reportage on this phenomenon, see Meghan Keneally, "Mega Millions Lottery: Where Does Lottery Money Go in Different States," ABC News, October 22, 2018, https://abcnews.go.com/US/mega-millions-lottery-lottery-money-states/story?id=58661412; and Peter O'Dowd, interview with Liberty Vittert, *Here & Now*, October 23, 2018, https://www.wbur.org/hereandnow/2018/10/23/where-do-lottery-profits-go.

68. Laurie H. Rubel, Vivian Y. Lim, Maren Hall-Wieckert, and Mathew Sullivan, "Teaching Mathematics for Spatial Justice: An Investigation of the Lottery," *Cognition and Instruction* 34, no. 1 (2016): 1–26.

69. Youth came up with scenarios to illustrate these probabilities. For example, you would have to drink 15,444 Arizona iced teas to consume four million calories; or wait sixty-one years for four million M subway trains to pass you on the platform.

70. As Rubel describes: "We brought them to Brooklyn College to present to faculty there. We had a group of New York City school kids, and each kid led a small group of faculty from lots of disciplines, showing them how to read the maps and how to interpret some of the data. That was neat." The students subsequently presented at Math for America (https://mathforamerica.org/our-model) and a national conference on math and social justice in San Francisco. Quote from Laurie Rubel, interview by Catherine D'Ignazio, July 25, 2018.

71. This is not to say that these women identified themselves as feminists; we note only that the identities of organizational leadership matter in ways that are both symbolic and material. They reflect how much an organization has prioritized the voices of minoritized bodies and, conversely, how much "privilege hazard" the organization faces by overvaluing the voices of majoritized bodies.

72. Rubel et al., "Teaching Mathematics for Spatial Justice."

73. Rubel et al., "Teaching Mathematics for Spatial Justice."

74. Donna J. Haraway, *Staying with the Trouble: Making Kin in the Chthulucene* (Durham, NC: Duke University Press, 2016).

## 3   On Rational, Scientific, Objective Viewpoints from Mythical, Imaginary, Impossible Standpoints

1. Alberto Cairo, "Emotional Data Visualization: Periscopic's 'U.S. Gun Deaths' and the Challenge of Uncertainty," Peachpit, April 3, 2013, http://www.peachpit.com/articles/article.aspx?p=2036558.

2. Nicole Amare and Alan Manning, *A Unified Theory of Information Design: Visuals, Text and Ethics* (New York: Routledge, 2016).

3. Theodore M. Porter, *Trust in Numbers: The Pursuit of Objectivity in Science and Public Life* (Princeton, NJ: Princeton University Press, 1996).

4. But is such distance possible? Pearson's own work, like that of so many important figures in statistical history, was influenced by his own deeply problematic beliefs. For more on Pearson and his support of the eugenics movement, in particular, see chapter 5. Pearson as quoted in Jonathan Gray, *The Data Epic: Visualisation Practices for Narrating Life and Death at a Distance,* in *Data Visualization in Society*, ed. H. Kennedy and M. Engebretsen (Amsterdam: Amsterdam University Press, 2019).

5. Adam Crymble, "The Two Data Visualization Skills Historians Lack," *Thoughts on Public & Digital History* (blog), March 13, 2013, http://adamcrymble.blogspot.com/2013/03/the-two-data-visualization-skills.html.

6. Haraway goes so far as to link the god trick of visualization with vision itself. She writes, "The eyes have been used to signify a perverse capacity—honed to perfection in the history of science tied to militarism, capitalism, colonialism, and male supremacy—to distance the knowing subject from everybody and everything in the interests of unfettered power." The redlining map in chapter 2 is a perfect example of such unfettered power: the pretense of distance and omniscience in the service of gender and race oppression. Donna Haraway, "Situated Knowledges: The Science Question in Feminism and the Privilege of Partial Perspective," *Feminist Studies* 14, no. 3 (1988): 575–599.

7. Edward R. Tufte, *The Visual Display of Quantitative Information*, 2nd ed. (Cheshire, CT: Graphics Press, 2015). For an example of how his ideas are used by contemporary practitioners, see "Maximiizing the Data-Ink Ratio in Dashboards and Slide Decks," *plotly,* December 11, 2017, https://medium.com/@plotlygraphs/maximizing-the-data-ink-ratio-in-dashboards-and-slide -deck-7887f7c1fab.

8. Mushon Zer-Aviv, "DataViz—The UnEmpathic Art," October 19, 2015, https://responsibledata .io/dataviz-the-unempathetic-art/.

9. Stephanie A. Shields, *Speaking from the Heart: Gender and the Social Meaning of Emotion* (Cambridge: Cambridge University Press, 2002).

10. Witney Battle-Baptiste and Britt Rusert, eds., *WEB Du Bois's Data Portraits: Visualizing Black America* (Hudson, NY: Chronicle Books, 2018); Laura Bliss, "The Hidden Histories of Maps Made by Women: Early North America," *CityLab*, March 21, 2016, https://www.citylab.com/ design/2016/03/women-in-cartography-early-north-america/471609/; and Lauren Klein, Caroline Foster, Adam Hayward, Erica Pramer, and Shivani Negi, "The Shape of History: Reimagining Elizabeth Palmer Peabody's Feminist Visualization Work," *Feminist Media Histories* 3, no. 3 (Summer 2017): 149–153. More about all of these visualizations and the contexts in which they were created can be found on Lauren's interactive book in progress, *Data by Design*, at http:// dataxdesign.io/.

11. Aristotle, *The Rhetoric and the Poetics* (New York: Random House, 1954).

12. Thanks go to M. Richard Zinman for helping us fact-check whether the men wore robes or tunics, as well as what their head garb consisted of.

13. Jessica Hullman and Nicholas Diakopoulos, "Visualization Rhetoric: Framing Effects in Narrative Visualization," *IEEE Transactions on Visualization and Computer Graphics* 17, no. 12 (December 2011): 2231–2240.

14. See Jonathan Stray, *The Curious Journalist's Guide to Data* (New York: Columbia Journalism School, 2016), https://legacy.gitbook.com/book/towcenter/curious-journalist-s-guide-to-data/ details; and Mike Bostock, Shan Carter, Amanda Cox, and Kevin Quealy, "One Report, Diverging Perspectives," *New York Times*, October 5, 2012, https://archive.nytimes.com/www.nytimes.com/ interactive/2012/10/05/business/economy/one-report-diverging-perspectives.html.

15. Since the 1950s, there has been a line of research focused on the important framing effects of titles of news articles on interpretation. More recently, scholars are showing that titles of visualizations are similarly important anchors for people to make sense of data graphics in popular media. For example, see Michelle A. Borkin, Zoya Bylinskii, Nam Wook Kim, Constance May Bainbridge, Chelsea S. Yeh, Daniel Borkin, Hanspeter Pfister, and Aude Oliva, "Beyond Memorability: Visualization Recognition and Recall," *IEEE Transactions on Visualization and Computer Graphics* 22, no. 1 (2016): 519–528.

16. See Stray, "The Curious Journalist's Guide to Data."

17. See Hullman and Diakopoulos, "Visualization Rhetoric."

18. Helen Kennedy, Rosemary Lucy Hill, Giorgia Aiello, and William Allen, "The Work That Visualisation Conventions Do," *Information, Communication & Society* 19, no. 6 (March 16, 2016): 715–735.

19. Haraway, "Situated Knowledges."

20. Sandra Harding, "'Strong Objectivity': A Response to the New Objectivity Question," *Synthese* 104, no. 3 (September 1995): 331–349.

21. Linda Alcoff, "Cultural Feminism versus Post-Structuralism: The Identity Crisis in Feminist Theory," *Signs: Journal of Women in Culture and Society* 13, no. 3 (1988): 405–436.

22. Nieca Goldberg, *Women Are Not Small Men: Life-Saving Strategies for Preventing and Healing Heart Disease in Women* (New York: Ballantine Books, 2002).

23. Most studies also continue to treat sex and gender as binary classifications, which they are not. (We address that in the next chapter.) And for more on what Carolina Criado-Perez calls the *gender data gap*, check out her book *Invisible Women: Data Bias in a World Designed for Men* (New York: Abrams, 2019).

24. Resisting binary thinking is a multipurpose tool in the feminist toolbox. We discuss the false gender binary in chapter 4. Feminist thinkers have demonstrated how other binaries also need a complete rethinking—like reason/emotion, nature/culture, subject/object, body/world, speaker/receiver, universal/particular, facts/values, and traditional/modern, among others. In short, beware binaries! They are probably hiding a hierarchy behind them.

25. Evelyn Fox Keller and Barbara McClintock, *A Feeling for the Organism: The Life and Work of Barbara McClintock*, 10th anniversary ed. (New York: Freeman, 1984).

26. Luke Stark, "Come on Feel the Data (and Smell It)," *Atlantic*, May 19, 2014, https://www.theatlantic.com/technology/archive/2014/05/data-visceralization/370899/.

27. If you're interested in appetites, Lauren's historical research deals with the cultural significance of appetite and eating in the early United States. See Lauren F. Klein, *An Archive of Taste: Race and Eating in the Early United States* (Minneapolis: University of Minnesota Press, 2020).

28. "Vision Impairment and Blindness," World Health Organization, October 11, 2018, http://www.who.int/news-room/fact-sheets/detail/blindness-and-visual-impairment.

29. Aimi Hamraie, "A Smart City Is an Accessible City," *Atlantic*, November 6, 2018, https://www.theatlantic.com/technology/archive/2018/11/city-apps-help-and-hinder-disability/574963/.

30. You can see the full performance of *A Sort of Joy (Thousands of Exhausted Things)* at https://vimeo.com/133815147.

31. This critique is at least as old as Linda Nochlin's canonical 1971 essay: "Why Have There Been No Great Women Artists?," in *Woman in Sexist Society: Studies in Power and Powerlessness*, ed. Vivian Gornick and Barbara Moran (New York: Basic Books, 1971), 344–366.

32. By Whitney Chadwick in Guerrilla Girls, *Confessions of the Guerilla Girls* (New York: Harper-Collins, 1995).

33. Feminist ruckuses with the Metropolitan Transit Authority continue today. In 2015, the menstruation start-up company THINX was also told their ads—featuring women with eggs and grapefruit—were too suggestive and that the word *period* wouldn't be allowed. After the public cried foul (MTA trains were already cluttered with cleavage due to easily approved advertisements for plastic surgery), the ads did run. In another subway-related incident, the woman-led sex toy company Dame sued the MTA for censorship in 2019 when it refused to run Dame's ads. See Rachel Krantz, "THINX Underwear Ads on NYC Subway Are Up—but the Company Has Another Big Announcement," *Bustle*, November 9, 2015; and Leila Ettachfini, "MTA Quietly Bans Sex Toys from Advertising on NYC Subway," *Vice*, January 10, 2019.

34. Wattenberg states: "A moment of insight, in which people see facts and patterns for themselves, can be rhetorically powerful" (2). See Robert Kosara, Sarah Cohen, Jérôme Cukier, and Martin Wattenberg, "Panel: Changing the World with Visualization," in *IEEE Visualization Conference Compendium* (Piscataway, NJ: IEEE, 2009).

35. In "The Eyes Have It," Ben Shneiderman writes about the design of graphic user interfaces to support data exploration for the purposes of analysis. There is clearly a different context and set of goals than an artistic performance, yet folks in the information visualization community have also showed how "overview first" doesn't necessarily apply for all analytic tasks in user interface design either. Still, it's useful to think about when "the whole picture" doesn't (and can't) provide the whole emotional picture and determine what strategies one might pursue to do so. See Shneiderman, "The Eyes Have It: A Task by Data Type Taxonomy for Information Visualizations," *The Craft of Information Visualization*, September 1996, 364–371; and Timothy Luciani, Andrew Burks, Cassiano Sugiyama, Jonathan Komperda, and G. Elisabeta Marai, "Details-First, Show Context, Overview Last: Supporting Exploration of Viscous Fingers in Large-Scale Ensemble Simulations," *IEEE Transactions on Visualization and Computer Graphics* 25, no. 1 (August 20, 2018): 1225–1235.

36. D'Ignazio and Sutton were struck by the similarity of Boston coastline maps from the past (seventeenth and eighteenth centuries) with the future predictions based on climate change (the year 2100 estimated with a seven-foot storm surge). The neighborhoods that Bostonians created

in the 1800s by trucking in gravel and dirt are the most vulnerable ones to rising sea levels in the future. In *Boston Coastline: Future Past*, the artists led a walking tour of the past/future coastline that was punctuated by microlectures from community members working on climate adaptation. Participants wore messages as they walked and then stenciled them into a timeline on the Boston Common to close the walk. Catherine D'Ignazio and Andi Sutton, "Boston Coastline: Future Past," kanarinka.com, 2018, accessed March 13, 2019, http://www.kanarinka.com/project/boston-coastline-future-past/.

37. Mikhail Mansion's project is called "Two Rivers." One chair sits on a platform in the Providence River and logs data about the currents and shifts in the water. Visitors to a gallery are invited to sit on a second chair and feel, in real-time, the motion of the Providence River at a distance. Mikhail Mansion, "Two Rivers (2011)," Vimeo, October 1, 2011, https://vimeo.com/29885745.

38. The Data Zetu project ("our data" in Swahili), in collaboration with a number of partners in Tanzania, ran the Data Khanga Design Challenge in which participants designed *khangas*—fabrics that traditionally carry social messages in East Africa. Designers worked to incorporate statistics about gender equality and health into their khangas. Models wore the winning designs in a fashion show to which all participants were invited. Maana Katuli, "Young Artists Use Fashion and Data to Promote Dialog on Sexual Health," Medium, March 28, 2018, https://medium.com/data-zetu/young-artists-use-fashion-and-data-to-promote-dialog-on-sexual-health-517429662ec2.

39. *Core Sample* is a GPS-based sound walk by Teri Rueb from 2007. Teri Reub, "Core Sample—2007," *Teri Rueb* (blog), 2007, accessed March 13, 2019, http://terirueb.net/core-sample-2007/.

40. *FM Radio Map* from 2006 is a paper map that plots the location of commercial and pirate radio stations in London. Viewers can use a modified radio to listen to each radio station by placing metal contacts on the station locations. The back of the map uses graphite to conduct electricity from the metal contacts to a small radio that tunes in to the station selected. Jo-Anne Green, "Simon Elvins' Silent London," *Networked_Music_Review* (blog), July 11, 2006, http://archive.turbulence.org/networked_music_review/2006/07/11/simon-elvins-silent-london/.

41. *A Piece of the Pie* by Annina Rüst leverages pie metaphors to make pie charts and literal, edible pies. Her robot also tweets its data about gender representation in technical fields. Annina Rüst, "A Piece of the Pie Chart", 2013, accessed March 13, 2019, http://www.anninaruest.com/pie/.

42. A 2010 study by Scott Bateman and colleagues in computer science at the University of Saskatchewan found that "embellished" charts—such as bar charts in the form of monsters—do not hinder people's ability to accurately read them—and in fact, they are actually easier to remember. When polled two to three weeks later, people were much more likely to recall the message of an embellished chart over a minimalist chart that displayed the same data. People also thought the "junk charts," decorated with monsters, were more attractive and enjoyed them more. (Duh. Who doesn't like monsters better than bar charts?!) Likewise, in 2016, Michelle Borkin and colleagues showed that visualizations that make use of novel presentation styles are more memorable. Relating the visual form to the topical content of a chart *can really work*. So, as data journalist Mona Chalabi says, "If it's about farts, draw a butt for god's sakes." See Scott Bateman,

Regan L. Mandryk, Carl Gutwin, Aaron Genest, David McDine, and Christopher Brooks, "Useful Junk?: The Effects of Visual Embellishment on Comprehension and Memorability of Charts," in *Proceedings of the SIGCHI Conference on Human Factors in Computing Systems* (New York: ACM, 2010), 2573–2582; Michelle A. Borkin, Zoya Bylinskii, Nam Wook Kim, Constance May Bainbridge, Chelsea S. Yeh, Daniel Borkin, Hanspeter Pfister, and Aude Oliva, "Beyond Memorability: Visualization Recognition and Recall," *IEEE Transactions on Visualization and Computer Graphics* 22, no. 1 (2015): 519–528; and Bryony Stone, "'If It's about Farts, Draw a Butt for God's Sakes': Mona Chalabi Tells Us How to Illustrate Data," *It's Nice That* (blog), March 8, 2018, https://www.itsnicethat.com/articles/mona-chalabi-illustration-internationalwomensday-080318.

43. You can read more about these chart forms and their purposes and functions at the Data Visualisation Catalogue, https://datavizcatalogue.com/.

44. According to work by Sarah Belia and colleagues, researchers themselves have a hard time understanding confidence intervals. See Sarah Belia, Fiona Fidler, Jennifer Williams, and Geoff Cumming, "Researchers Misunderstand Confidence Intervals and Standard Error Bars," *Psychological Methods* 10, no. 4 (2005): 389–396.

45. Take, for example, the weather report. Forecasts such as "There's a 30 percent chance of rain tomorrow" are generally interpreted by the public to mean "It will rain 30 percent of the time" or "It will rain in 30 percent of my area," and not as a 30 percent probability of it raining. The standard meteorological measure is *probability of precipitation* (PoP), which takes both time and geography into account. PoP is calculated by multiplying a confidence measure (that rain will occur somewhere in a geographic area in a given time period) by an area measure (the percentage of the geographic area that will receive any rain in a given time period). In an installment of her series Just the Facts, data journalist Mona Chalabi detailed how the weather industry has an acknowledged "wet bias," meaning forecasters consistently overpredict rain so as not to make people angry that they didn't bring umbrellas. Mona Chalabi, "Is the National Weather Service Lying to You?," *Guardian*, March 17, 2017, https://www.theguardian.com/us-news/2017/mar/17/national-weather-service-forecasting-temperatures-storms.

46. Hullman and colleagues did a study on hypothetical outcome plots, which animate different simulated outcomes for a single quantity. By viewers seeing where the outcomes tended to land animated over time, they were able to infer more about the probability of variation than in standard violin plots and error bars. Jessica Hullman, Paul Resnick, and Eytan Adar, "Hypothetical Outcome Plots Outperform Error Bars and Violin Plots for Inferences about Reliability of Variable Ordering," *PLOS ONE* 10, no. 11 (2015): e0142444.

47. Richard Porczak (@tsiro), "Straight up: the NYT needle jitter is irresponsible design at best and unethical design at worst and you should stop looking at it," Twitter, November 8, 2016, 9:58 p.m., https://twitter.com/tsiro/status/796185282718511104. J. K. Trotter, "*The New York Times* Live Presidential Election Meter Is Fucking with Me," *Gizmodo*, November 8, 2016, https://gizmodo.com/the-new-york-times-live-presidential-meter-is-fucking-w-1788732314.

48. Gregor Aisch, "Why We Used Jittery Gauges in Our Live Election Forecast," Vis4.net, November 14, 2018, https://www.vis4.net/blog/2016/11/jittery-gauges-election-forecast/.

49. Email to Catherine D'Ignazio and Lauren Klein, January 7, 2019.

50. Margaret Wickens Pearce, "'Coming Home' Map," Canadian-American Center, 2018, accessed March 13, 2019, https://umaine.edu/canam/publications/coming-home-map/.

51. Margaret Pearce, interview by Catherine D'Ignazio, March 15, 2019.

52. As Pearce and Hornsby write, "To be 'coming home' is itself a kind of reconciliation, a moving away from settler time and moving toward Indigenous time." M. Pearce and S. Hornsby, "The Making of Coming Home," *Canadian Geographer* 64, no. 1 (2020).

53. No god trick reveals *everything* because that would be impossible at any scale. The trick part is that it gives the viewer the impression that everything is revealed.

54. Pearce and Hornsby, "The Making of Coming Home."

55. Elizabeth Grosz, in "Architectures of Excess," explains: "Communities, which make language, culture, and thus architecture their modes of existence and expression, come into being not through the recognition, generation, or establishment of universal, neutral laws and conventions that bind and enforce them, but through the remainders they cast out, the figures they reject, the terms that they consider unassimilable, that they attempt to sacrifice, revile and expel" (152). Grosz, "Architectures of Excess," in *Architecture from the Outside: Essays on Virtual and Real Space* (Cambridge, MA: MIT Press, 2006), 151–166. Also see Shaowen Bardzell, "Feminist HCI: Taking Stock and Outlining an Agenda for Design," in *Proceedings of the SIGCHI Conference on Human Factors in Computing Systems* (New York: ACM, 2010), 1301–1310.

56. Michaelanne Dye, Neha Kumar, Ari Schlesinger, Marisol Wong-Villacres, Morgan G. Ames, Rajesh Veeraraghavan, Jacki Oneill, Joyojeet Pal, and Mary L. Gray, "Solidarity across Borders: Navigating Intersections towards Equity and Inclusion," in *Companion of the 2018 ACM Conference on Computer Supported Cooperative Work and Social Computing—CSCW 18* (New York: ACM, 2018), 487–494.

57. Collins, *Black Feminist Thought*.

58. As more designers and illustrators enter the field, a new generation of data visualizers is challenging the antiemotion and antiembellishment dogma. These include Jessica Bellamy, Giorgia Lupi, Stefanie Posavec, Federica Fragapane, and Kelli Anderson, among many others. On a practical level, engineering productive collisions between data science people (sophisticated in analytic methods and abstraction) and artists, designers, media folks, and humanists (sophisticated in rhetoric, form, and embodiment) might be the surest way to overcome the false binary of reason versus emotion.

## 4  "What Gets Counted Counts"

1. For coverage of Munir's remarks, see Nadia Khomami, "I Thought It's Now or Never, Says Student Who Came Out as Non-binary to Obama," *Guardian*, April 24, 2016, https://www .theguardian.com/world/2016/apr/24/now-or-never-says-student-who-came-out-as-non-binary

-to-obama. Munir, who has since graduated, is now an activist, writer, and public speaker. Their website is https://mariamunir.com.

2. See, for example, Rena Bivens and Oliver L. Haimson, "Baking Gender into Social Media Design: How Platforms Shape Categories for Users and Advertisers," *Social Media and Society* (October–December 2016), 1–12, which documents how five of the eight most popular social media sites (as of 2016) required new users to input a gender as part of the signup process, and of those five, "all but one (Google+) conceptualized gender as a binary" (4).

3. Email to Lauren Klein, July 18, 2018.

4. In this chapter and throughout the book, we attempt to use "man" instead of "male," and "woman" instead of "female," in our discussions of gender. Here, however, our use of "male" and "female" reflects the terminology most commonly employed in online forms. As for the assertion that there are "millions of nonbinary people in the world," how many millions of nonbinary people actually are there? We don't really know. In their 2015 "Non-Binary Gender Identities Fact Sheet," the American Psychological Association explains, "Because there is limited research on individuals with non-binary gender identities, it is difficult to estimate the exact number of people who identify as non-binary. ... From the limited research that has [included nonbinary as a response category when asking about gender], it is estimated that non-binary individuals make up 25–35% or more of *transgender* populations. However, these studies sampled only transgender populations and did not capture non-binary individuals who do not identify as transgender." The Williams Institute has asserted that 0.3 percent of US adults are transgender. But this figure is likely an underestimate, both for the reasons described by the APA and for the additional personal reasons that might impact an individual's decision to self-disclose. See Mona Chalabi, "Why We Don't Know the Size of the Transgender Population," *FiveThirtyEight*, July 29, 2014, https://fivethirtyeight.com/features/why-we-dont-know-the-size-of-the-transgender-population/.

5. There are some exceptions. Countries that allow some form of nonbinary or third-gender designations on passports include Australia, Canada, Denmark, India, the Netherlands, New Zealand, and the United Kingdom, though some restrictions (like medical or legal documentation) apply.

6. Email to Lauren Klein, July 18, 2018.

7. Joni Seager, "Missing Women, Blank Maps, and Data Voids: What Gets Counted Counts," talk at the Boston Public Library, March 22, 2016, https://civic.mit.edu/2016/03/22/missing-women-blank-maps-and-data-voids-what-gets-counted-counts/.

8. Seager, "Missing Women."

9. The Calling the Shots project, from which this example is drawn, is exemplary in its attention to the categories of data collection, as well as the processes by which categories are ascribed to individual records. See the project website at https://www.southampton.ac.uk/cswf/, as well as the discussion offered by two of the project team members, Natalie Wreyford and Shelley Cobb, in "Data and Responsibility: Toward a Feminist Methodology for Producing Historical Data on Women in the Contemporary UK Film Industry," *Feminist Media Histories* 3, no. 3 (Summer 2017): 107–132.

10. Quoted in Natalie Wreyford and Shelley Cobb, in "Data and Responsibility: Toward a Feminist Methodology for Producing Historical Data on Women in the Contemporary UK Film Industry," *Feminist Media Histories* 3, no. 3 (Summer 2017): 108.

11. Ann Oakley, "Paradigm Wars: Some Thoughts on a Personal and Public Trajectory," *International Journal of Social Research Methodology* 2, no. 3 (1999): 247–254.

12. According to Bivens and Haimson, Google+ was the first of the major social media sites to offer "other" as a gender category, which it included as early as 2011. But Facebook was the first to receive wide media coverage for the decision. See Will Oremus, "Here Are All the Different Genders You Can Be on Facebook," *Slate*, February 13, 2014, http://www.slate.com/blogs/future _tense/2014/02/13/facebook_custom_gender_options_here_are_all_56_custom_options.html.

13. In another essay, Bivens observes that the options vary according to the user's language choice, and in some languages a binary choice remains the only option. See Rena Bivens, "The Gender Binary Will Not Be Deprogrammed: Ten Years of Coding Gender on Facebook," *New Media and Society* 19, no. 6 (2017): 880–898.

14. Note that these choices vary according to language and geography.

15. Bivens, "The Gender Binary Will Not Be Deprogrammed."

16. The canonical work on the politics of classification systems is Geoffrey C. Bowker and Susan Leigh Star, *Sorting Things Out: Classification and Its Consequences* (Cambridge, MA: MIT Press, 2000), which we discuss later in this chapter. On gender as a social construct, see Judith Butler, *Gender Trouble: Feminism and the Subversion of Identity* (New York: Routledge, 1990), which we also discuss later in this chapter. More recently, work in the field of transgender studies has shown how an insistence on the social construction of gender inadvertently (or, some argue, quite intentionally) reinforces another false binary between gender and "biological" sex. While we touch on this issue later in the chapter as well, interested readers may wish to consult, for example, Julian Gill-Peterson's *Histories of the Transgender Child* (Minneapolis: University of Minnesota Press, 2018).

17. In addition, all of these structures must be maintained over time in order to maintain their structural integrity, as Oliver Haimson noted in a comment on the draft version of this manuscript. The Golden Gate Bridge must be reinforced and repaired. The Facebook Ads API must be patched and versioned. And the gender binary must also be actively maintained if it is to endure. This maintenance takes the form of the many small acts that reinforce the gender binary, such as the M/F checkboxes we routinely encounter on forms, as well as those that reinforce the roles those genders should play (e.g., gifts of dolls for girls and trucks for boys). We would argue that this structure is well past due for an upgrade. For a more in-depth discussion of maintenance, see chapter 7.

18. Aristotle saw women as inferior to free men but higher than enslaved men, and he gave written form to many of the stereotypes that exist to this day, such as women being more sentimental and emotional than men (see chapter 3). He also illogically claimed that only paler women had orgasms and that men had more teeth than women (and did not bother to check; all genders have the same number of teeth). Charlotte Witt and Lisa Shapiro, "Feminist History

of Philosophy," in *The Stanford Encyclopedia of Philosophy* (Fall 2018 Edition), ed. Edward N. Zalta, https://plato.stanford.edu/archives/fall2018/entries/feminism-femhist/.

19. Thomas Laqueur, *Making Sex: Body and Gender from the Greeks to Freud* (Cambridge, MA: Harvard University Press, 1992).

20. In fact, in the United States, the sex binary took much longer to solidify than the black/white divide. While many state laws already limited voting rights to men, it was not until the passage of the Fourteenth Amendment, in 1868, which granted voting rights to "male citizens," that sex difference entered the Constitution for the first time.

21. Ibram X. Kendi, *Stamped from the Beginning* (New York: Bold Type Books, 2016).

22. To learn more about Linnaeus's *Systemae Naturae*, first published in 1735, visit *Linné Online*, Upsala University, accessed July 31, 2019, http://www2.linnaeus.uu.se/.

23. Julia Angwin, Jeff Larson, Lauren Kirhner, and Surya Mattu. "Minority Neighborhoods Pay Higher Car Insurance Premiums than White Areas with the Same Risk," ProPublica, April 5, 2017, https://www.propublica.org/article/minority-neighborhoods-higher-car-insurance -premiums-white-areas-same-risk.

24. Blaise Agüera y Arcas, Margaret Mitchell, and Alexander Todorov, "Physiognomy's New Clothes," Medium Artificial Intelligence, May 6, 2017, https://medium.com/@blaisea/physiognomys-new -clothes-f2d4b59fdd6a.

25. Miriam Posner and Lauren F. Klein, "Editor's Introduction—Data as Media," *Feminist Media Histories* 3, no. 3 (Summer 2017): 1–8.

26. Geoffrey C. Bowker and Susan Leigh Star, *Sorting Things Out: Classification and Its Consequences* (Cambridge, MA: MIT Press, 2000).

27. Representative Carrie Meek (D-FL) of the Black Caucus stated that she was "very troubled" by the recommendation and reminded Congress that the purpose of counting race was to permit enforcement of antidiscrimination laws and the equal protection provisions of the Fourteenth Amendment of the Constitution. See Alice Robbin, "Classifying Racial and Ethnic Group Data in the United States: The Politics of Negotiation and Accommodation," *Journal of Government Information* 27, no. 2 (March 2000): 129–156.

28. For an excellent visual timeline of the evolution of racial categories on the US census, see "What Census Calls Us: A Historical Timeline," created by Pew Research: https://www .pewsocialtrends.org/interactives/multiracial-timeline/.

29. Lizette Alvarez, "Meet Mikey, 8: U.S. Has Him on Watch List," *New York Times*, January 13, 2010, https://www.nytimes.com/2010/01/14/nyregion/14watchlist.html.

30. See Joe Sharkey, "With Hair Pat-Downs, Complaints of Racial Bias," *New York Times*, August 15, 2001, https://www.nytimes.com/2011/08/16/business/natural-hair-pat-downs-warrant -a-rethinking.html. For a more scholarly exploration of this same issue, see Simone Browne,

"'What Did TSA Find in Solange's Fro': Security Theater at the Airport," in *Dark Matters: Race and the Surveillance of Blackness* (Durham, NC: Duke University Press, 2015), 131–160.

31. See, for example, Haroon Moghul, "The Unapologetic Racial Profiling of Muslims Has Become America's New Normal," *Quartz*, April 20, 2016, https://qz.com/665317/the-unapologetic-racial -profiling-of-muslims-has-become-americas-new-normal/; and Assia Boundaoui's film, *The Feeling of Being Watched* (2018), which documents the government surveillance experienced by an Arab-American community in Bridgeview, Illinois: http://www.feelingofbeingwatched.com/.

32. Sasha Costanza-Chock, "Design Justice, A.I., and Escape from the Matrix of Domination," *Journal of Design and Science*, last updated July 26, 2018, https://jods.mitpress.mit.edu/pub/ costanza-chock.

33. Dean Spade, *Normal Life: Administrative Violence, Critical Trans Politics, and the Limits of the Law* (Boston: South End Press, 2011). Security pat-downs in particular have long been the focus of criticism and scholarly critique. Angela Davis, in her work on pat-downs in prison, has argued that they are a form of state-sponsored assault: *Are Prisons Obsolete?* (New York: Seven Stories Press, 2013). Shoshana Magnet and Tara Rodgers extend Davis's critique to airport screenings in "Stripping for the State: Whole Body Imaging Technologies and the Surveillance of Othered Bodies," *Feminist Media Studies* 12 (2012): 101–118. Poet Stacey Waite's "On the Occasion of Being Mistaken for a Man by Security Personnel at Newark International Airport," in *Love Poem to Androgyny* (Mint Hill, NC: *Main Street Rag*, 2006), offers a personal meditation on this experience. Paisley Currah and Tara Mulqueen, in "Securitizing Gender: Identity, Biometrics, and Transgender Bodies at the Airport," *Social Research* 78, no. 2 (Summer 2011): 557–582, provide a scholarly analysis of the experience that Waite describes.

34. The philosopher Michel Foucault has described this state of affairs as "the power to make live and let die." Foucault is also responsible for the concept of *biopower*, also known as the *distribution of life chances*, which we reference in chapter 2 in our analysis of redlining maps.

35. Although we do not discuss this particular issue here, "bathroom bills" and other attempts to police public bathroom use along the lines of binary gender is perhaps the most pervasive and most publicized instance of this form of administrative (and sometimes physical) violence. The National Conference of State Legislatures maintains the web page, "Bathroom Bill Legislative Tracking," which provides links to updates (and legal challenges) to the numerous states that attempted to pass such bills. See http://www.ncsl.org/research/education/-bathroom-bill -legislative-tracking635951130.aspx.

36. Jan Diehm and Amber Thomas, "Someone Clever Once Said Women Were Not Allowed Pockets," *Pudding*, August 2018, https://pudding.cool/2018/08/pockets/.

37. Victoria and Albert Museum, "A History of Pockets," accessed July 31, 2019, http://www .vam.ac.uk/content/articles/a/history-of-pockets/.

38. Butler, *Gender Trouble*.

39. American Medical Association, "AMA Adopts New Policies at 2018 Interim Meeting," November 13, 2018, https://www.ama-assn.org/press-center/press-releases/ama-adopts-new-policies -2018-interim-meeting.

40. On the former, see Aniruddha Dutta and Raina Roy, "Decolonizing Transgender in India: Some Reflections," *TSQ: Transgender Studies Quarterly* 1, no. 3 (August 2014): 320–337. On the latter, see Qwo-Li Driskill, "Doubleweaving Two-Spirit Critiques: Building Alliances between Native and Queer Studies," *GLQ: A Journal of Gay and Lesbian Studies* 16, no. 1–2 (April 2010): 69–92.

41. There are some exceptions. For example, states like Oregon and California allow nonbinary gender markers on identification documents.

42. Aliya Saperstein, "Gender Identification," in *Pathways: The Poverty and Inequality Report*, 2018, 5–8, https://inequality.stanford.edu/sites/default/files/Pathways_SOTU_2018_gender-ID.pdf.

43. This is an extension of the nation's originary attempt to assert control over portions of its population, as well as the populations of Indigenous peoples, who did not conform to its racist, colonialist, cis-sexist, and heteronormative ideas about who should be allowed to thrive. There is a particularly grim tradition of the weaponization of these terms and the categories that underlie them in the United States, where normative categories of gender and sexuality have long been imposed on Indigenous populations—first by European colonial powers, and then by the US government. These terms and others push back against what Two-Spirit scholar Qwo-Li Driskill characterizes as "the ways colonial projects continually police sexual and gender lines." In the United States, in particular, we must remain aware of how categories of gender and sexuality were once deployed on behalf of European colonial powers to subjugate and dispossess the Indigenous populations they encountered and whose land and resources they sought; and we might view the Trump administration's attack on the rights of transgender people as an extension of this originary attempt to consolidate power and control. See Ann Laura Stoler, *Race and the Education of Desire: Foucault's* History of Sexuality *and the Colonial Order of Things* (Durham, NC: Duke University Press; 1995); Deborah Miranda, "Extermination of the Joyas: Gendercide in Spanish California," in *The Transgender Studies Reader 2*, ed. Susan Stryker and Aren Z. Aizura (New York: Routledge, 2013), 347–360; and Susan Stryker and Paisley Currah, "Introduction," *TSQ: Transgender Studies Quarterly* 1, no. 1–2 (May 2014): 1–18.

44. See Oliver Haimson (@oliverhaimson), commenting on the Positive Voices survey from the United Kingdom, "This is the best 2-step gender measure I've seen. ... I think asking the second question is invasive if you don't actually need that data," Twitter, April 5, 2019, 5:25 a.m., https://twitter.com/oliverhaimson/status/1114142113007009792.

45. When is sex classification actually necessary? Heath Fogg Davis argues that it isn't in many cases. See *Beyond Trans: Does Gender Matter?*, vol. 2 (New York: NYU Press, 2018).

46. FollowBias (not available any longer) was a Twitter application circa 2017 that would give you a breakdown of what percentage of men you were following on Twitter and what percentage of women (it stuck to the binary). It would attempt to detect gender based on people's names on

Twitter, and show the user aggregated statistics like "80% of the people you follow are men," with the idea that that might nudge them toward following more women—particularly important for the journalists that it was trying to influence. See J. Nathan Matias, Sarah Szalavitz, and Ethan Zuckerman, "FollowBias: Supporting Behavior Change toward Gender Equality by Networked Gatekeepers on Social Media," in *Proceedings of the 2017 ACM Conference on Computer Supported Cooperative Work and Social Computing* (New York: ACM, 2017), 1082–1095. It is part of a not-uncontroversial class of technology known as *automated gender detection* (in which applications try to infer gender from names or handwriting or blogging style) or *automated gender recognition* (in which applications try to infer gender from photographs, video, or audio). On the one hand, these automated systems have been used to survey vast archives of research papers to quantify gender bias in scientific publishing. On the other hand, these same systems have been used to attempt to create bathroom security systems based on binary, heteronormative, scientifically incorrect notions of gender. A 2018 study found that transgender individuals see high risks for harm stemming from the use of these technologies. See Foad Hamidi, Morgan Klaus Scheuerman, and Stacy M. Branham, "Gender Recognition or Gender Reductionism?: The Social Implications of Embedded Gender Recognition Systems," in *Proceedings of the 2018 CHI Conference on Human Factors in Computing Systems* (New York: ACM, 2018), 8. And also Os Keyes, "The Misgendering Machines: Trans/HCI Implications of Automatic Gender Recognition," in *Proceedings of the ACM on Human-Computer Interaction* 2, no. CSCW (2018): article 88.

47. Brittney Cooper and Margaret Rhee, "Introduction: Hacking the Black/White Binary," *Ada: A Journal of Gender, New Media, and Technology* 6 (January 2015), https://adanewmedia.org/2015/01/issue6-cooperrhee/.

48. Claire Cohen, Patrick Scott, and Ellie Kempster, "Born Equal. Treated Unequally." *Telegraph*, March 8, 2018, https://www.telegraph.co.uk/women/business/women-mean-business-interactive/.

49. Lisa Charlotte Rost, "An Alternative to Pink & Blue: Colors for Gender Data," *Datawrapper*, July 10, 2018, https://blog.datawrapper.de/gendercolor/.

50. Sam Morris, Juweek Adolphe, and Erum Salam, "Does the New Congress Reflect You?," *Guardian*, updated June 7, 2019, https://www.theguardian.com/us-news/ng-interactive/2018/nov/15/new-congress-us-house-of-representatives-senate.

51. Visualization expert Andy Kirk has an excellent talk about creative strategies to represent absences, nulls, and zeros—what he calls the *design of nothing*. Andy Kirk, "The Design of Nothing: Null, Zero, Blank," YouTube, May 28, 2014, https://www.youtube.com/watch?v=JqzAuqNPYVM.

52. See note 16 for a discussion of the false binary between sex and gender. For more on Montañez's design process, see her blog post, "Visualizing Sex as a Spectrum," *Scientific American Blog Network*, August 29, 2017, https://blogs.scientificamerican.com/sa-visual/visualizing-sex-as-a-spectrum/.

53. Montañez, "Visualizing Sex as a Spectrum."

54. Anne Fausto-Sterling, *Sexing the Body: Gender Politics and the Construction of Sexuality* (New York: Basic Books, 2000). For a distillation of this argument, see Fausto-Sterling, "Why Sex Is Not Binary," *New York Times*, October 25, 2018, https://www.nytimes.com/2018/10/25/opinion/sex-biology-binary.html.

55. The website of the Intersex Campaign for Equity cites Fausto-Sterling as the source of this statistic, and further notes that this statistic "makes being intersex about as common as having red hair (1%-2%)." Hida Viloria, "How Common is Intersex? An Explanation of the Stats," April 1, 2015, Intersex Campaign for Equality, https://www.intersexequality.com/how-common-is-intersex-in-humans/.

56. Oliver L. Haimson and Anna Lauren Hoffmann, "Constructing and Enforcing 'Authentic' Identity Online: Facebook, Real Names, and Non-normative Identities," *First Monday* 21, no. 6 (June 10, 2016), https://doi.org/10.5210/fm.v21i6.6791.

57. Haimson and Hoffman write that we must imagine "alternative conceptions of safety that may hinge on obscurity or invisibility—strategies that are often employed by those with marginalized or non-normative identities (like trans people) or those managing risks of physical and emotional danger if found (like abuse survivors)." "Constructing and Enforcing 'Authentic' Identity Online."

58. To learn more about the Colored Conventions Project, visit http://coloredconventions.org/. An anthology of essays about the Colored Conventions and the Colored Conventions Project, *Colored Conventions in the Nineteenth Century and the Digital Age*, ed. P. Gabrielle Forman, Jim Casey, and Sarah Patterson, is in preparation.

59. "CCP Corpus and Word Trends," accessed July 31, 2019, http://coloredconventions.org/intro-corpus.

60. As the document explains, "This is our shared commitment to recovering a convention movement that includes women's activism and presence—even though it's largely written out of the minutes themselves." See "Teaching Partner Memo of Understanding," accessed July 31, 2019, http://coloredconventions.org/memo-of-understanding.

61. A *breast pump* is a machine that helps postpartum mothers, nonbinary parents, and trans dads remove breastmilk from their breasts when they are not with their nursing baby or unable to nurse them. Breast pumps can be lifesaving in the case of premature babies and helpful for parents that work outside the home. They suck, literally and figuratively, and many parents tell stories of pumping in closets, server rooms, and bathrooms. In the US context, breast pumps are a mediocre workaround for a larger social problem, which is the lack of a comprehensive paid family leave policy that would support parents to care for their new babies. The co-organizers of the 2014 breast pump hackathon were Tal Achituv, Catherine D'Ignazio, Alexis Hope, Taylor Levy, Alexandra Metral, David Raymond, and Che-Wei Wang.

62. Catherine D'Ignazio, Alexis Hope, Becky Michelson, Robyn Churchill, and Ethan Zuckerman, "A Feminist HCI Approach to Designing Postpartum Technologies: 'When I First Saw a

Breast Pump I Was Wondering if It Was a Joke,'" CHI'16 (May 2016). More information on the hackathons can be found at https://www.makethebreastpumpnotsuck2018.com.

63. It is a product of the hegemonic domain of the matrix of domination that so many stigmatized cultural topics have to do with women's bodies. Along with breastfeeding, there is abortion, miscarriage, birth, menstruation, sex, sexual assault, domestic violence, fertility, and more. The cultural taboos around speaking publicly on these topics leads individuals to thinking of these shared experiences as being purely personal—and blaming themselves for any trauma they experience. You may have heard the phrase, "The personal is political." This comes from an essay of the same name by activist Carol Hanisch in 1970 about the practice of "consciousness-raising" in the 1960s and 1970s. She wrote of these gatherings, "One of the first things we discover in these groups is that personal problems are political problems. There are no personal solutions at this time. There is only collective action for a collective solution." Although wholly underexplored in data science (but explored in other fields), there is an opening here to use story-sharing—qualitative data collection in a community—as a collective healing process and the basis for political action. This was one of the main goals of the breast pump hackathon, as well as the Make Family Leave Not Suck Policy Summit that accompanied it.

64. Our reflection process and reflections themselves are described in Alexis Hope, Catherine D'Ignazio, Josephine Hoy, Rebecca Michelson, Jennifer Roberts, Kate Krontiris, and Ethan Zuckerman, "Hackathons as Participatory Design: Iterating Feminist Utopias," in *Proceedings of the 2019 CHI Conference on Human Factors in Computing Systems* (New York: ACM, 2019), 61.

65. Jenn Roberts, of Versed Education Group, suggested many of the key accountability structures, including drafting a values statement, convening an advisory board, and thinking about participation numbers that truly make a space welcoming versus tokenizing for marginalized groups.

66. An interesting team interaction occurred during this process. When establishing metrics for racial diversity, Catherine (the executive director) proposed, "How about 50 percent people of color?" Jenn Roberts, the equity and inclusion lead, reminded the team that a space equally filled by white people and people of color will still feel like a white-dominated space, so we increased our goal to 70 percent. For more information on the demographics of the event and the process, see "Building Community at our Hackathon," Medium, April 3, 2018, https://medium.com/make-the-breast-pump-not-suck-hackathon/building-community-at-our-hackathon-a08a76bb5ea6.

## 5   Unicorns, Janitors, Ninjas, Wizards, and Rock Stars

1. Vincent Del Giudice and Wei Lu, "America's Rich Get Richer and the Poor Get Replaced by Robots," *Bloomberg.com*, April 26, 2017, https://www.bloomberg.com/news/articles/2017-04-26/america-s-rich-poor-divide-keeps-ballooning-as-robots-take-jobs.

2. "Wages: Median (Old) San Francisco-Oakland-Fremont, CA Metro Area," National Equity Atlas, 2018, accessed March 19, 2019, http://nationalequityatlas.org/indicators/Wages:_Median_(old)/Over_time:7616/San_Francisco-Oakland-Fremont,_CA_Metro_Area/false/. "Snapshot of

Poverty: San Francisco County," United Way, 2017, accessed March 19, 2019, https://uwba.org/wp-content/uploads/2017/10/SanFrancisco-Snapshot.pdf.

3. "About/Acerca De Nosotros," Anti-Eviction Mapping Project, 2016, accessed March 19, 2019, https://www.antievictionmap.com/about.

4. *Counterpoints: Bay Area Data and Stories for Resisting Displacement* will be published by PM Press in 2020.

5. Eviction Defense Collaborative, "City of Change: Fighting for San Francisco's Vanishing Communities," Evictiondefense.org, 2016, accessed March 19, 2019. http://www.antieviction mappingproject.net/EDC_2016.pdf.

6. Eviction Defense Collaborative, "City of Change."

7. Anti-Eviction Mapping Project, *Tech Bus Stops and No-Fault* evictions, accessed August 6, 2019, http://www.antievictionmappingproject.net/techbusevictions.html.

8. See Rebecca Solnit, "Diary," *London Review of Books* 35, no. 3 (2013): 34–35, https://www.lrb.co.uk/v35/n03/rebecca-solnit/diary.

9. Anti-Eviction Mapping Project, *Narratives of Displacement and Resistance*, accessed August 6, 2019, http://www.antievictionmappingproject.net/narratives.html.

10. In her book *Gentrification and Resistance*, Laura Naegler, a professor of cultural criminology, links cleanliness with gentrification: "Dirt—or as antonym, cleanliness—becomes a symbolic means to demarcate the actors legitimately appropriating and using a certain space." Laura Naegler, *Gentrification and Resistance: Cultural Criminology, Control, and the Commodification of Urban Protest in Hamburg*, vol. 50 (Berlin: LIT Verlag, 2012).

11. So-called data dashboards are increasingly a trend in cities attempting to demonstrate how they are keeping with the times. In "Mission Control: A History of the Urban Dashboard," Shannon Mattern shows how these dashboards also enact a particular form of control. *Places Journal* (March 2015), https://placesjournal.org/article/mission-control-a-history-of-the-urban-dashboard/.

12. Katie Lloyd Thomas, "Lines in Practice: Thinking Architectural Representation through Feminist Critiques of Geometry," *Geography Research Forum* 21 (2016): 57–76.

13. Hadley Wickham, "Tidy Data," *Journal of Statistical Software* 59, no. 10 (2014): 1–23.

14. See Thomas H. Davenport and D. J. Patil, "Data Scientist: The Sexiest Job of the 21st Century," *Harvard Business Review*, October 2012, https://hbr.org/2012/10/data-scientist-the-sexiest-job-of-the-21st-century.

15. See Davenport and Patil, "Data Scientist."

16. Steve Lohr, "For Big-Data Scientists, 'Janitor Work' Is Key Hurdle to Insights," *New York Times*, August 17, 2014, https://www.nytimes.com/2014/08/18/technology/for-big-data-scientists-hurdle-to-insights-is-janitor-work.html.

17. *Eugenics* describes the view that humans should control the evolution of their species by encouraging the reproduction of "superior" kinds of people (namely, white ones) and discouraging the reproduction of all others (where "discouraging" took the form of forced sterilization and, in its worst instantiations, murder and death). Karl Pearson held the first Eugenics chair at the University of London, where he developed many of the statistical concepts and methods still in use today. For more on this history, see Spade and Rohlfs, "Legal Equality, Gay Numbers and the (After?)Math of Eugenics."

18. Banu Subramaniam, *Ghost Stories for Darwin: The Science of Variation and the Politics of Diversity* (Urbana: University of Illinois Press, 2014).

19. While we can no longer point to examples like the state-level eugenics program that prompted World War II, scholars point out—rightly—that certain eugenic assumptions have never gone away. There is a long history of the United States employing sterilization programs in prisons, for example. See David M. Perry, "Our Long, Troubling History of Sterilizing the Incarcerated," *The Marshall Project* (blog), July 26, 2017, https://www.themarshallproject.org/2017/07/26/our-long-troubling-history-of-sterilizing-the-incarcerated. For a data-based project that attempts to reckon with this history's roots, see Jacqueline Wernimont and Alexandra Minna Stern, "The Eugenic Rubicon: California's Sterilization Stories" (2017), accessed August 6, 2019, http://scalar.usc.edu/works/eugenic-rubicon-/index.

20. That being said, a point about the inherent racism of certain statistical techniques could be made. For a detailed critique along these lines, see Wendy H. K. Chun, "On Patterns and Proxies, or the Perils of Reconstructing the Unknown," *Accumulation*, September 25, 2018, https://www.e-flux.com/architecture/accumulation/212275/on-patterns-and-proxies/.

21. Katie Rawson and Trevor Muñoz, "Against Cleaning," in *Debates in the Digital Humanities 2019*, ed. Matthew K. Gold and Lauren F. Klein (Minneapolis: University of Minnesota Press, 2019), 290.

22. Yanni A. Loukissas, *All Data Are Local: Thinking Critically in a Data-driven Society* (Cambridge, MA: MIT Press, 2019); see also Anne Luther, "Local Data Design: An Interview with Professor Yanni Loukissas," *Data Matters*, July 13, 2017, https://data-matters.nyc/?p=18579.

23. Reuben S. Rose-Redwood, "Indexing the Great Ledger of the Community: Urban House Numbering, City Directories, and the Production of Spatial Legibility," *Journal of Historical Geography* 34, no. 2 (2008): 286–310.

24. This is a point that Rose-Redwood makes with respect to the nineteenth century as well. It wasn't just the mail carrier who needed these signs. Rather, the street names were for the "stranger, merchant, or businessman"—this from an 1838 Philadelphia business directory—who came from elsewhere to operationalize a new landscape in the service of their own profit. See Rose-Redwood, "Indexing the Great Ledger of the Community," 295.

25. "Front End Engineer," Amazonjobs, accessed March 19, 2019, https://www.amazon.jobs/en/jobs/700028/front-end-engineer.

26. Gayatri Chakravorty Spivak, *Can the Subaltern Speak? Reflections on the History of an Idea* (New York: Columbia University Press, 2010). The idea that the data cannot speak for themselves is something we will return to at length in chapter 6.

27. We exclude the unicorn from this assessment, as it is a mythical creature that does not have a readily apparent gender.

28. "Janitors & Building Cleaners: Race & Ethnicity," Data USA, accessed March 19, 2019, https://datausa.io/profile/soc/37201X/#demographics. See also Andrew Ti, "Team, We Have to Give up on Ninja," Angry Asian Man (blog), September 21, 2016, http://blog.angryasianman.com/2016/09/team-we-have-to-give-up-on-ninja.html.

29. See Daniela Aiello, Lisa Bates, Terra Graziani, Christopher Herring, Manissa Maharawal, Erin McElroy, Pamela Phan, and Gretchen Purser, "Eviction Lab Misses the Mark," *ShelterForce*, August 22, 2018, https://shelterforce.org/2018/08/22/eviction-lab-misses-the-mark/.

30. Matthew Desmond, interview by Catherine D'Ignazio, April 5, 2019.

31. Manissa M. Maharawal and Erin McElroy. "The Anti-Eviction Mapping Project: Counter Mapping and Oral History toward Bay Area Housing Justice," *Annals of the American Association of Geographers* 108, no. 2 (December 1, 2016): 380–389.

32. As information studies professor Johanna Drucker has stated, "Knowledge is partial ... Knowledge is situated ... Knowledge is historical." From her talk at the Unflattening and Enacting Visualization Workshop at the City University of New York, June 9, 2016.

33. Feminist standpoint theory, developed by Sandra Harding and elaborated by later feminist thinkers, asserts that there are group-based experiences, *standpoints*, that go beyond individual, specific experiences. As Maggie Walter and Chris Anderson, the authors of the book *Indigenous Statistics*, write, "It is our position in social space, our capital relationalities, that shapes our life chances, and while we experience relationalities and life chances as individuals, we share this position with those of similar social, economic, cultural, *and* racial capitals" (emphasis theirs). Note how this is different than a call for simple diversity in individual perspectives—what people in the tech industry characterize as *thought diversity* or *cognitive diversity*. This means explicitly acknowledging and taking steps to address the unjust structural forces at play in our work, including racism, sexism, and more. Walter and Andersen, *Indigenous Statistics*.

34. See David Weinberger, "Transparency: The New Objectivity," *KMWorld*, August 28, 2009, http://www.kmworld.com/Articles/Column/David-Weinberger/Transparency-the-new-objectivity-55785.aspx.

35. See Eric Roston and Blacki Miglozzi, "What's Really Warming the World?," *Bloomberg Businessweek*, June 24, 2015, https://www.bloomberg.com/graphics/2015-whats-warming-the-world.

36. At other times, reflexivity can be more difficult to implement. Databases and charts, like the data themselves, are often very good at obscuring the perspectives of their human creators, as we discuss in chapter 3. But this is a technical problem as much as an ideological one and could be addressed through the development of different file formats and metadata standards.

For example, spreadsheets are described as "broken" when they have notes and metadata at the top. But they are only broken because this is a convention that has been encoded into software and implemented in most data-processing scripts. It would be very easy to develop a different format in which authors, participants, and process notes were expected to appear at the top of the file. Or a standard in which each field in a database is annotated with its business purpose and the person who requested it. These additions to metadata norms and standards, like the idea of pluralism itself, would help to further elucidate a project's human actors, research questions, methods, and results, and allow them to be more fully explored by others seeking to build on them. In other words: reflexivity values scientific reproducibility and takes it one step further on the path to knowledge and discovery.

37. Additional examples of visualizing and crediting data workers are discussed in chapter 7.

38. In the interest of self-disclosure: Catherine D'Ignazio is an organizer in the Public Lab community. Public Lab creates open-source, low-cost, environmental-monitoring technologies that include aerial imaging, devices to log water-quality parameters, and a cardboard spectrometer to detect contaminants. The organization has grassroots chapters around the world. As Catherine and her colleagues wrote in a paper about the community, "The goal of Public Lab projects is to create or curate live archives of data, collected and produced in a decentralized manner. On the one hand, these data support scientific investigation, advocacy, and, in some cases, regulatory actions. On the other hand, they foster diverse participation and collaboration of concerned residents and local organizations in the development of techno-scientific tools and methods." Pablo Rey-Mazon, Hagit Keysar, Shannon Dosemagen, Catherine D'Ignazio, and Don Blair, "Public Lab: Community-based Approaches to Urban and Environmental Health and Justice," *Science and Engineering Ethics* 24, no. 3 (May 3, 2018): 971–997.

39. This helpful summary comes in the context of Nash's own critique of the limits of intersectionality and its cooptation by the academy. See Jennifer C. Nash, "Re-Thinking Intersectionality," *Feminist Review* 89, no. 1 (June 2008): 1–15; see also her recent book, *Black Feminism Reimagined: After Intersectionality* (Durham, NC: Duke University Press, 2019).

40. Michaelanne Dye, Neha Kumar, Ari Schlesinger, Marisol Wong-Villacres, Morgan G. Ames, Rajesh Veeraraghavan, Jacki Oneill, Joyojeet Pal, and Mary L. Gray, "Solidarity across Borders," in *Companion of the 2018 ACM Conference on Computer Supported Cooperative Work and Social Computing—CSCW 18* (New York: ACM, 2018), 487–494.

41. "Design Justice Network Principles," Design Justice, accessed March 19, 2019, http://designjusticenetwork.org/network-principles/. Professor of informatics Shaowen Bardzell makes a similar call for design that starts first and foremost with the perspective of the "marginal user." See Shaowen Bardzell, "Feminist HCI," in *Proceedings of the 28th International Conference on Human Factors in Computing Systems—CHI 10* (New York: ACM, 2010), 1301–1310.

42. "A Step Change: DataKind Raises $20M Investment to Support the Data Science for Social Good Ecosystem," DataKind, January 22, 2019, https://www.datakind.org/blog/a-step-change-datakind-raises-20m-investment-to-support-the-data-science-for-social-good-ecosystem.

43. "Data Science for Social Good Summer Fellowship," Data Science for Social Good, 2018, accessed March 19, 2019, https://dssg.uchicago.edu/.

44. Sara Hooker, "Why 'Data for Good' Lacks Precision," *Towards Data Science* (blog), July 22, 2018, https://towardsdatascience.com/why-data-for-good-lacks-precision-87fb48e341f1.

45. Hooker, "Why 'Data for Good' Lacks Precision."

46. This shift is particularly necessary in work involving partnerships across knowledge systems, in which the data—as a form of knowledge—are fundamentally connected to issues of power. For example, in many Indigenous communities, data relating to the community constitutes sacred knowledge and cannot (and should not) be easily shared. Platforms like Murkutu, designed by Kim Cristen and Craig Dietrich, which enable fine-grained privacy controls for sharing digital cultural heritage materials, and classification systems like Ngā Upoko Tukutuku/Māori Subject Headings, developed by LIANZA, Te Rōpū Whakahau, and the National Library of New Zealand, represent thoughtful attempts to bridge issues of data access with cultural concerns. See http://mukurtu.org/about/ and https://natlib.govt.nz/nga-upoko-tukutuku.

47. Emily and Rahul Bhargava, interview by Catherine D'Ignazio, February 7, 2018.

48. Interview with Emily and Rahul Bhargava, 2018. For an insider account, see Rahul Bhargava's description, "Mural-ing Our Way to Data Literacy," MIT Center for Civic Media, August 6, 2013, https://civic.mit.edu/2013/08/06/mural-ing-our-way-to-data-literacy/.

49. Data murals like the one in Somerville are becoming a more common practice to tell a data-driven public story about an important issue. Detroit Future Schools undertook a series of data murals in 2015. See "Detroit Future Schools Data Murals Project: What Stories Can We Tell from Data?," August 7, 2015, https://www.alliedmedia.org/news/2015/08/07/detroit-future -schools-data-murals-project-what-stories-can-we-tell-data. In Dar Es Salaam, the Data Zetu project ("our data" in Swahili) ran a listening campaign in four low-income districts. They compiled the residents' concerns, as well as statistical data, into a data mural about teenage pregnancy and sexual health (https://datazetu.or.tz/). And there are many examples of participatory mapping that combine data collection and analysis with explicit acknowledgement of oppression with explicit goals around building community capacity and solidarity. For example, The Morris Justice Project is a collaboration between neighborhood residents in the South Bronx and academic institutions in New York City. Founded by a group of mothers in 2011, the project has been documenting policing behaviors in its members' neighborhood around Yankee Stadium, with a focus on New York's stop and frisk policy (http://morrisjustice.org/).

50. Digital Democracy home page, accessed August 6, 2019, https://www.digital-democracy.org/.

51. Westside Atlanta Land Trust Program home page, accessed August 6, 2019, https://www .helporginc.org/walt-2015.html.

52. In the process, they partnered with scholars Ellen Zegura, Amanda Meng, and Carl DiSalvo at the Georgia Institute of Technology, who wrote about the collaborative undertaking as a case of data activism and as care work. See Amanda Meng and Carl Disalvo, "Grassroots Resource

Mobilization through Counter-data Action," *Big Data & Society* 5, no. 2 (September 14, 2018); and Ellen Zegura, Carl DiSalvo, and Amanda Meng, "Care and the Practice of Data Science for Social Good," in *Proceedings of the 1st ACM SIGCAS Conference on Computing and Sustainable Societies* (New York: ACM, 2018), 34.

53. Erhardt Graeff, "Digital Democracy: Participatory Mapping & Tool-building in the Amazon," MIT Center for Civic Media, September 14, 2017, https://civic.mit.edu/2017/09/14/digital-democracy-participatory-mapping-tool-building-in-the-amazon/.

54. Meng and DiSalvo make a strong case that these more *affective* outcomes of community data projects are understudied. Empowerment does not only come through achieving a specific policy win with data, but also through strengthening the social infrastructure of the community. They draw on Black feminist scholar Audré Lorde, who advocates for examination and self-actualization to precede social transformation: a community cannot articulate a more just world without understanding the oppressive conditions of the current one. Meng and Disalvo, "Grassroots Resource Mobilization through Counter-data Action."

55. The "data as a campfire" analogy is from Denise Ross, cofounder of the White House Police Data Initiative under Barack Obama. In her book, *Data | Action* (Cambridge, MA: MIT Press, 2020), urban science expert Sarah Williams makes a compelling case that collecting data can strengthen communities around their shared interests as well as build affective ties that are crucial to achieving civic goals.

56. Note the underlying paternalism of much of the "data for good" work: Who is in a position to do good for whom? Too often, data service work ends up being framed as charity work from benevolent experts for poor victims who can't help themselves. This framing reinscribes existing hierarchies, denies agency to people and communities, and does harm. For more on deficit narratives, see chapter 2.

57. María Isabel Casas-Cortés, "Social Movements as Sites of Knowledge Production: Precarious Work, the Fate of Care and Activist Research in a Globalizing Spain" (PhD diss., University of North Carolina at Chapel Hill, 2009).

58. Here we draw from discussions of feminist empiricism as theorized by Maitree Wickramasinghe and Donna Haraway, among others. As researcher Ambika Tandon helpfully summarizes, "Feminist empiricism is cognizant of the multiplicity of local knowledges, as well as their 'unequal translation and exchange' with dominant knowledge. It then looks at the translatability of knowledge between different social groups and communities in the matrix of oppression and privilege, with the aim of conducting and disseminating research in ways that speaks to a range of such groups." See Tandon, *Feminist Methodology in Technology Research* (Bangalore: Center for Internet and Society, 2018). In other words, while feminist empiricism acknowledges the value and dignity of local knowledge, it also asserts that there are strong reasons, grounded in social justice, to understand macrophenomena at the regional, country, or global level and beyond. On this point, postcolonial feminist theorist Chandra Mohanty explains, "These arguments are not against generalization as much as they are for careful, historically specific complex

generalizations." Chandra Talpade Mohanty, *Feminism without Borders* (Durham, NC: Duke University Press, 2003).

59. See "Mapping Environmental Justice," Environmental Justice Atlas, accessed March 19, 2019, https://ejatlas.org/. See also the scholarly papers that the team has written about its work: Leah Temper, D. Del Bene, and J. Martinez-Alier, "Mapping the Frontiers and Front Lines of Global Environmental Justice: The EJAtlas," *Journal of Political Ecology* 22, no. 1 (2015): 255–278; and Leah Temper, F. Demaria, A. Scheidel, D. Del Bene, and J. Martinez-Alier, "The Global Environmental Justice Atlas (EJAtlas): Ecological Distribution Conflicts as Forces for Sustainability," *Sustainability Science* 13, no. 3 (2018): 573–584. They discuss the challenges of such a large-scale mapping process in Leah Temper and D. Del Bene, "Transforming Knowledge Creation for Environmental and Epistemic Justice," *Current Opinion in Environmental Sustainability* 20 (2016): 41–49.

60. Grettel Navas and Daniela Del Bene, "Proyecto Hidroeléctrico Agua Zarca, Honduras," Environmental Justice Atlas, March 3, 2018, https://ejatlas.org/conflict/proyecto-hidroelectrico -agua-zarca-honduras.

61. Nina Lakhani, "Berta Cáceres: Seven Men Convicted of Murdering Honduran Environmentalist," *Guardian*, November 30, 2018, https://www.theguardian.com/world/2018/nov/29/ berta-caceres-seven-men-convicted-conspiracy-murder-honduras.

62. Partners include the Latin American Observatory of Mining Conflicts (OCMAL), the World Rainforest Movement (WRM), and the Brazilian Network of Environmental Justice (RBJA), among many others. Many of these organizations carry out their work using methods from participatory action research (PAR), which explicitly includes frontline communities in the process of knowledge production. PAR is an established methodological approach to academic research that values community-based knowledge, commits to producing knowledge in the interest of social change, and centers the leadership of affected communities. To learn more about the intersection of PAR and feminist approaches, see M. Brinton Lykes and Erzulie Coquillon, "Participatory Action Research and Feminisms," in *Handbook of Feminist Research: Theory and Praxis*, ed. Sharlene Nagy Hesse-Biber (Thousand Oaks, CA: SAGE, 2007), 297–326.

63. "Protect Kok-Zhailau," Ile-Alatau State National Nature Park, Kazakhstan, Environmental Justice Atlas, January 29, 2016, https://ejatlas.org/conflict/protect-kok-zhailau-ile-alatau-state -national-nature-park-kazakhstan.

64. Begüm Özkaynak, Beatriz Rodríguez-Labajos, Cem İskender Aydın, Ivonne Yanez, and Claudio Garibay, *Towards Environmental Justice Success in Mining Resistances: An Empirical Investigation*, Environmental Justice Organisations, Liabilities and Trade (EJOLT) Report No. 14, April 2015, http://www.ejolt.org/2015/04/towards-environmental-justice-success-mining-conflicts/.

65. Özkaynak et al., "Towards Environmental Justice Success in Mining Resistances."

## 6   The Numbers Don't Speak for Themselves

1. See Mona Chalabi, "Kidnapping of Girls in Nigeria Is Part of a Worsening Problem," *FiveThirtyEight*, May 8, 2014, https://fivethirtyeight.com/features/nigeria-kidnapping/.

2. You can see the whole thread on the archived version of Storify at https://web.archive.org/web/20140528062637/https://storify.com/AthertonKD/if-a-data-point-has-no-context-does-it-have-any-me, as well as on Simpson's account directly: Erin Simpson (@charlie_simpson), "So if #GDELT says there were 649 kidnappings in Nigeria in 4 months, WHAT IT'S REALLY SAYING is there were 649 news stories abt kidnappings," Twitter, May 13, 2014, 4:04 p.m., https://twitter.com/charlie_simpson/status/466308105416884225.

3. Kalev Leetaru, "The GDELT Project," GDELT, accessed May 12, 2018, https://www.gdeltproject.org/.

4. Our association of Big Dick Data with cis-masculinism is intended to call attention to how cis-heteropatriarchy currently dominates data studies. It is not intended to erase the experiences of gender non-conforming people. On the range of expressions of masculinity and the dicks that accompany them, see Amanda Phillips, "Dicks Dicks Dicks: Hardness and Flaccidity in (Virtual Masculinity)," *Flow: A Critical Forum on Media and Culture*, March 23, 2017. As an example of a critique of big data that does not rely upon the dick as signifier, see Jen Jack Gieseking, "Size Matters to Lesbians, Too: Queer Feminist Interventions into the Scale of Big Data," *Professional Geographer* 70, no. 1 (2018): 150–156.

5. APIs allow a little program one writes to talk to other computers over the internet that are ready to receive data queries. Twitter, Zillow, and MOMA are some examples of large entities that have APIs available to programatically download data.

6. Here are some of our favorites: Dogs of Zurich (https://www.europeandataportal.eu/data/en/dataset/https-data-stadt-zuerich-ch-dataset-pd_stapo_hundenamen); UFO sightings (https://www.kaggle.com/NUFORC/ufo-sightings); all of the cartoon-based murals of Brussels (https://opendata.brussels.be/explore/dataset/comic-book-route/images/); Things Lost on the New York City subway system (http://advisory.mtanyct.info/LPUWebServices/CurrentLostProperty.aspx); and a list of abandoned shopping carts in Bristol (https://data.gov.uk/dataset/abandoned-shopping-trolleys-bristol-rivers). Some of the best of the newsletters include *Data Is Plural*, curated by Jeremy Singer-Vine, who is the data editor for Buzzfeed; and *Numlock News*, a daily email newsletter by Walt Hickey, which tries to provide some context around the numbers we see in the news.

7. "Scottish Witchcraft," Data.world, May 18, 2017, https://data.world/history/scottish-witchcraft.

8. Trevor Muñoz and Katie Rawson, "Data Dictionary," Curating Menus, 2016, accessed April 23, 2019, http://curatingmenus.org/data_dictionary/.

9. Haraway uses the phrase "unlocatable, and so irresponsible, knowledge claims." Donna Haraway, "Situated Knowledges: The Science Question in Feminism and the Privilege of Partial Perspective," *Feminist Studies* 14, no. 3 (Autumn 1988): 575–599, https://doi.org/10.2307/3178066.

10. For example, philosopher Lorraine Code argues that connecting knowledge to its specific biographic, historical, and geographic locations leads to "more responsible knowings." Code, *Ecological Thinking: The Politics of Epistemic Location* (New York: Oxford University Press, 2006).

11. Christine L. Borgman, *Big Data, Little Data, No Data: Scholarship in the Networked World* (Cambridge, MA: MIT Press, 2015).

12. "Open Knowledge International." Open Knowledge International. Accessed March 27, 2019. https://okfn.org/. The *Guardian* newspaper, out of the United Kingdom, launched the Free Our Data campaign in 2006 to petition government agencies to make public data available to taxpayers and companies for free. Among other things, they focused on geographic data collected by the Royal Ordnance Survey which had restrictive licenses on reuse by citizens. The campaign was largely successful: in 2010, the United Kingdom created the Open Government License and launched data.gov.uk, one of the first national data portals in the world. See Charles Arthur and Michael Cross, "Give Us Back Our Crown Jewels," *Guardian*, March 9, 2006, https://www.theguardian.com/technology/2006/mar/09/education.epublic.

13. See Peter R. Orszag, "Memorandum for the Heads of Executive Departments and Agencies Re: Open Government Directive," Washington, DC, Executive Office of the President, December 8, 2009, https://obamawhitehouse.archives.gov/sites/default/files/omb/assets/memoranda_2010/m10-06.pdf.

14. Although the movement under Obama was toward openness (Orszag, "Memorandum for the Heads of Executive Departments and Agencies Re: Open Government Directive"), the current administration has retreated from this position, according to a Sunlight Foundation audit, which found that "the Open Government Initiative, Open Government Partnership, and related programs, initiatives and partnerships across the federal government are being ignored, neglected or even forgotten in federal agencies." Briana Williams, "Under Trump, U.S. Government Moves from /Open to /Closed," Sunlight Foundation, January 24, 2018, https://sunlightfoundation.com/2018/01/24/under-trump-u-s-government-moves-from-open-to-closed/.

15. "The International Open Data Charter," Open Data Charter, accessed March 27, 2019, https://opendatacharter.net/principles/.

16. Tim Davies, "Exploring Participatory Public Data Infrastructure in Plymouth," *Public Sector Blogs*, September 11, 2017, https://www.publicsectorblogs.org.uk/2017/09/exploring-participatory-public-data-infrastructure-in-plymouth-tim-davies/.

17. *Zombie data* was named by Daniel Kaufmann, an economist with the Revenue Watch Institute. Joel Gurin, "Open Governments, Open Data: A New Lever for Transparency, Citizen Engagement, and Economic Growth," *SAIS Review of International Affairs* 34, no. 1 (Winter 2014): 71–82. While the name is certainly evocative, it's also important to acknowledge the history of zombies, which can be traced to seventeenth-century Haiti as a response to the incursion of slavery. As Mike Mariani helpfully summarizes, enslaved Haitians "believed that dying would release them back to *lan guinée*, literally Guinea, or Africa in general, a kind of afterlife where they could be free." But "those who took their own lives wouldn't be allowed to return to *lan guinée*. Instead, they'd be condemned to skulk the Hispaniola plantations for eternity, an undead slave at once denied their own bodies and yet trapped inside them—a soulless zombie." See Mariani, "The Tragic, Forgotten History of Zombies," *Atlantic*, October 28, 2015, https://www.theatlantic.com/entertainment/archive/2015/10/how-america-erased-the-tragic-history-of-the-zombie/412264/.

18. See Chris Anderson, "The End of Theory: The Data Deluge Makes the Scientific Method Obsolete," *Wired*, June 23, 2008, https://www.wired.com/2008/06/pb-theory/.

19. Fulvio Mazzocchi makes the connection between Bacon and big data in "Could Big Data Be the End of Theory in Science?," EMBO reports 16, no. 10 (2015): 1250–1255. While Bacon's *Novum Organum* (1620) was indeed a masterful work that influenced centuries of scientists, he was not alone in his promulgation of a (proto) scientific method. Margaret Cavendish (1623–1717), for example, was an author of both natural philosophy (as scientific theory was known at the time) and science fiction. In fact, her scientific treatise, *Observations upon Experimental Philosophy*, was published alongside her science fiction text, *The Blazing World* (1666), and together they worked to challenge the domination of science by men—a reality even in the seventeenth century.

20. Historian Matthew Jones has written an intellectual history of this line of thinking and demonstrates how it has led to a computational "culture of predictive utility" in which prediction is prized above other possible measures of success. See Jones, "How We Became Instrumentalists (Again)," *Historical Studies in the Natural Sciences* 48, no. 5 (November 5, 2018): 673–684.

21. Safiya Umoja Noble, *Algorithms of Oppression: How Search Engines Reinforce Racism* (New York: NYU Press, 2018), 80–81.

22. See https://clerycenter.org/policy-resources/the-clery-act/. The data include separate and specific numbers on sexual assault, dating violence, domestic violence, and stalking. It includes sexual assault incidents experienced by women, men, and nonbinary people.

23. The term *rape culture* was coined by second-wave feminists in the 1970s to denote a society in which male sexual violence is normalized and pervasive, victims are blamed, and the media exacerbates the problem. Rape culture includes jokes, music, advertising, laws, words, and images that normalize sexual violence. In 2017, following the election of a US president who joked about sexual assault on the campaign trail and the exposé of Harvey Weinstein's predatory behavior in Hollywood, high-profile women began speaking out against rape culture with the #MeToo hashtag. #MeToo, a movement started over a decade ago by activist Tarana Burke, encourages survivors to break their silence and build solidarity to end sexual violence.

24. In 2012, two members of BU's hockey team were charged with sexual assault, and a report by the university found that the team had created a "culture of sexual entitlement." See Mary Carmichael, "Graphic Details Emerge from BU Hockey Panel Reports," *Boston Globe*, September 6, 2012, https://www.boston.com/news/local-news/2012/09/06/graphic-details-emerge-from-bu-hockey-panel-reports.

25. The students' full story is excellent. You can read it here: Patrick Torphy, Michaela Halnon, and Jillian Meehan, "Reporting Sexual Assault: What the Clery Act Doesn't Tell Us," *Atavist*, April 26, 2016, https://cleryactfallsshort.atavist.com/reporting-sexual-assault-what-the-clery-act-doesnt-tell-us.

26. Sixteen staff members at the US Department of Education are devoted to monitoring the more than seven thousand higher-education institutions in the country, so it is unlikely that underreporting by an institution would be discovered, except in very high-profile cases. See Michael Stratford, "Clery Fines: Proposed vs. Actual," *Inside HigherEd*, July 17, 2014, https://www

.insidehighered.com/news/2014/07/17/colleges-often-win-reduction-fines-federal-campus-safety -violations. For example, the Sandusky Case at Penn State involved systematic sexual abuse of young boys by a football coach, and the university was subsequently fined $2.4 million for failing to properly report and disclose these crimes.

27. In this context, one might consider the decision of Christine Blasey Ford to testify about her assault by (now) US Supreme Court Justice Brett Kavanaugh. Coming forward involved relinquishing her privacy and reliving her trauma multiple times over, on a national stage.

28. Abigail Golden, "Is Columbia University Mishandling LGBT Rape Cases?," *Daily Beast*, April 30, 2014, https://www.thedailybeast.com/is-columbia-university-mishandling-lgbt-rape-cases?ref =scroll.

29. Sara Ahmed has written powerfully on the violent effects of this silencing of assault victims. "Silence enables the reproduction of the culture of harassment and abuse. When we don't speak about violence we reproduce violence. Silence about violence is violence," she explains. Ahmed, "Speaking Out," *Feministkilljoys* (blog), June 2, 2016, https://feministkilljoys.com/2016/06/02/ speaking-out/.

30. Lisa Gitelman and Virginia Jackson, "Introduction," in *"Raw Data" Is an Oxymoron* (Cambridge, MA: MIT Press, 2013), 2. Here they are following a statement from information studies scholar Geoffrey Bowker, "Raw data is both an oxymoron and a bad idea; to the contrary, data should be cooked with care," as quoted in *Memory Practices in the Sciences* (Cambridge, MA: MIT Press, 2005). The dichotomy between "raw" and "cooked," in turn, owes its source to the renowned structural anthropologist Claude Levi-Strauss. His famous book, *The Raw and the Cooked* (1964), analogizes the process of transforming nature into culture as akin to the process of transforming raw food into cooked. Your false binary and hidden hierarchy alarm bells should already be going off; and indeed, much of the work of the feminist theory of the early 1970s was to challenge this false dichotomy, as well as the assumptions (and examples) that it rested upon. See Lévi-Strauss, trans. John Weightman and Doreen Weightman, *The Raw and the Cooked: Introduction to a Science of Mythology*, vol. 1 (New York: Harper & Row, 1969).

31. "Google Flu Trends," accessed August 6, 2019, https://www.google.org/flutrends/about/.

32. Sally Kestin and John Maines, "Cops among Florida's Worst Speeders, Sun Sentinel Investigation Finds," *Sun Sentinel*, February 11, 2012.

33. A brilliant idea—to try to link searches for flu symptoms to actual cases of the flu to see if one could predict where the next outbreak would be—Google Flu Trends seemed to work for the first couple years. Then, in the 2012–2013 flu season, Google estimated more than double the flu cases that the CDC did. This discrepancy was possibly due to media panic about swine flu, to Google updating its technology to include recommendations, or perhaps to something else. These are the dangers of prioritizing prediction and utility over causation and context: it all works temporarily, until something in the environment changes. See David Lazer, Ryan Kennedy, Gary King, and Alessandro Vespignani, "The Parable of Google Flu: Traps in Big Data Analysis," *Science* 343, no. 6176 (2014): 1203–1205.

34. In the paper "Tampering with Twitter's Sample API," Jürgen Pfeffer, Katja Mayer, and Fred Morstatter demonstrate how the opacity of sampling done by platforms makes the data vulnerable to manipulation. Pfeffer, Mayer, and Morstatter, "Tampering with Twitter's Sample API," *EPJ Data Science* 7, no. 1 (December 19, 2018).

35. Gaffney and Matias found that the supposedly complete corpus is missing at least thirty-six million comments and twenty-eight million submissions. At least fifteen peer-reviewed studies have used the dataset for research studies on topics like politics, online behavior, breaking news, and hate speech. Depending on what the researchers used the corpus for, the missing data may have affected the validity of their results. Devin Gaffney, and J. Nathan Matias, "Caveat Emptor, Computational Social Science: Large-Scale Missing Data in a Widely-Published Reddit Corpus," *PLOS ONE* 13, no. 7 (July 6, 2018).

36. Lauren F. Klein, "The Image of Absence: Archival Silence, Data Visualization, and James Hemings," *American Literature* 85, no. 4 (Winter 2013): 661–688.

37. The title of a book by Dave Dewitt, *The Founding Foodies: How Washington, Jefferson, and Franklin Revolutionized American Cuisine* (Naperville, IL: Sourcebooks, 2010).

38. The subject of Adrian Miller's *The President's Kitchen Cabinet: The Story of the African Americans Who Have Fed Our First Families, from the Washingtons to the Obamas* (Chapel Hill: University of North Carolina Press, 2017) and of Lauren's more academic book on the subject, *An Archive of Taste: Race and Eating in the Early United States* (Minneapolis: University of Minnesota Press, 2020).

39. See Nikhil Garg, Londa Schiebinger, Dan Jurafsky, and James Zou, "Word Embeddings Quantify 100 Years of Gender and Ethnic Stereotypes," *Proceedings of the National Academy of Sciences* 115, no. 16 (2018): E3635–E3644.

40. In 1970, Daniel Halloran and colleagues wrote, "Events will be selected for news reporting in terms of their fit or consonance with pre-existing images—the news of the event will confirm earlier ideas." James Dermot Halloran, Philip Ross Courtney Elliott, and Graham Murdock, *Demonstrations and Communication: A Case Study* (London: Penguin Books, 1970).

41. Desmond Patton, interview by Catherine D'Ignazio, August 30, 2018.

42. Patton, interview by D'Ignazio.

43. This method is described in detail in their paper: William R. Frey, Desmond U. Patton, Michael B. Gaskell, and Kyle A. McGregor, "Artificial Intelligence and Inclusion: Formerly Gang-Involved Youth as Domain Experts for Analyzing Unstructured Twitter Data," *Social Science Computer Review* (2018).

44. Patton, interview by D'Ignazio.

45. Collins, *Black Feminist Thought*.

46. See "Tressie McMillan Cottom—Upending Stereotypes of Black Womanhood with 'Thick,'" *The Daily Show with Trevor Noah*, video, 7:20, January 21, 2019, https://www.youtube.com/watch?v=EYNu6yvv8HU.

47. Context is crucial for understanding social media conversations. This becomes a particularly fraught problem once we start automating meaning-making with techniques like sentiment analysis and quantitative text analysis. Language and image meanings shift and change quickly, often based on local knowledge, culture, and circumstances. Mariana Giorgetti Valente, director of the Brazilian nonprofit InternetLab, gives the example of a 2010 attack on a gay man in São Paulo in which he was hit on the head with a neon lamp. The image of a lamp then became used in hate speech online. When somebody would subsequently speak out in support of gay rights on Brazilian social media, trolls would post a lamp to communicate a threat of violence. But how would a machine-learning classifier understand that an image of a lamp is a threat without knowing this local context? Valente and InternetLab are collaborating with IT for Change in India to see how they can incorporate context into the detection of hate-speech and anti-hate-speech practices online. Mariana Valente, interview by Catherine D'Ignazio, March 11, 2019.

48. One of the principles of this code is that "social workers recognize the central importance of human relationships." As new codes of ethics are developed for emerging work in machine learning and artificial intelligence, it may be useful to look toward those fields, like social work, that have long-standing histories and specific language for navigating social inequality. In a blog post, Catherine adapted the National Association of Social Workers Code of Ethics and replaced *social worker* with *data scientist* as a way of speculating about whether design and technical fields might ever be able to deal so explicitly with concepts of justice and oppression. Catherine D'Ignazio, "How Might Ethical Data Principles Borrow from Social Work?," Medium, September 2, 2018, https://medium .com/@kanarinka/how-might-ethical-data-principles-borrow-from-social-work-3162f08f0353.

49. Patton, interview by D'Ignazio.

50. Fatos Kaba, Angela Solimo, Jasmine Graves, Sarah Glowa-Kollisch, Allison Vise, Ross Macdonald, Anthony Waters, et al., "Disparities in Mental Health Referral and Diagnosis in the New York City Jail Mental Health Service," *American Journal of Public Health* 105, no. 9 (September 2015): 1911–1916.

51. Prison reform advocate and formerly incarcerated person Eddie Ellis states that terms like *prisoner*, *inmate*, *convict*, and *felon* "are no longer acceptable for us and we are asking people to stop using them." See Eddie Ellis, "Language," Prison Studies Project, accessed July 29, 2019, http://prisonstudiesproject.org/language/.

52. Pete Vernon, "Dancing around the Word 'Racist' in Coverage of Trump," *Columbia Journalism Review*, September 25, 2017, https://www.cjr.org/covering_trump/trump-racism.php.

53. As a nonhypothetical example of this, see the recent interactive feature from the *New York Times*, "Extensive Data Shows Punishing Reach of Racism for Black Boys," which models much of this advice in both naming racism and reflecting the findings of the study that served as the basis for the report. See https://www.nytimes.com/interactive/2018/03/19/upshot/race-class-white-and -black-men.html.

54. Kimberly Seals Allers, "What Privileged Kids—and Parents—Can Learn from Low-Income Youth," *Washington Post*, March 2, 2018.

55. How might we focus less attention on minoritized groups' disadvantages and more attention on dominant groups' unearned privileges? For example, instead of focusing on the women that are "missing" from data science and AI, perhaps we should be focusing on the overabundance of men in data science and AI who don't see it as a problem worth their time and energy (because the system works for them).

56. See Gisele S. Craveiro and Andrés M. R. Martano, "Caring for My Neighborhood: A Platform for Public Oversight," in *Agent Technology for Intelligent Mobile Services and Smart Societies* (Berlin: Springer, 2014), 117–126.

57. See Heather Krause, "Data Biographies: Getting to Know Your Data," Global Investigative Journalism Network, March 27, 2017, https://gijn.org/2017/03/27/data-biographies-getting-to -know-your-data/.

58. Timnit Gebru, Jamie Morgenstern, Briana Vecchione, Jennifer Wortman Vaughan, Hanna Wallach, Hal Daumeé III, and Kate Crawford, "Datasheets for Datasets," ArXiv.org, July 9, 2018.

59. Likewise, James Zou and Londa Schiebinger advocate for standardized metadata to accompany AI training datasets that spells out demographics, geographic limitations, and relevant definitions and collection practices. Zou and Schiebinger, "AI Can Be Sexist and Racist—It's Time to Make It Fair," *Nature*, July 18, 2018, https://www.nature.com/articles/d41586-018-05707-8.

60. "Data User Guides," accessed August 6, 2019, http://www.wprdc.org/data-user-guides/.

61. Emerson Engagement Lab, "Civic Data Ambassadors: Module 2 Video 3—Civic Data Guides," video, 6:25, March 18, 2018, https://vimeo.com/260650894.

62. "School Segregation Data," ProPublica, December 2017, https://www.propublica.org/ datastore/dataset/school-segregation-charter-district-data.

63. WomanStats.org is solely focused on the status of women and does not collect any indicators on nonbinary people.

64. Valerie Hudson, interview with Catherine D'Ignazio, January 31, 2019.

65. "Codebook," WomanStats, accessed March 27, 2019, http://www.womanstats.org/new/codebook/.

66. "Codebook," http://www.womanstats.org/new/codebook/.

67. Moreover, if one of the goals is transparency and accountability, the institutions in power often have strong incentives to *not* provide context, so the data setting is rife with conflicts of interest. Indeed, Gebru and colleagues foresee challenges to publishers specifying ethical considerations on their datasheets because they may perceive it as exposing themselves to legal and public relations risks. See Gebru et al., "Datasheets for Datasets."

68. Ricardo Ramírez, Balaji Parthasarathy, and Andrew Gordon, "From Infomediaries to Infomediation at Public Access Venues: Lessons from a 3-Country Study," in *Proceedings of the Sixth International Conference on Information and Communication Technologies and Development: Full Papers*, vol. 1 (New York: ACM, 2013), 124–132.

69. Shannon Mattern, "Public In/Formation," *Places Journal*, November 2016, https://placesjournal.org/article/public-information/.

70. "ProPublica Data Store," ProPublica, accessed August 6, 2019, https://www.propublica.org/datastore/.

71. See https://measuresforjustice.org/.

72. Aaron Brenner et al., "Civic Switchboard," accessed August 6, 2019, https://civic-switchboard.github.io/.

## 7  Show Your Work

1. The authors do not appear to have considered nonbinary genders, although they do maintain a category for "gender-neutral" usernames that cannot be categorized as either men's or women's names. It is also worth noting that this study led to Mozilla, an open-source software company, undertaking an experiment in gender-blind code reviews, in which the gender of a project contributor remains hidden. The results are yet to be published. On the initial study, see J. Terrell, A. Kofink, J. Middleton, C. Rainear, E. Murphy-Hill, C. Parnin, and J. Stallings, "Gender Differences and Bias in Open Source: Pull Request Acceptance of Women versus Men," *PeerJ Computer Science* 3 (2017): e111. On the Mozilla experiment, see, Judy McConnell, "Mozilla Experiment Aims to Reduce Bias in Code Reviews," *Mozilla Blog*, March 8, 2018, https://blog.mozilla.org/blog/2018/03/08/gender-bias-code-reviews/.

2. Claire Cain Miller, "GitHub Founder Resigns after Investigation," *New York Times*, April 21, 2014, https://bits.blogs.nytimes.com/2014/04/21/github-founder-resigns-after-investigation/.

3. Ethan Baron, "GitHub Paid Executive Less because She's Asian and Female, Fired Her for Complaining: Lawsuit," *Mercury News*, October 1, 2018, https://www.mercurynews.com/2018/10/01/github-paid-executive-less-because-shes-asian-and-female-fired-her-for-complaining-lawsuit/.

4. Coraline Ada Ehmke, "Antisocial Coding: My Year at GitHub," *Coraline Ada Ehmke* (blog), July 5, 2017, https://where.coraline.codes/blog/my-year-at-github/.

5. To explore the interactive version of the *Ship Map*, visit https://www.shipmap.org/.

6. For a summary of the cognitive demands required by this type of work, see, for instance, John Levi Martin, "Life's a Beach but You're an Ant, and Other Unwelcome News for the Sociology of Culture," *Poetics* 38, no. 2 (April 2010): 229–244.

7. Miriam Posner, "See No Evil," *Logic Magazine*, April 2018, https://logicmag.io/04-see-no-evil/.

8. Silvia Federici's *Wages against Housework* (Bristol, UK: Power of Women Collective and Falling Water Press, 1975) is the most in depth articulation of the goals of the campaign. The ephemera collection of the Barnard Center for Research on Women contains many of the flyers associated with the campaign and has made them available online at http://bcrw.barnard.edu/archive/workforce.htm. Louise Toupin's *Wages for Housework: A History of an International Feminist*

*Movement, 1972–1977* (Vancover: University of British Columbia Press, 2018) provides a history of the movement.

9. On the positions and impact of the Wages for Housework movement and the International Feminist Collective, see *Wages for Housework: The New York Committee, 1972–1977*, ed. Silvia Federici and Arlen Austin (New York: Autonomedia, 2017).

10. Angela Davis, "The Approaching Obsolescence of Housework: A Working-Class Perspective," in *Women, Race, & Class* (New York: Penguin Random House, 1983), 222–244. In other work, Davis traces this phenomenon back to the institution of slavery. See "Reflections on the Black Woman's Role in the Community of Slaves," *Massachusetts Review* 13, no. 1–2 (Winter/Spring 1972): 81–100. In the years since that foundational essay, scholars such as Stephanie Camp, Jennifer Morgan, and Thavolia Glymph have contributed additional historical and theoretical lenses on the racialization of domestic labor. See Camp, *Closer to Freedom: Enslaved Women and Everyday Resistance in the Plantation South* (Chapel Hill: University of North Carolina Press, 2004); Morgan, *Laboring Women: Reproduction and Gender in New World Slavery* (Philadelphia: University of Pennsylvania Press, 2004); and Glymph, *Out of the House of Bondage: The Transformation of the Plantation Household* (Cambridge: Cambridge University Press, 2008). Mignon Duffy, in *Making Care Count: A Century of Gender, Race, and Paid Care Work* (New Brunswick: Rutgers University Press, 2011), takes this history back up to the present.

11. Other organizations worked in concert with Wages for Housework in the late 1970s and early 1980s to tackle race and other intersectional differences. For example, Black Women for Wages for Housework had a mission to "see to it that challenging racism and challenging sexism, which reinforce each other, are prioritized equally." The group's members called for not only remunerating housework but all of their unwaged work over time, including reparations for slavery, imperialism, and neocolonialism. Wages Due Lesbians lobbied for additional funds to compensate the social discrimination faced by lesbians. WinVisibility fought for economic independence and recognition for disabled women. And the English Collective of Prostitutes sought to decriminalize sex work and fight poverty so that no woman would be forced into unwanted sex. See the listing of "Autonomous Organizations" included in the flyer for "The International Wages for Housework Campaign," Freedom Archives, accessed May 13, 2019, http://freedomarchives.org/Documents/Finder/DOC500_scans/500.020.Wages.for.Housework.pdf.

12. One of the first essays defining the term *invisible labor* was "Invisible Work" by Arlene Kaplan Daniels (*Social Problems* 34, no. 5 [1987]: 403–415), in which she writes that "gender expectations, and separation between the public and private worlds are mixed together with paid work to create a special type of problem for which women are expected to take responsibility." Since then it has become a broad, umbrella term for the many kinds of labor (often gendered) that go uncompensated, undervalued, or unseen.

13. Tiziana Terranova, "Free Labor: Producing Culture for the Digital Economy," *Social Text* 18, no. 2 (2000): 33–58.

14. "Netflix Prize," Netflix, accessed August 1, 2019, https://www.netflixprize.com/.

15. "MPs Expenses: The Guardian Launches Major Crowdsourcing Experiment," *Guardian*, June 23, 2009, https://www.theguardian.com/gnm-press-office/crowdsourcing-mps-expenses.

16. "EEBO-TCP: Early English Books Online," accessed August 1, 2019, https://www.textcreation partnership.org/tcp-eebo/.

17. When he coined the term in the mid-2000s, Jeff Howe argued that *crowdsourcing* was a powerful way to tap networked knowledge for corporate use. "The labor isn't always free, but it costs a lot less than paying traditional employees," he wrote in *Wired*. See Jeff Howe, "The Rise of Crowdsourcing," *Wired*, January 6, 2006, https://www.wired.com/2006/06/crowds/.

18. Ashe Dryden, "Programming Diversity," talk at Mix-IT, Lyon, France, April 29, 2014. Video and transcript available online at https://www.ashedryden.com/mixit-programming-diversity.

19. Benjamin Mako Hill and Aaron Shaw, "The Wikipedia Gender Gap Revisited: Characterizing Survey Response Bias with Propensity Score Estimation," *PLOS ONE* 8, no. 6. (June 2013): 1–5; and "Editor Survey 2011/Executive Summary," WikiMedia, accessed August 1, 2019, https://meta.wikimedia.org/wiki/Editor_Survey_2011/Executive_Summary. Neither of the studies consider nonbinary genders.

20. "#VisibleWikiWomen 2019," *Whose Knowledge?*, 2019, accessed April 12, 2019, https://whoseknowledge.org/initiatives/visiblewikiwomen-2019/.

21. Heather Ford and Judy Wajcman, "'Anyone Can Edit', Not Everyone Does: Wikipedia's Infrastructure and the Gender Gap," *Social Studies of Science* 47, no. 4 (2017): 511–527.

22. Veerle Miranda, "Cooking, Caring and Volunteering: Unpaid Work around the World," OECD Social, Employment and Migration Working Papers no. 116, March 3, 2011, https://www.oecd.org/berlin/47258230.pdf. The study did not consider nonbinary genders.

23. For more on the attention-leaching effects of capitalism, see McKenzie Wark, "Cognitive Capitalism," *Public Seminar*, February 19, 2015, http://www.publicseminar.org/2015/02/cog-cap/.

24. Mechanical Turk, also known as MTurk, is a marketplace for crowdsourced human labor—individuals and businesses that have "human intelligence tasks" can set up jobs for online workers to complete. Often repetitive, like describing images or taking surveys, the jobs are compensated in tiny financial increments that don't add up to a minimum hourly wage—not to mention a living wage. See Amazon Mechanical Turk, accessed April 22, 2019, https://www.mturk.com.

25. Amazon Mechanical Turk, accessed April 22, 2019, https://www.mturk.com.

26. Paul Hitlin, "Research in the Crowdsourcing Age, a Case Study," Pew Research Center: Internet & Technology, July 11, 2016, http://www.pewinternet.org/2016/07/11/research-in-the-crowdsourcing-age-a-case-study/.

27. Joel Ross, Lilly Irani, M. Six Silberman, Andrew Zaldivar, and Bill Tomlinson, "Who Are the Crowdworkers? Shifting Demographics in Mechanical Turk," in *CHI 2010: Imagine All the People* (New York: ACM, 2010), 2863–2872.

28. To view the documentary *Workers Leaving the Googleplex*, visit https://vimeo.com/15852288.

29. Lucy Yang, "13 Incredible Perks of Working at Google, According to Its Employees," *Insider*, July 11, 2017, https://www.thisisinsider.com/coolest-perks-of-working-at-google-in-2017-2017 -7#perhaps-one-of-googles-most-well-known-perks-employees-can-eat-every-meal-at-work-for -free-and-save-a-ton-of-money-1.

30. Lilly Irani, "Justice for 'Data Janitors,'" *Public Books*, January 15, 2015, https://www .publicbooks.org/justice-for-data-janitors/.

31. Scholars have uncovered many examples of "hidden figures" in the history of data work that have not yet (but probably should) become major motion pictures. For example, Lisa Nakamura has recently exposed the below-minimum-wage pay of the Navajo women who, in the early days of digital computing, were tapped to assemble integrated circuits for the largest electronics supplier in the country, Fairchild Semiconductor. Or consider the female line workers, called "little old ladies" by their colleagues in engineering, who were employed by the US military defense contractor Raytheon to weave the delicate core rope memory that was used on the computers that accompanied the Apollo astronauts to the moon—a contribution that Daniela Rosner and her team have explored. See Lisa Nakamura, "Indigenous Circuits: Navajo Women and the Radicalization of Early Electronic Manufacture," *American Quarterly* 66, no. 4. (2014): 919–941; and Daniela K. Rosner, Samantha Shorey, Brock R. Craft, and Helen Remick, "Making Core Memory: Design Inquiry into Gendered Legacies of Engineering and Craftwork," in *CHI 2018* (New York: ACM, 2018), 1–13.

32. "About," Turkopticon, December 28, 2018, https://turkopticon.ucsd.edu/.

33. At the time of writing, Turkopticon had announced on Twitter that the project was shuttering. Since then, however, they have regrouped, organizing events for MTurk workers and developing new co-designed methods for worker governance and maintenance of the platform.

34. Andrea Chen, "The Laborers Who Keep Dick Pics and Beheadings Out of Your Facebook Feed," *Wired*, October 23, 2014, https://www.wired.com/2014/10/content-moderation/. Chen went on to collaborate with filmmaker Ciaran Cassidy on a short documentary called *The Moderators* (https://fieldofvision.org/the-moderators). Other work on the labor of content moderation includes Sarah T. Roberts, *Behind the Screen: Content Moderation in the Shadows of Social Media* (New Haven: Yale University Press, 2019), and Tarleton Gillespie, *Custodians of the Internet: Platforms, Content Moderation, and the Hidden Decisions that Shape Social Media* (New Haven: Yale University Press, 2018).

35. Mary Gray and Siddarth Suri, *Ghost Work: How to Stop Silicon Valley from Building a New Global Underclass* (New York: Houghton Mifflin Harcourt, 2019).

36. Filmmaker and author Astra Taylor has a word for this phenomenon: *fauxtomation*. Taylor, "The Automation Charade," *Logic*, October 2, 2018, https://logicmag.io/05-the-automation-charade/.

37. For an account of this episode in the context of capitalism, see Ian Baucom, *Specters of the Atlantic: Finance Capital, Slavery, and the Philosophy of History* (Durham, NC: Duke University Press, 2005).

38. Christina Sharpe, *In the Wake: On Blackness and Being* (Durham, NC: Duke University Press, 2016).

39. Marlene NourbeSe Philip, *Zong!* (Middletown, CT: Wesleyan University Press, 2011).

40. Philip's poem has become a touchstone of contemporary information studies scholarship. Miriam Posner, in "Seeing Like a Supply Chain" (forthcoming), connects *Zong!* to the history and present of capitalism. Jeffrey Moro connects the poem to digital data in "Want of Water, Want of Data: The Trans-Atlantic Slave Database and Oceanic Computing" (presented at SIGCIS 2018, Stored in Memory, St. Louis, Missouri, October 14, 2018).

41. Due to the systemic racism of the United States, those in prison are disproportionately Black and Latinx; those groups constitute around 32 percent of the US population, but comprised 56 percent of all incarcerated people in 2015 ("Criminal Justice Fact Sheet," NAACP, 2019, accessed April 23, 2019, https://www.naacp.org/criminal-justice-fact-sheet/). Scholars such as Michelle Alexander, who we mentioned in chapter 1, as well as filmmakers such as Ava DuVernay, have demonstrated how mass incarceration is a contemporary, legally permissible extension of slavery. And who benefits? Corporations. The Corrections Accountability Project released a comprehensive study in 2018 that maps and exposes all the commercial interests in the prison system: "The Prison Industrial Complex: Mapping Private Sector Players," Urban Justice Center: Corrections Accountability Project, April 2018. On the Victoria's Secret connection in particular, see Caroline Winter, "What Do Prisoners Make for Victoria's Secret?," *Mother Jones*, July/August 2008, https://www.motherjones.com/politics/2008/07/what-do-prisoners-make-victorias-secret/. On the BP oil spill, see Abe Louise Young, "BP Hires Prison Labor to Clean Up Spill while Coastal Residents Struggle," *Nation*, July 21, 2010, https://www.thenation.com/article/bp-hires-prison-labor-clean-spill-while-coastal-residents-struggle/. On incarceration as a new form of enslavement, see Ava DuVernay, *13th*, Netflix, documentary (Sherman Oaks, CA: Kandoo Films, 2016).

42. Siddharth Kara, "Is Your Phone Tainted by the Misery of the 35,000 Children in Congo's Mines?," *Guardian*, October 12, 2018, https://www.theguardian.com/global-development/2018/oct/12/phone-misery-children-congo-cobalt-mines-drc.

43. Jacopo Ottaviani, "E-waste Republic," *Aljazeera*, accessed June 10, 2018, http://www.spiegel.de/international/tomorrow/electronic-waste-in-africa-recycling-methods-damage-health-and-the-environment-a-1086221.html. The narratives in the popular press that focus on the toxicity of these sites often overlook the ingenuity of those who work there. For scholarship on Ghana in particular, see Stephen J. Jackson, "Rethinking Repair," in *Media Technologies: Essays on Communication, Materiality and Society*, ed. Tarleton Gillespie, Pablo Boczkowski, and Kirsten Foot (Cambridge, MA: MIT Press, 2014), 221–239; for other geographic contexts, see Cuban artist Ernesto Orozo's concept of the "architecture of necessity," as elaborated in Alex Gil, "Interview with Ernesto Oroza," in *Debates in the Digital Humanities 2016*, ed. Matthew K. Gold and Lauren F. Klein (Minneapolis: University of Minnesota Press, 2016), 184–193; or Daniela Rosner's concept of "critical fabulations," in *Critical Fabulations: Reworking the Methods and Margins of Design* (Cambridge, MA: MIT Press, 2018).

44. Safiya Umoja Noble and Robert Mejia, among others, have demonstrated these links in their essays in *The Intersectional Internet: Race, Sex, Class, and Culture Online*, ed. Safiya Umoja Noble and Brendesha M. Tynes (New York: Peter Lang International Academic Publishers, 2016).

45. Miranda Banks, "Production Studies," *Feminist Media Histories* 4, no. 2 (Spring 2018): 157–161. Also see Banks, "Gender Below-the-Line: Defining Feminist Production," in *Production Studies Cultural Studies of Media Industries*, ed. Vicki Mayer, Miranda J. Banks, and John T. Caldwell (London: Taylor and Francis Group, 2009), 87–98.

46. This work can also be connected to the area of research known as *maintenance studies*, as popularized by groups such as the Maintainers, which we discuss in more detail ahead. See "The Maintainers," accessed August 1, 2019, http://themaintainers.org/.

47. Kate Crawford and Vladan Joler, "Anatomy of an AI System: The Amazon Echo as an Anatomical Map of Human Labor, Data and Planetary Resources," AI Now Institute and Share Lab, September 7, 2018, https://anatomyof.ai.

48. Sara Ahmed, "Making Feminist Points," *Feministkilljoys* (blog), September 11, 2013, https://feministkilljoys.com/2013/09/11/making-feminist-points/.

49. "Collaborators' Bill of Rights," in *Off the Tracks: Laying New Lines for Digital Humanities Scholars*, MediaCommons, 2011, http://mcpress.media-commons.org/offthetracks/part-one-models-for-collaboration-career-paths-acquiring-institutional-support-and-transformation-in-the-field/a-collaboration/collaborators%E2%80%99-bill-of-rights/. More recently, at UCLA, a team of eleven students and faculty members worked together to author the Student Collaborators' Bill of Rights. Supplementing the original document with ten additional principles, the student version of the document emphasizes the importance of empowering students to "make critical decisions about the intellectual design of a project or a portion of a project" and credit them accordingly. See Haley Di Pressi, Stephanie Gorman, Miriam Posner, Raphael Sasayama, and Tori Schmitt, with contributions from Roderic Crooks, Megan Driscoll, Amy Earhart, Spencer Keralis, Tiffany Naiman, and Todd Presner, "A Student Collaborators' Bill of Rights," June 8, 2018, https://humtech.ucla.edu/news/a-student-collaborators-bill-of-rights/.

50. See J. K. Gibson-Graham and the Community Economies Collective, *Cultivating Community Economies*, Next System Project, February 27, 2017, https://thenextsystem.org/cultivating-community-economies. Thanks go to Kate Diedrick for telling us about this work.

51. As a hack, the Drupal development community has begun to create "issues" in its bug-tracking system to document in-person meetings. When the issues are resolved at the close of each meeting, people who attended the meeting are entered into the system so that their participation is formally logged. See Angela Byron, "Proposal: Add a New 'Meetings' Component for Initiative (and Other?) Meetings," Drupal.org, June 22, 2018, https://www.drupal.org/project/drupal/issues/2976614.

52. Benjamin M. Schmidt, "A Brief Visual History of MARC Cataloging at the Library of Congress," *Sapping Attention* (blog), May 16, 2017, http://sappingattention.blogspot.com/2017/05/a-brief-visual-history-of-marc.html.

53. For more on Loukissas's work and his idea of the *data setting*, see chapter 5.

54. For a survey of this form of labor, see Amy S. Wharton, "The Sociology of Emotional Labor," *Annual Review of Sociology* 35 (August 2009): 147–165.

55. Arlie Russell Hochschild, *The Managed Heart: Commercialization of Human Feeling*, 3rd ed. (1979; Berkeley: University of California Press, 2012).

56. As an entry point, see Guillermina Altomonte, "Affective Labor in the Post-Fordist Transformation," Public Seminar, May 8, 2015, http://www.publicseminar.org/2015/05/affective-labor -in-the-post-fordist-transformation/. See also Melissa Gregg and Gregory J. Seigworth, eds., *The Affect Theory Reader* (Durham, NC: Duke University Press, 2010).

57. Enda Brophy and Jamie Woodcock, "The Call Centre Seen from Below: Issue 4.3 Editorial," *Notes from Below*, February 14, 2019, https://notesfrombelow.org/article/call-centre-seen -below-issue-43-editorial.

58. An increasing amount of scholarship exists on this phenomenon. See, for instance, Sara Ahmed, *On Being Included: Racism and Diversity in Institutional Life* (Durham, NC: Duke University Press, 2012).

59. The invisible labor of domestic work has long been the subject of scholarship in feminist geography. See, for example, Mei-Po Kwan, "Gender, the Home-Work Link, and Space-Time Patterns of Nonemployment Activities," *Economic Geography* 75, no. 4 (1999): 370–394, as well as her important essay, "Feminist Visualization: Re-envisioning GIS as a Method in Feminist Geographic Research," *Annals of the Association of American Geographers* 92, no. 4 (2002): 645–661. See also Marianna Pavlovskaya and Kevin St. Martin, "Feminism and Geographic Information Systems: From a Missing Object to a Mapping Subject," *Geography Compass* 1, no. 3 (2007): 583–606. Thanks to Shannon Mattern for alerting us to this work.

60. See "The Value of Self-tracking to Caregivers," in V. Rajiv Mehta and Dawn Nafus, "Atlas of Caregiving Pilot Study Report," 2016, 30–31, https://atlasofcaregiving.com/wp-content/uploads/2016/03/Study_Report.pdf.

61. See Giorgia Lupi, "Bruises—the Data We Don't See," Medium: Neuroscience, January 31, 2018, https://medium.com/@giorgialupi/bruises-the-data-we-dont-see-1fdec00d0036. Complete work viewable at https://www.youtube.com/watch?v=QvxVWukROTw.

62. Lupi, "Bruises."

63. Ellen Samuels, "Six Ways of Looking at Crip Time," *Disability Studies Quarterly* 37, no. 3 (2017), https://doi.org/10.18061/dsq.v37i3.5824.

64. Samuels, "Six Ways of Looking at Crip Time."

65. Mierle Laderman Ukeles, "Manifesto for Maintenance Art, 1969," Arnolfini, 1969, accessed April 23, 2019, https://www.arnolfini.org.uk/blog/manifesto-for-maintenance-art-1969.

66. In domestic work, for example, women represent 80 percent of the sixty-seven million workers across the globe. See Abigail Hunt and Fortunate Machingura, "A Good Gig? The Rise of On-Demand Domestic Work," *Overseas Development Institute* (December 2016): 1–43, https://www.odi.org/publications/10658-good-gig-rise-demand-domestic-work.

67. See, for example, Andrew L. Russell and Lee Vinsel, "Make Maintainers: Engineering Education and an Ethics of Care," in *Does America Need More Innovators?*, ed Matthew Wisnioski, Eric S. Hintz, and Marie Stettler Kleine (Cambridge, MA: MIT Press, 2019), 249–272; and Bethany Nowviskie, "Capacity through Care," in *Debates in the Digital Humanities 2019*, ed. Matthew K. Gold and Lauren F. Klein (Minneapolis: University of Minnesota Press, 2019), 424–426.

68. Hunt and Machingura, "A Good Gig?"

69. See "Alia," NDWA Labs, 2015, accessed April 23, 2019, https://www.ndwalabs.org/alia.

## Conclusion: Now Let's Multiply

1. Google Walkout for Real Change, "Google Employees and Contractors Participate in Global 'Walkout for Real Change,'" Medium: Technology, November 2, 2018, https://medium.com/@GoogleWalkout/google-employees-and-contractors-participate-in-global-walkout-for-real-change-389c65517843.

2. The women that are hired into Google are not compensated equally to their male counterparts. The corporation is facing a class-action lawsuit about gender pay equity after the Department of Labor found "systemic compensation disparities against women pretty much across the entire workforce." The 31 percent figure is from Alex Morris, "When Google Walked: Rage Drove the Protests Last Year, but Can It Bring About Lasting Change at Tech Companies?," *New York Magazine*, February 5, 2019, http://nymag.com/intelligencer/2019/02/can-the-google-walkout-bring-about-change-at-tech-companies.html. See also Sam Levin, "Google Gender Pay Gap: Women Advance Suit that Could Affect 8,300 Workers," *Guardian*, October 26, 2018, https://www.theguardian.com/technology/2018/oct/26/google-gender-pay-gap-women-class-action-lawsuit.

3. Daisuke Wakabayashi and Katie Benner, "How Google Protected Andy Rubin, the 'Father of Android,'" *New York Times*, October 25, 2018, https://www.nytimes.com/2018/10/25/technology/google-sexual-harassment-andy-rubin.html.

4. Wakabayashi and Benner, "How Google Protected Andy Rubin."

5. Morris, "When Google Walked."

6. Morris, "When Google Walked."

7. Mar Hicks, "The Long History Behind the Google Walkout: Tech Companies Ignore Labor History at Their Own Peril," *Verge*, November 9, 2018, https://www.theverge.com/2018/11/9/18078664/google-walkout-history-tech-strikes-labor-organizing.

8. Moira Weigel, "Coders of the World, Unite: Can Silicon Valley Workers Curb the Power of Big Tech?," *Guardian*, October 31, 2017, https://www.theguardian.com/news/2017/oct/31/coders-of-the-world-unite-can-silicon-valley-workers-curb-the-power-of-big-tech.

9. Alex Press, "Code Red: Organizing the Tech Sector," *n+1*, no. 31 (Spring 2018), https://nplusonemag.com/issue-31/politics/code-red/.

10. See https://twitter.com/hashtag/TechWontBuildIt?src=hash.

11. The change was short-lived and caused an uproar in the open-source software development community, with prominent open-source advocates claiming that this constituted discrimination (against US Immigration and Customs Enforcement—which, to remind you, was separating young children from their parents at the border and putting them in jails). The case is a fascinating collision between the rising political consciousness of the tech sector and the libertarian values that drove the 1990s internet and rise of open-source software. See Daniel Oberhaus, "Open Source Devs Reverse Decision to Block ICE Contractors From Using Software," *Motherboard*, August 30, 2018, https://motherboard.vice.com/en_us/article/pawnwv/open-source-devs-reverse-decision-to-block-ice-contractors-from-using-software.

12. See https://www.techworker.coop/, and a partial list of global tech cooperatives can be found here: "Hng/tech-coops," GitHub, April 11, 2019, https://github.com/hng/tech-coops. The Design Action Collective's Points of Unity are available at https://designaction.org/about/points-of-unity/.

13. The Toronto Declaration was written by a coalition of technology and human rights groups. It advocates for standards to ensure that machine learning systems respect basic principles of human rights, including equality and non-discrimination. It was published on May 16, 2018, by Amnesty International and Access Now, and launched at RightsCon 2018 in Toronto, Canada. See https://www.accessnow.org/cms/assets/uploads/2018/08/The-Toronto-Declaration_ENG_08-2018.pdf.

14. Canada's principles for open government include commitments to gender equity, accessibility, and reconciliation. The last of these resolves to design all services and programs in collaboration and consultation with Indigenous communities. See Government of Canada, "Canada's 2018–2020 National Action Plan on Open Government," accessed July 31, 2019, https://open.canada.ca/en/content/canadas-2018-2020-national-action-plan-open-government#toc3-5.

15. See http://designjusticenetwork.org/network-principles/.

16. Una Lee, "Generating Shared Principles for Design Justice," *And Also Too*, July 13, 2016. Costanza-Chock is writing a longer book-length work called *Design Justice* (Cambridge, MA: MIT Press, 2020).

17. Una Lee et al., "Design Justice Network Principles," *Design Justice*, accessed April 23, 2019, http://designjusticenetwork.org/network-principles/.

18. UCIBrenICS, "Abolish Big Data—Yeshimabeit Milner," video, 58:09, March 8, 2019, https://www.youtube.com/watch?v=26lM2RGAdlM.

19. UCIBrenICS, "Abolish Big Data."

20. As quoted in Morris, "When Google Walked."

21. Mark Bergin and Josh Eidelson, "Inside Google's Shadow Workforce," *Bloomberg News*, July 25, 2018, https://www.bloomberg.com/news/articles/2018-07-25/inside-google-s-shadow-workforce.

22. Leah Fessler, "Google CEO Sundar Pichai's Full Memo Outlining Updated Sexual-Harassment Policies," *Quartz at Work*, November 8, 2019, https://qz.com/work/1456530/google-ceo-sundar-pichais-full-memo-to-employees-on-sexual-harassment.

23. Nick Bastone, "Google Has Publicly Stated That It Supports Employee Walkouts, But Its Latest Actions Say Otherwise," *Inc.*, January 24, 2019, https://www.inc.com/business-insider/google-asked-us-to-reconsider-rules-that-allow-employees-to-organize-without-fear-of-punishment.html.

24. See Kyle Wiggers, "How Google Treats Meredith Whittaker Is Important to Potential AI Whistleblowers," *Venture Beat*, April 24, 2019, https://venturebeat.com/2019/04/24/how-google-treats-meredith-whittaker-is-important-to-potential-ai-whistleblowers/.

25. See Julia Carrie Wong, "'I've Paid a Huge Personal Cost': Google Walkout Organizer Resigns over Alleged Retaliation," *Guardian*, June 7, 2019, https://www.theguardian.com/technology/2019/jun/07/google-walkout-organizer-claire-stapleton-resigns; Sugandha Lahoti, "#NotOk-Google: Employee-Led Town Hall Reveals Hundreds of Stories of Retaliation at Google," *Packt* (blog), April 27, 2019, https://hub.packtpub.com/notokgoogle-employee-led-town-hall-reveals-hundreds-of-stories-of-retaliation-at-google/; and Meredith Whittaker, "Onward! Another #GoogleWalkout Goodbye," Medium, July 16, 2019, https://medium.com/@GoogleWalkout/onward-another-googlewalkout-goodbye-b733fa134a7d.

26. In her 2019 talk "Abolish Big Data," Yeshimabeit Milner discusses how certain applications of AI and big data need to be outright abolished. As she defines it, "to abolish Big Data would mean [to] put data in the hands of people who need it most," as well as to name the "data-industrial complex." From UCIBrenICS, "Abolish Big Data."

27. See Dean Spade, *Normal Life: Administrative Violence, Critical Trans Politics, and the Limits of Law* (Durham, NC: Duke University Press, 2015); and Shaka McGlotten, "Black Data," in *No Tea, No Shade: New Writings in Black Queer Studies*, ed. E. Patrick Johnson (Durham, NC: Duke University Press, 2016), 262–286.

28. Walter and Andersen, *Indigenous Statistics*, and Kukutai, Tahu, and John Taylor, eds., *Indigenous Data Sovereignty: Toward an Agenda* (Canberra: ANU Press, 2016).

29. Lilly Irani, Janet Vertesi, Paul Dourish, Kavita Philip, and Rebecca E. Grinter, "Postcolonial Computing: A Lens on Design and Development," in *Proceedings of the SIGCHI Conference on Human Factors in Computing Systems* (New York: ACM, 2010), 1311–1320. Mustafa Ali, "Towards a Decolonial Computing," in *Ambiguous Technologies: Philosophical Issues, Practical Solutions, Human Nature* (Lisbon: International Society of Ethics and Information Technology, 2014), 28–35.

30. Payal Arora, "Bottom of the Data Pyramid: Big Data and the Global South," *International Journal of Communication* 10 (March 14, 2016), https://ijoc.org/index.php/ijoc/article/view/4297; Stefania Milan and Emiliano Treré, "Big Data from the South(s): Beyond Data Universalism," *Television & New Media* 20, no. 4 (2019): 319–335; and a workshop, "Big Data from the South: Decolonization, Resistance and Creativity," Universiteit Van Amsterdam, December 3, 2018, https://www.spui25.nl/spui25-en/events/events/2018/12/big-data-from-the-south-decolonization-resistance-and-creativity.html.

31. See Margaret Mitchell, Simone Wu, Andrew Zaldivar, Parker Barnes, Lucy Vasserman, Ben Hutchinson, Elena Spitzer, Inioluwa Deborah Raji, and Timnit Gebru, "Model Cards for Model Reporting," in *Proceedings of the Conference on Fairness, Accountability, and Transparency* (New York: ACM, 2019), 220–229.

32. Clarity should not always be the goal of data communication. Confusion and disorientation can be extremely effective emotions for producing a new seeing of the world. For more about the *Decoding Possibilities* project, see https://elegantcollisions.com/decoding-possibilities.

## Our Values and Our Metrics for Holding Ourselves Accountable

1. For more information on Jenn Roberts and Versed Education, see https://www.versededucationgroup.com/. To read the AADHum Initiative's "Statement of Our Values," see https://aadhum.umd.edu/conference/values/. To read the Colored Convention Project's project principles, see http://coloredconventions.org/ccp-principles.

# Name Index

The letter *f* following a page number denotes a figure.

# Subject Index

The letter *f* following a page number denotes a figure.